David Frey is Associate Professor of History and Director of the Center for Holocaust and Genocide Studies, United States Military Academy at West Point.

'In his wonderful study of the Hungarian film industry which, among other things, provided Hollywood with some of its greatest directors, producers and actors, David Frey demonstrates the corrosive effect of German National Socialist culture and politics on the Hungarian film industry. Hungarian antisemites and German Nazis caused the Hungarian government to put the "Jewish Question" in the forefront of pre-World War II and wartime Hungarian filmmaking; still, both the country's government and Hungarian movie industry preserved enough autonomy to dominate the Balkan market and even to influence German filmmaking. David Frey shows us an artistic world dominated by extraordinary talent and one that did not always succumb to evil.'

István Deák,
Seth Low Professor Emeritus, Columbia University

'This well-researched book is a much-needed reminder of the significant position Hungary's cinema occupied on the European film market in the 1930s and the first years of World War II. One of the particular strengths of this work lies in David Frey's ability to analyze and describe complex national and transnational processes in clear prose, linking his study to important debates and research strands in historical and media studies. This is a mandatory read, not just for film scholars interested in the production and distribution of sound cinema in Europe (1929–45) but also for students of "national cinemas" and cultural industries.'

Roel Vande Winkel,
KU Leuven, Belgium

'*Jews, Nazis, and the Cinema of Hungary* is a wonderful and thoughtful study of the relationship between Hungarian cinema and politics in one of the darkest periods of the history of Hungary. David Frey's powerful approach and his serious scholarship makes his work essential reading for everyone interested in the impact of ideological dictatorship on culture, and a precious contribution to Hungarian film historiography.'

András Bálint Kovács,
ELTE University, Budapest

JEWS, NAZIS, AND THE CINEMA OF HUNGARY

THE TRAGEDY OF SUCCESS, 1929-44

DAVID FREY

I.B. TAURIS
LONDON · NEW YORK

Every attempt has been made to gain permission for the use of the images in this book. Any omissions will be rectified in future editions.

References to websites were correct at the time of writing.

International Library of Twentieth Century History 93

ISBN: 978 1 78076 451 1
eISBN: 978 1 78672 061 0
ePDF: 978 1 78673 061 9

A full CIP record for this book is available from the British Library
A full CIP record is available from the Library of Congress

Library of Congress Catalog Card Number: available

Typeset by Newgen Knowledge Works Pvt Ltd
Printed and bound by CPI Group (UK) Ltd, Croydon, CR0 4YY

FSC www.fsc.org
MIX
Paper from responsible sources
FSC® C013604

Contents

List of Figures

Notes on citation abbreviations in figure captions

MaNDA, Budapest = Hungarian National Digital Archive and Film Institute, Budapest, Hungary.
MFA = Hungarian Film Archive Photo Collection
MKT CD 1 = *MozgóKépTár. Magyar filmtörténeti sorozat,* Compact Disk, volume 1: Játékfilmek a kezdetektől 1944-ig. Magyar Filmintézet, 1996.

List of Tables

Acknowledgements

I am grateful to the many, many people who assisted me with this project and helped me see it through to completion. In Hungary, Germany, and the United States, I was aided by a bevy of helpful and insightful scholars and archivists. My Columbia University advisors were of course irreplaceable. István Deák stirred me to study Central Europe. His guidance and assistance on everything, from translations to contacts in Hungary, were invaluable, and his example as a scholar and mentor inspiring. Victoria de Grazia challenged me as no one else can, and this book bears her imprint. Volker Berghahn supported me throughout the dissertation process, and I believe he, more than any other, was responsible for my first teaching position. Dr. Attila Pók of the Institute of History at the Hungarian Academy of Sciences, as he has for many scholars, provided inestimable assistance and guidance, from introductions to help with housing. Gyöngyi Balogh and Tibor Sándor, both film experts and film historians, receive my thanks for their research guidance and vast knowledge of the interwar and wartime periods. Vera Gyürey, the former Director of the Hungarian National Film Institute and Archive, allowed me access to the treasure trove of the Archive's film and photo collection. Márika Paldy, the Institute's librarian, made my research effective and enjoyable through her warmth and knowledge. Dr. László Karsai directed me toward my most important archival finds. To the many archivists at the various Hungarian National Archives and Budapest City Archives, the librarians at the Ervin Szabó and Széchenyi Libraries and in the Széchenyi's special collections, I am in your debt. The librarians at the Margaret Herrick Library and the Academy of Motion Picture Arts and Sciences Archive made my research in southern California a pleasure. David Langbart at the National Archives and Records Administration, who although his role in this book's research was limited, played a critical role helping me link Hungarian film émigrés and postwar State Department activities. Finally, the librarians at Columbia University and West Point, particularly those in Interlibrary Loan, never get enough credit. They did more than they can imagine in helping me write this book.

My friends and colleagues on both sides of the Atlantic deserve special recognition. The Hegedüs family in Budapest, the Juhász family in Stuttgart, and the Helmchen family in Berlin were of inestimable aid throughout. Without a room at the Helmchen house and my dear friend Jane Helmchen's ability to navigate German bureaucracies, my research in the *Bundesarchiv* would never have been so successful. Monika Biró-Wise corrected my imperfect written Hungarian whenever necessary, which was often. Roger Spiller, Robert Citino, Denise Youngblood, and Thomas Sakmyster, all read my early drafts and provided wonderful, if often contradictory, guidance! The entire Kennebunkport Circle, specifically Paul Hanebrink, Eagle Glassheim, Melissa Feinberg, and Cynthia Paces have my eternal gratitude. They, along with Jonathan Gumz, provided feedback on my work at different stages and I'm privileged to count them as friends. Another friend, Michael Miller, has provided encouragement, contacts, insight on drafts, camaraderie, and an apartment in Budapest. I am also indebted to Zsolt Nagy and Anna Manchin, whose research pushed me in novel directions and towards different sources. And to Gene Garthwaite, who helped me with my proposal and was instrumental arranging the book's publication, for this and much more I am in your debt.

I am especially grateful to my colleagues in the Department of History at West Point, particularly Brigadier General (R) Lance Betros; Colonels Ty Seidule, Gail Yoshitani, and Beth Behn; Professors Greta Bucher and Steve Waddell, and retired Colonels Greg Daddis, Kevin Farrell, and Pilar Ryan. These leaders all helped protect my time and believed not only in this project, but in my work as Director of West Point's Center for Holocaust and Genocide Studies. My former office mates and interlocutors Ray Kimball, Brian Schoellhorn, Jon Due, Kevin Scott, Charlie Thomas, Brian Drohan, and J.P. Clark all forced me to think deeply about issues of identity and nation, and taught me what true service to the nation entails.

My family, of course, played the largest role in this project. My brothers Howard and Daniel have spent their lifetimes requiring me to support arguments with fact. Howard was especially helpful by providing critical and constructive guidance early in the writing process, while Daniel's optimism and consistent ribbing helped sustain my motivation. My in-laws,

Ben and Joan Wood, and parents, Michael and Eleanor Frey, have made so many things possible in my life. In their very different ways, they have enabled the completion of this work. Their love, encouragement, and support were unending, as was that of my immediate family. My children Anna, Sophie, and Katie have all grown up too fast, and they have all inherited their uncles' frankness and smarts. I look forward to when they write their first books!

Finally, there is no way to avoid the cliché, nor should it be avoided. My wife Beth is a fantastic person. She should be listed as an author and copy-editor, and she should receive a halo. She put in untold hours of discussion, made countless personal sacrifices, read an unfortunate number of drafts and endured way too many years of waiting for this book to be finished. For this and many other reasons, I'm deeply indebted to her, thankful for her, and profoundly in love with her.

Fellowships and the support of a number of outstanding institutions made this project possible. Grants from the German Academic Exchange Service (DAAD), the International Research and Exchanges Board (IREX), the Fulbright-Hays Doctoral Dissertation Research Abroad program administered by the US Department of Education, and the American Council of Learned Societies (ACLS) supported my early research and writing. While working at West Point, I received several Department of Defense grants to which allowed me to continue research at various American and Hungarian archives. Without a subvention from the Littauer Foundation, this book might never have been realized. The Ernst Galéria provided permission to use several period film posters, including the cover of the book. The Hungarian National Digital Archives granted me the rights to multiple stills and photographs from its extensive archives. To all of the individuals who and organizations which supported this project, you have my deepest gratitude.

List of Important Organizations and Abbreviations

AA	*Auswärtiges Amt* German Foreign Ministry
BM	*Belügyminisztérium* Hungarian Interior Ministry
Film Council	*Filmtanács*
FIF	*Filmipari Alap* Film Industry Fund
FK	*Filmkultúra* The trade journal *Film Culture*
GsB	*Gesandtschaft Budapest* German Embassy Budapest
IFK	*Internationale Filmkammer/Nemzetközi Filmkamara* International Film Chamber
KKM	*Kereskedelem- és Közlekedésügyi Minisztérium* Trade and Transport Ministry
KM or KÜM	*Külügyminisztérium* Hungarian Foreign Ministry
KMI	*Kozma Miklós iratai* Miklós Kozma papers
ME	*Miniszterelnökség* Prime Minister's Office
MF	*Magyar Film* The trade journal *Magyar Film*
MFI	*Magyar Film Iroda* Hungarian Film Office
MFSzSz	*Magyar Filmalkalmazottak Szabad Szakszervezet* Hungarian Film Employees' Free Union

MKK	*Magyar Királyi Konsultátus* Hungarian Royal Consulate
MKT	*MozgóKépTár. Magyar filmtörténeti sorozat.* Moving Picture Gallery CD
MLVs	Multiple Language Version Films
MMOE	*Magyar Mozgóképüzemengedélyesek Országos Egyesülete* Hungarian National Association of Movie Theater Licensees
Nb.	*Népbíróság* People's Court
OMB	*Országos Mozgóképvizsgáló Bizottság* National Committee for the Examination of Motion Pictures, a.k.a. the Censorship Committee
OMF	*Országos Magyar Filmegyesület* National Hungarian Film Union
OMME	*Országos Magyar Mozgóképipari Egyesület* Hungarian National Association of Film Producers
ONFB or ONF	*Országos Nemzeti Filmbizottság* National Film Committee
PM	*Pénzügyminisztérium* Finance Ministry
RFK	*Reichsfilmkammer* Reich's Film Chamber
RMVP	*Reichsministerium für Volksaufklärung und Propaganda* Reich's Ministry for Enlightenment and Propaganda
VKM	*Vallás- és Közoktatásügyi Minisztérium* Hungarian Ministry of Religion and Education

Archives – Abbreviations

AAA	*Auswärtiges Amt Archiv – Berlin* German Foreign Ministry Archive – Berlin
BA	*Bundesarchiv Berlin* German National Archives
BADH	*Bundesarchiv Dahlwitz-Hoppegarten* German National Archives, Temporary Archive

BFL	*Budapest Fővárosi Levéltár* Budapest Municipal Archives
Herrick Library	Margaret Herrick Library, Academy of Motion Picture Arts and Sciences, Beverly Hills
MaNDA	*Magyar Nemzeti Digitális Archívum és Filmintézet* Hungarian National Digital Archive and Film Institute, Budapest, Hungary
MOL	*Magyar Országos Levéltára* Hungarian National Archives
MOL-Ó	*Magyar Országos Levéltára – Óbuda* Hungarian National Archives – Óbuda branch
NARA	National Archives and Records Administration (US) – College Park, Maryland
OSzK	*Országos Széchényi Könyvtár* Széchényi Library/Hungarian National Library
SdK	*Stiftung deutsche Kinematek* Foundation for German Cinema Studies, Berlin
TH	*Történeti Hivatal* Historical Office, Budapest

Introduction

> [He] sat in the office of a newly-founded institute whose task it was to create new national cultures for some small peoples... by furnishing them with a new alphabet, primers and primitive newspapers....It emerged that the people were not learning fast enough. They had to be helped with films.[1]
>
> Joseph Roth, 1927

The Trailer

When István Székely, fresh from Berlin, cruised down Pasareti Street in Budapest in his brand new Buick in 1931, he was the embodiment of the optimism that was soon to reinvigorate Hungarian filmmaking.[2] In a world buffeted by the Depression, the head of the state-run Hunnia studio, János Bingert, took a flyer. He and producer Albert Samek invited Székely, who was earning his directorial spurs abroad at the time, to make the film *Hyppolit, the Butler* [*Hyppolit, a lakáj*]. While several other Hungarian sound films preceded *Hyppolit,* it was this film, written by the famed Károly Nóti and starring a laundry list of stars from the Hungarian stage, that was the first box office smash made in Hungary and, most importantly, in Hungarian.[3] *Hyppolit* was proof that a Hungarian sound cinema was viable, and that it could, and should, play a leading role in shaping a regenerative national culture.

If the early history of Hungary's sound film industry were itself a movie, it would have all the elements of a blockbuster. The script might read as follows. Sandwiched between the fanaticism of Nazi Germany and the fantasy world of Hollywood, a poor, upstart industry strives to make its mark. There are cigar-chomping bureaucrats, backroom deals, and bankruptcy; demagogues and danger; smashing successes and fabulous failures. Wonderful plot twists abound as the industry's quest for self-discovery nearly results in self-destruction. Within the industry, antisemites do battle with Jewish plutocrats but protect those same Jews from Nazi pressures in their struggle for sovereignty. The industry experiences tumultuous highs and lows as it seeks an identity of own and an ability to reproduce the nation, endeavors complicated by political radicalization and the outbreak of war. This war brings not only deprivation but also unprecedented good fortune, catapulting the industry to its greatest successes. Yet in the final scene it dies a sudden death, just months after reaching the height of its power. This death, or perhaps it was murder, was enabled by internal weaknesses and hastened by the machinations of Hungary's depraved next-door neighbor.

The mystery, comedy, and tragedy that was the evolution of the early Hungarian sound film industry can provide important insights into how small states devised the institutions of nation and survived the buffeting from the powerful ideological, political, and commercial winds battering the world from the late 1920s through the mid-1940s. In the cauldron that was the motion picture world, the relationship of Jews to the nation; Hungary's position in the Nazi 'New Order'; the tortured relations among the Central European states; Hollywood's extensive European influence; and the roles of capitalism, urban modernity, and populist politics all boiled together. Through the lens of film, this study demonstrates how these transnational trends, international relations, and local influences impacted the development of an industry explicitly imagined by its progenitors to be the clarion of a new national culture.

Prologue

To understand the film industry's special task in the early age of sound and its proverbial rise and fall, one must have a sense of the narrative arc of Hungarian history. Hungarians trace the origins of their cultural presence in Central Europe to the late 800s and 900s. During this time, the Magyar tribes established their kingdom's initial outlines, which Pope Sylvester II and the Holy Roman Emperor Otto III recognized in the year 1000 when they officially accepted the Christian Magyar chieftain István as the legitimate ruler of the plains of Central Europe. István's successors ranked among the most powerful rulers of medieval and early modern Europe until the Battle of Mohács in 1526, when the Ottomans defeated the Hungarian forces.

With this loss began 350 years of mythologized subservience both to the victorious Ottomans and to the Habsburgs, to whom desperate Hungarian nobles offered the Hungarian crown after the Mohács debacle. Revolutions in Transylvania in the early 1700s and throughout Hungary in 1848 failed to bring about the desired restoration of sovereignty. By 1867, however, conditions had changed. Weakened Ottoman Turkish rulers had lost much of their European territory. As this foe receded, Hungary's gentry leadership took advantage of the sudden leverage they gained when their own empire lost the Austro-Prussian War. The prostrate Habsburg rulers feared the Hungarians would break away, and to preempt this possibility they invited Hungary's leaders to negotiate a new political structure. A Hungarian contingent of nobles and politicians won the sovereignty within the Hungarian crown lands their compatriots had sought through revolution less than 20 years earlier. Creating this new system, however, was no simple matter. The byzantine compromise they hashed out established a dual empire, ruled by the Habsburg Emperor Franz Joseph as a sovereign of two equal halves.

While he maintained powers over budgets, armies, and foreign policy, the Emperor surrendered to the Hungarian parliament extensive rights and privileges. Hungary's leaders now reigned virtually autonomously over Transleithia, a region east of the Leitha River. This included Transylvania in today's Romania, the southern corner of today's Ukraine, all of today's Slovakia, the northern portions of today's Serbia, and all of today's Croatia. Stretching from the Adriatic to the Carpathians, this landmass, equivalent to about 60 percent of today's France, contained nearly 21 million inhabitants by 1910.[4] To celebrate their regained power, Hungary's paladins created a new capital out of three riverside cities – Buda, Pest, and Óbuda – and built themselves an enormous parliament abutting the majestic Danube River.

In fits and starts Hungary's parliamentarians established the rule of law in 'their' half of the Empire, reserving true autonomy for aristocrats, gentry, and some of the middle classes of Hungarian ethnicity. The Dualist era that began with the 1867 Compromise soon transformed into a triumphal 'golden age' of Hungarian liberalism. In law, Hungary's oligarchs enshrined the concept of nation, proclaiming in 1868 that 'All Hungarian citizens constitute a nation in the political sense, the one and indivisible Hungarian nation'.[5] By the turn of the twentieth century, the Kingdom of Hungary's capital Budapest was rapidly becoming the 'greatest financial and media centre of Europe east of Vienna'.[6] Although the pace of industrialization lagged compared to Western Europe, Hungary's civil society was vibrant. By 1900, Budapest accommodated 39 newspapers, easily outpacing Berlin, Vienna, and London.[7] The number of primary schools tripled between 1868 and 1914, and new universities opened in many major cities.[8] Hungarian culture, particularly in urban centers, prospered. This was in no small measure due to the significant numbers of Jews who 'assimilated' into Hungarian culture in the second half of the 1800s and early 1900s, scores of whom, such as Leo Szilard, John von Neumann, Eugene Wigner, Michael Polanyi, Imre Bródy, Robert Capa, and Arthur Koestler, went on to win Nobel prizes in the sciences or make contributions to the arts and popular culture which still resonate today.[9]

Included in this *fin-de-siècle* renaissance was the birth of the motion picture, followed soon after by the onset of war. Hungary emerged from the Great War wounded and radically changed, unclear about what it should become once healed: pastoral and idyllic or dynamic and modern. Newly independent after 1918, Hungary, like most of its neighbors, was a peculiar mix of contrasts. Hungarians relished their nation's independence from the Habsburg dynasty, yet complete autonomy came at a price. Hungary was severely punished for its role in the conflagration of 1914–18. The Treaty of Trianon stripped the state of nearly 70 percent of its pre-war territory, and these border revisions resulted in massive demographic shifts and collective trauma. Hungary's population declined to fewer than eight million, and millions of Hungarian speakers now lived outside the country's borders. Its economy, one of the more advanced in Central Europe prior to 1918, was devastated.[10] After a brief Bolshevik interregnum in 1919, the government that emerged included a group of aristocratic conservatives and violent radicals, some of whom tended toward oligarchic 'conservative liberalism' while others favored mass-based proto-fascist ideas.[11] These counterrevolutionaries united behind Miklós Horthy, who eventually took the title Regent of Hungary and ruled as a surrogate monarch. The government of this oft-ridiculed Admiral-without-a-navy built its rule upon sweeping anti-Communist, antisemitic, and anti-democratic principles. Surprisingly, Horthy's Hungary supported a stable, well-functioning parliament which was characterized by real political heterogeneity into the early 1940s.

Yet the rump state Horthy took over in 1920 was in shambles. The Great War, occupations by Romanian and Czechoslovak troops, Hungary's Bolshevik revolution, and Horthy's own bloody counterrevolution had exacted a heavy price. Hungarians were universally wounded, victims, they believed, of unwarranted humiliation at the hands of Great Powers who misunderstood them. Hungary's truncation transformed it from a multinational partner in one of Europe's ruling empires, to a small, far less ethnically heterogeneous state in search of a new identity. Thus, at a time when Hungary's leaders agreed that nothing was more important than revision of the hated Trianon 'diktat', they also faced serious challenges. Hungary lacked the tools of power. Its navy was gone, its army heavily restricted, its diplomatic leverage and prestige non-existent, and with its loss of land and people, it was truly impoverished. In the Central European maelstrom of empire dissolution and the birth of multiple new nation-states, Hungary needed a new, persuasive *raison d'être*.

Hungary's elites concluded that their desire to modify Trianon, regain regional hegemony, reestablish Hungary's place in Europe, and re-root and revitalize what was essentially a new country depended on their ability to prove the existence, uniqueness, and superiority of the Hungarian nation to themselves and the Great Powers. Scholars have shown that they had an extensive nationalizing palette from which to choose. Language standardization, mapping, law, censuses, pledges and oaths, newspapers, commemorations, food, music, histories, and novels – virtually anything could be a tool to nationalize individuals and naturalize national identities. The process could occur almost anywhere, in theaters, armies, sports stadia, museums, churches, beer halls, homes, and the public square. The nationalizing practice could take many forms, whether economic incentives for in-groups such as land reform or trade freedoms; access to political power and/or education; symbolic actions ranging from fashion or monument construction; infrastructure building; warfare and diplomacy; or even recasting the nation along racial-biological lines.[12]

In parts of Europe, nationalism became a method for widening national groupings to include people heretofore ignored. Yet many of these paths, if they involved expanding the electorate, sharing economic or political power, or significant state investment, did not suit Hungary. Its political class, dominated by great magnates and landless gentry, adopted a strategy of nationalization without fully incorporating the majority as full-fledged members of the national polity. Hungary's leaders consented to only superficial land reform, thus forfeiting the opportunity to tighten communal bonds through socio-economic transformation.[13] They also rolled back the franchise so that as of 1925, less than 30 percent of adult men and women could vote.[14] For a country whose political class dismissed 'the blind rule of the masses' and held that 'true democracy…assured the leadership of the intelligent classes', national integration through either biology alone, progressive political/social change, or some form of inclusive patriotism was off the table.[15] During the interwar era, which Eric Hobsbawm termed the 'Apogee of Nationalism', the interrelated beliefs that cultures reflected systems of irreducible values and that culture was a function of nation became axiomatic and widespread, particularly in Central Europe.[16] Eschewing other alternatives, the pathology of cultural nationalism became feverish in Hungary. Its elites grasped at any and all cultural forms in order to delineate the essence of Hungarian identity. This would be the surest way to promote domestic unity, hold fast against destabilizing historical change, and build a case for Hungary's return to the top of Europe's hierarchy.

This task of creating an authentic, legible, and exportable concept of a nation and its culture was daunting, but highly appealing.[17] As the sound revolution hit the motion picture industry in the later 1920s, proponents of a national culture discovered a new and powerful tool. The dream was enticing for Hungary's patricians: create and institutionalize a national film industry which would then reveal the unifying, distinct national aesthetic or identity which was latent in Hungarian society. This, in turn, would legitimize Hungary's place in Europe, suppress class division, set a future trajectory, and ultimately contribute to the efforts to overturn Hungary's Great War defeat. Speaking of radio as well as film, a leading Hungarian statesman encapsulated the expectations of the time, and the weight on

István Székely's shoulders as his Buick wended through Budapest: 'Our culture is not only there as a guardian of our national existence, but also as a weapon in achieving our future goals'.[18]

Plot

Not only in Hungary, but throughout the developed world during the late 1920s, 30s, and 40s, motion pictures became a boon for groups of elites envisioning their nations' cultures. These elites began to imagine film as the 'vehicle for cultural processes through which nations established traditions and inhabitants of states became citizens'.[19] So imagined, the cinema, scholars argue, played a critical role in negotiating, establishing, and affirming 'the boundaries of the national community'.[20] Nearly every European state that could afford to construct studios – from Britain to Denmark to Poland – did so, and across the political spectrum in Hungary there was agreement that Hungary could not be left out, lest it risk the 'nation's purity' and 'honor'.[21] With the advent of sound, the stakes rose further, and the cinema industry became a central collision point in the quest to define, locate, and discover Hungary's national nature.

This study traces how self-appointed cultural and political vanguards specifically discussed and took action to forge and regulate an industry whose products they assumed would shape representations of individuality and group belonging for millions of 'Hungarians'.[22] It analyzes the elites' beliefs that they could configure their industry as a nationalizing tool which would delineate the boundaries of, and bind together, the national community. Their industry, later re-christened the Christian national film industry, would inform the people of Hungary and audiences abroad about what 'Hungarianness' looked like, how it sounded, what behaviors it entailed, what pasts it referenced, what places it occupied.[23] By generating a limited number of collective representations for mass consumption, their industry would become the most important element of the didactic nationalizing process.

The twist in this tale is that Hungary not only engaged in this national filmmaking odyssey, it unexpectedly thrived. Between 1929 and 1942, the Hungarian sound film industry experienced a meteoric rise, a remarkable achievement given the constraints of the Hungarian language. Hungary progressed from a country that did not produce features to the European continent's third most prolific motion picture maker by 1942, trailing only Germany and Italy and exporting around the globe. Film became an integral part of Hungarian culture, surpassing literature, music, theater, and the visual arts as it assumed a 'privileged position' in markets abroad.[24] In this book, I explain these astonishing feats and the contributions that sound film and the institutions that created it made to Hungarian nation-building efforts in the 1930s and early 1940s. Simultaneously, I question the nation-building project itself. I examine the complex blend of forces that made and unmade the Hungarian national film industry and account for its tragic success.

In both historical and film studies, decades of texts have focused on the nation state or used the term 'national cinema' to provide a coherent narrative for a particular territory or to rescue cultural producers or traditions from obscurity. Most authors now recognize how artificial these studies can be.[25] Over a half-century of nation and nationalism scholarship has taught us that national groups do not arise organically from some long-repressed collective desire and that nations are not primordial or natural entities. Rather, the consensus is that nations are constructed and modern political and cultural forms, processed entities built through explicit manipulation of public symbols, land ownership, education, social institutions, legal and political systems, and popular and material culture.[26] Within the last decade, the best scholarship on Central Europe has exposed the teleologies in accounts of the rise of nation-states, fore-grounded resistance to or indifference in the face of nationalist projects, warned us about accepting national categories as givens, and utilized alternative regional, interzonal, transnational, gendered or theoretical methodologies to explain historical developments.[27] Film historians and cinema studies scholars have reached similar conclusions.[28] They continually caution that one must challenge categories and essentialism. They demand sensitivity to hybridity and the countervailing, divisive forces embedded within so-called national cinemas and to the centrifugal forces pulling from without. They implore us to recognize that national cinemas 'do not simply represent or express stable features of a national culture, but are themselves one of the loci of debates about a nation's governing principles, goals, heritage, history' and mythology.[29] Combined, this scholarship has so destabilized the concepts of the nation and national culture that there are those who question whether either the vocabulary of nationalism or the category of national cinema possesses any intellectual integrity at all.[30]

In *Jews, Nazis, and the Cinema of Hungary,* I analyze why contemporaries believed a national 'character', 'essence', or 'spirit' existed.[31] I examine their pursuit of this holy grail of nationhood and I detail their convoluted attempts to naturalize it through film. By historicizing the evolution of Hungarian sound cinema from its birth in 1929 through its abrupt demise after the Nazi occupation in 1944, we learn about the complexities of the ephemeral processes of group formation.[32] We also discover the sometimes surprising and paradoxical reasons why certain visions of national coherence failed to coalesce in the ways their advocates intended.[33] Confounding their hopes for unity, film and political elites frequently found their 'nation-building' schemes and actions in conflict once they set about defining their nationalist and nationalizing missions. Their disagreements were so extensive and subversive, even within state organizations normally assumed to be monolithic, one might question whether there even was a dominant 'national' film culture in interwar and wartime Hungary. Their attempts at crafting, regulating, and reproducing the nation expose the limits of the nationalizing state. They also reveal a Hungary whose nature was unstable and contingent, a more complex cluster of cultural, political, and economic strategies than most studies of interwar Hungary acknowledge. Rather than efface or ignore this competition, this study explains the persistence of this diversity by highlighting multidirectional movements, particularly resistance within the Hungarian film industry and state institutions to prevailing rightist and Nazi pressures. I detail why attempts to imprint in celluloid specific nationalist ideologies and beliefs about Hungary's national culture failed, despite the significant forces arrayed within the Hungarian government, on the Hungarian street, and on Hungary's borders in support of those visions. My study ultimately reveals that while segments of Hungary fully endorsed a nationality of exclusion, the existence of a single 'official' or agreed upon concept of nation is extremely difficult to find in the early industry and its products. It shows, ironically, that the closest Hungary's national film industry came to producing a distinctive form may have been the cosmopolitan, progressive, class-crossing, Jewish-produced vision of nation, precisely the opposite of that normally ascribed to the Hungary of this period.[34]

As Hungary began its sound film making venture, it was not without advantages. It had great power pretensions, a filmmaking legacy tracing back through World War I, and for a short time in 1931 possessed one of the most state-of-the-art studios in Europe. With the return of experienced film professionals from America and throughout Europe in the early 1930s, especially those trained in Germany, Hungary gathered together an expert supply of labor. Hungary's location seemed to make it an ideal pivot point, a crossroads where the great European and American film producers could set up shop, make 'talkies' cheaply, and distribute their wares to the rest of Central Europe and the Balkans.

Hungary also faced obstacles, some embedded in the geopolitical realities of the age, others of its own making.[35] In a regional sense, none of the aforementioned characteristics set Hungary apart from Czechoslovakia or Austria, whose filmmaking pedigrees were just as refined as Hungary's. From a domestic perspective, interwar Hungary was plagued by a 'culture of étatism' – a reliance upon the state to resolve problems that industry itself found were too difficult to handle on its own.[36] Hungarian cinema could not escape this bug. Every film constituency lobbied state authorities to intercede in film matters. Even when private industry did not invite them, Hungary's political leaders and government functionaries still believed they had the bureaucratic and nation-building mandates to intervene. Similar to their counterparts in other states with fledgling motion picture industries, Hungary's officials meddled in haphazard and sometimes destructive ways. They did not shy away from pressuring the industry to use its products for purported 'national cultural' purposes, while only grudgingly offering the industry financial support for these endeavors. Short-sighted bureaucrats censored features post-production, treated motion picture theater ownership as a retirement perk, and insured conflicts of interest by awarding themselves stock in Hungary's studios. Far from creating uniform practices of power or enduring officially sanctioned imagery, they often purposely disaggregated power and undermined their own productions. It would be wrong, however, to blame all the film industry's shortcomings on bureaucratic corruption, incompetence, and polycratic intra-governmental competition. The country's language proved isolating and its domestic market too small and too undercapitalized to support self-sustaining production. Divisions between producers, distributors, and exhibitors thwarted all aspirations to unify and centralize the industry. Contradictory demands – should Hungary produce for its domestic audiences or for audiences beyond its borders; is filmmaking primarily a material, ideological, or artistic venture – coexisted and clashed within the motion picture community until feature filmmaking ceased soon after Nazi armies occupied Hungary in 1944. And not only was the industry irreconcilably split, but contributors to discourses about nation-sustaining film believed audiences to be just as divided. By the mid-point of World War II, one of Hungary's top film regulators argued that rural and urban audiences had developed such disparate tastes that the best solution for Hungary would be to produce two entirely different types of films.[37] So much for the unifying magic of cinema and its ability to construct national subjects.

This book will show that ironically, it was also these internal contradictions and imbalances that prevented the Hungarian motion picture industry from ruining itself. The archetypical example of these imbalances that this project considers in depth is the industry's treatment of its 'Jewish question'.[38] The Jewish question, a catch-all matter with many permutations, forcefully emerged in interwar Hungary for a variety of reasons. Hungary's ruling elite had welcomed the assimilation of its Jewish citizens during the Dual Monarchy period of 1867–1918. Government officials considered 'assimilated' Jews as ethnic Hungarians, as least in terms of statistics, and their numbers were crucial to maintaining a Magyar majority in the Transleithanian half of the Austro-Hungarian Empire.[39] By the 1920s, the Jewish community in Hungary had become, by many measures, highly successful. Despite legal impediments limiting their access to higher education, Jews averaged around ten percent of Hungary's university students, double the approximately five percent of the overall Hungarian population Jews represented in 1920.[40] They were extraordinarily over-represented in most of the educated liberal professions.[41] Hungarians of Jewish origin thus played an enormous role in their country's economy. By the 1920s they were undoubtedly the core, if not the bulk of the Hungarian middle and upper middle classes.[42] In the mid-1930s, 70 percent of the board members of Hungary's 20 largest industrial enterprises were Jewish.[43]

After World War I, however, Hungary's Jews lost their swing constituency status as Hungary became an ethnically homogenous nation-state. No longer necessary for the project of justifying Hungarian rule, they found themselves blamed for a laundry list of catastrophes, including the ills of capitalism, the Hungarian Bolshevik experience, and Hungary's postwar punishment. Multiple social groups began to see continued Jewish overrepresentation in the liberal professions and among Hungary's economic vanguard as undesirable. Hungarian society was reliant on but resentful of Jewish success. Collective attempts to come to terms with this jealousy and need became the kernel of what was known as the Jewish question.

The idea of 'the Jew', the roles of Jews in filmmaking, and the Jewish place in the national polity are central themes that I trace through this text.[44] From the film industry's origins, Jews built and sustained it. Film distribution was almost entirely the domain of Jewish Hungarians; the largest and most influential cinemas were owned or managed by Jews; nearly all of Hungary's top producers and most of its top directors were of Jewish origin; and the corpus of 1930s film has been interpreted as presenting an identifiably Jewish vision of Hungarian modernity.[45] This of course made the film industry a target, and it could not insulate itself from prevailing social and political currents. Antisemitism seeped through the cracks that already fractured certain industry segments. The spread of a more virulent form of antisemitism in the later 1930s led Hungary's political and right-wing cultural elites to attempt to excise everything they deemed 'Jewish' from the movie business. They portrayed 'film Jews' as invaders, unwelcome interlopers in Hungarian vital space and distorters of the national heritage. Capitalist commerce, escapist entertainment, urban middle-class culture, and the people who produced these things were what Hungary's non-Jewish elite and most of the country's nationalist right decided they could not stomach. All had to go. What Hungary's more radical elites failed to realize was rather than make the cinema industry more potent, stripping it of its Jewish components emasculated it. Film is a commercial art that succeeds because it entertains. Removing capitalism and the cosmopolitan forms and genres contemporary Hungarians associated with Jews – especially middle-class comedy – left little more than artless ideological pabulum that few would pay to see. The story of early sound film in Hungary will show, as some of Hungary's more moderate film elites realized, that if Hungary were to define its industry as being absolutely 'non-Jewish', it would essentially be mandating the destruction of the motion picture business. Attempts to purge the industry of its Jewish elements and 'Jewish nature' were damaging, and for some Jews proved fatal as World War II came to a close. Surprisingly, however, the industry remained a relative haven for Jews until the Nazi takeover. We must concede the irony, as contemporaries did, that without Hungarian 'film Jews', there was no possibility of constructing a 'Christian national' Hungarian film industry.

Counter-intuitively, the phenomenon of Hungarian étatism at times exerted a moderating influence vis-à-vis the Jewish question. Using a strategy employed in several other cases, Hungary's government created structures to appease antisemitic radicals, including a film chamber based on the Nazi model, without endowing them with the requisite power to purge all Jews. In part this was due to misgivings shared by some leading government figures about the wisdom of enacting an extreme antisemitic agenda. It was also likely the result of rampant equivocation among the film elite concerning the centralization of control over the industry. But the end result was the presence of bulkheads which prevented extreme antisemitism from flooding the entire film endeavor. Further, these actions created opportunities for businesses, professional associations, and even moderate film regulators to allow some Jews to remain active in the moving picture business. Their continued participation, technical know-how, and international contacts, in turn, permitted Hungary to manufacture and distribute products which had allure both domestically and abroad, contributing to the wartime flowering of the film industry and the persistence of cinematic diversity well into the 1940s.

The complications of the Jewish question and other domestic considerations, however, only partially determined how the Hungarian motion picture industry evolved and the nationalist visions its progenitors created. It is also imperative we understand the international and transnational[46] contexts out of which nationalisms emerge.[47] Small states and nations do not develop in vacuums. They are battered and pressured by the Great Powers, yet also possess the agency to exploit the spaces great power competition creates. In this sense, *Jews, Nazis, and the Cinema of Hungary* looks beyond Hungary's territorial borders. I propose the interwar and wartime cinema of Hungary is best analyzed in regional, European and global contexts, as a 'minor cinema' – a cash-poor, internally riven, language-limited entity – fighting for recognition and validation.[48] Part of its nationalist project involved protecting its sovereignty in a rough-and-tumble international arena which forced it to define itself in ways and in situations over which it sometimes had limited control. Nazi power, Hollywood heft, cross-border flows of capital and talent, and Central European markets greatly influenced the evolution of Hungary's film institutions and styles. These facts created a dynamic whereby nationalist imperatives were frequently rent by internal inconsistencies and competing objectives. All of this took place in an era and space dominated by Nazi Germany. Notions of the national, and questions of sovereignty and authenticity, thus became nested in and complicated by the debates over Europeanness and discourses about the new Europe and Hungary's place in it.[49]

Sound filmmaking in Hungary began as a project to differentiate Hungary and its 'superior' culture from its less advanced, less modern neighbors to its south and east. Although viewing them as inferior, Hungary's film establishment became ecstatic when the people of the Balkans developed a voracious appetite for Hungary's light film fare. Without the 'backward' Balkans, Hungary's film business never would have transformed as it did. In terms of its relations with the United States, with the exception of the first few years of the 1930s, Hollywood products dominated the programs of most Hungarian theaters, and Hungarian audiences, like much of the rest of the world in the 1930s, became attached to American features. Hollywood officials raided Hungarian talent pools and objected to every change in Hungarian film policy that contained any scent of protectionism. While appreciating this attention from across the ocean, much of the Hungarian film establishment preferred a more regional focus, either by cooperative arrangements within Europe or by soliciting a more intimate relationship with Germany. Preferable to all other options were German-language motion picture production in Hungarian studios and Hungarian access to Nazi Germany's film markets. Through 1944, Hungary remained the unrequited lover. When the Nazis did become seriously involved with Hungary, the relationship was not what either party expected. Germany's war brought early Hungarian sound film creation to a climax, allowing it to reach unprecedented levels of production and sales abroad. But this produced tensions, which were aggravated when the Nazis snatched the dominant role from Hollywood in the early 1940s. Rather quickly, segments of Hungary's film elite and its audiences began pining for old times, as German domination proved oppressive. Treating Hungary as a pest, not a partner, the Nazis stripped Hungary of the sovereignty Hungary's leaders thought they deserved. Germany controlled Hungarian production by rationing raw film, restricting Hungarian exports, and intervening inside Hungary. Ultimately, Germany's war brought about the destruction of Hungary's motion picture business, and with it, the murder of some of Hungary's greatest talents.

While Central European historians have scoured the interwar and wartime era, Central European cinema histories pay far less attention to the coming of sound. There are a number of reasons for this phenomenon. First, for much of Europe, scholarship has painted the interwar era with what I term the brush of the right. Fascism, integral nationalism, and antisemitism pervaded much of Europe. Many scholars have assumed that the end of political pluralism meant the concomitant decline of cultural pluralism.[50] In addition, film historians generally agree that the early years of sound marked Hollywood's golden age. This combination of assumed political/cultural homogenization and Hollywood hegemony results in many authors summarily dismissing the early sound period as one of cultural stagnation or 'extreme insignificance'.[51] Exceptions to this general narrative do exist in which scholars have challenged the ideas of historical unity and ideological conformity and have discovered diversity where earlier accounts found none.[52] This work will add to their number.

In light of the innovations of the wartime and postwar eras, *auteurs* in Poland, Czechoslovakia, Hungary, Italy, France, and other states all developed distinct art cinema styles or genres. It is certainly understandable why film scholars became attracted to the study of these edgy and experimental alternatives and their living creators, and why they framed them in national terms. Of course, one ignores at great peril the ideological lenses through which both the Western and Soviet Blocs viewed the world in the postwar era. The Cold War age hardened readings of the pre-war era, and for years historians and cultural commentators alike interpreted the interwar era as a historical anomaly, the period where the march of progress went wrong. For Communists, the interwar era was fascist, pure and simple. For the West, it was totalitarian or at least anti-liberal and state dominated, and thus uniform. It has been only two decades since these ideological shackles have broken. These factors have certainly influenced studies of Hungarian cinema. While there are extensive publications on post-1945 Hungarian films and filmmakers, there are only a handful of books written on the pre-war period, and many focus on Hungarians who emigrated during the silent era.[53] The most comprehensive studies of 1930s film remain available only in Hungarian, and some remain unpublished.[54] Almost every general overview of Hungarian cinema, including John Cunningham's *Hungarian Cinema: From Coffeehouse to Multiplex*, give far more space to the 65 years of cinema after World War II than the 50 years prior, and all rely heavily on the work of a single excellent author, István Nemeskürty.[55] On the other hand, general histories of Hungary or works on Hungarian cultural politics covering the 1919 to 1945 period rarely devote more than a few pages to film.[56]

This study returns to the early sound era, not to argue it deserves a larger pedestal in the pantheon of Hungarian and film history but to challenge convention and widen the aperture through which we see the processes and practices of cultural and national construction. The example of the evolution of the Hungarian national film industry during the interwar and wartime years should give pause to those who subscribe to the beliefs that cultures precede nations, nationalisms reify common cultures, or that state-led processes of nation building can be insulated from external forces. The one issue Hungarian interwar and wartime cultural critics, politicians, and film industry elites alike agreed upon was the film industry's inability to locate and reproduce the authentic national culture. Right wing nationalists in particularly believed that the film industry categorically failed, to paraphrase film historian Susan Hayward, to put the hyphen between nation and state.[57] Despite its explicit task and in part because of its unprecedented success abroad, the national film industry ineffectively reinforced the racist, integral nationalist ideology that postwar scholars agree pervaded interwar and especially wartime Hungary. While much of Hungarian society may have adopted a 'national identity of excuse', accepting backwardness as a badge of honor and placing the purported national culture at the center of its value system, the Hungarian film establishment was unable or unwilling to wholeheartedly commit to projecting this notion of distinction onto the silver screen.[58] For the radical cultural commentators and the film elite's immoderate elements, it was easy to define what the nation was not: not Jewish, not bourgeois, and not urban. However, defining what the nation was not did not mean the remainder was, *ipso facto*, a coherent alternative. Neither were the emphases on racial hatred, radical nationalism, Christian custom, and cultural despair in the early 1940s sufficient bases upon which to construct a new national film style or articulate a national identity transposable to film. Perhaps most maddeningly for the right and their Nazi ideological compatriots, international audiences happily consumed *as Hungarian* precisely the Jewish-produced goods they wished to let rot. When it came time to define the national industry and its 'deep Hungarian' products in detail, the contingency of the process became clear, and Hungary's political leaders and its guardians of culture floundered. They spoke endlessly of a 'Christian national cinema' but never reached agreement on its contours or even who could make it.

The goal of delineating the nation and its identity through movies inexorably collided with the reality of the aesthetic, material, political, temporal, mass and transnational bases of culture. If films seek to project society, then they also reflect society, including all its unrealized and changing aspirations and fissures. Interwar and wartime Hungarian society was wracked by division, and the problems experienced by the film industry were manifestations of those splits. Hungary's ambivalence about the role of the state; its troubled relationship with capitalism, the forces of modernization, and its Jews; its attempts to balance profit and cultural-political goals; and finally the fact that it had to redefine itself while situated in the tumultuous center of Europe – all were projected upon the screen of the film industry. Hungarian audiences' attraction to Hollywood's fantasies and the industry's preference for the West were counter-balanced by the industry's orientation first toward Central Europe, then specifically the Balkans, and the fact that throughout the early sound era Hungary was a satellite orbiting Nazi Germany. The country's development of a flourishing cinema industry, film culture, and robust exports helped torpedo its own attempts to formulate a distinctive national style reflective of an immutable national identity. Little Hungary's national cinema – its industry, output, culture, and elites – was stretched, tugged, and torn by internal and external currents. It ran into walls, at almost every turn, many of which it constructed itself, others built by its Nazi neighbors. Still Hungarian filmmakers, many of Jewish background, managed for a time to attain a degree of quantitative, and in some ways qualitative, success not matched since.[59] *Jews, Nazis, and the Cinema of Hungary* explains how this morass of clash, competition, cultural plasticity, and ideological dissonance came to pass.[60]

Notes

1. Roth, *Flight without End* (New York, 2003 [1927]), 30.
2. Székely's arrival is recounted in M. Cenner, et al. (eds), *Kabos Gyula 1887–1941* (Budapest, 1987), 199.
3. The film starred Gyula Kabos, Pál Jávor, Éva Fenyvessy, Gyula Csortos, Gyula Gózon, Mici Haraszti, and Mici Erdélyi.
4. Paul Lendvai, *The Hungarians: A Thousand Years of Victory in Defeat,* trans. A. Major (Princeton, NJ, 2003), 286.
5. Law XLIV of 1868, translated in Paul Hanebrink, *In Defense of Christian Hungary* (Ithaca, NY, 2006), 10–11.
6. Lendvai, *The Hungarians*, 331.
7. Ibid.
8. László Kontler, *A History of Hungary* (New York, 2002), 306.
9. On this incredible cohort, see T. Frank, *Double Exile: Migrations of Jewish-Hungarian Professionals through Germany to the United States, 1919–1945* (Oxford, UK, 2009) and K. Marton, *The Great Escape. Nine Jews Who Fled Hitler and Changed the World* (New York, 2006). I use the term assimilate to include those of Jewish origin who adopted a majority culture but maintained Jewish identity and/or practice, and those who converted to Christianity.
10. Hungary signed the Treaty of Trianon on 4 June 1920. Hungarian territory, some 282,000 square kilometers in 1914, was reduced to 93,000 square kilometers. With its population loss, Hungary became one of the most ethnically homogenous states in Europe, with nearly 90 percent of its population defined as Magyar. Hungary's economic infrastructure was crippled by the war and its settlement. Hungary lost 89 percent of its iron production, 84 percent of its forests, 62 percent of its rail network, and 44 percent of its food processing industries. See Tibor Hajdú & Zsuzsa L. Nagy, 'Revolution, Counterrevolution, Consolidation', in P. Sugar, et al. (eds), *A History of Hungary* (Bloomington, IN, 1994), 314 and Ignác Romsics, *Magyar Sorsfordulók* (Budapest, 2012), 40–3. Hungary was strapped with a greater per capita foreign debt than any other East Central European state in 1930. C.A. Macartney, *October Fifteenth: A History of Modern Hungary, 1929–44,* vol. I (Edinburgh, 1957), 91.
11. On 'conservative liberalism' see A. C. Janos, *The Politics of Backwardness in Hungary 1825–1945* (Princeton, NJ, 1982), Chapters 3–6; and András Gerő, *Modern Hungarian Society in the Making* (Budapest, 1995), Chapters 7–10.
12. On race, studies of Central European eugenics make this point. See Maria Bucur, *Eugenics and Modernisation in Interwar Romania* (Pittsburgh, PA, 2002); Marius Turda & Paul Weindling (eds), *'Blood and Homeland': Eugenics and Racial Nationalism in Central and Southeast Europe* (Budapest, 2007); and Tudor Georgescu, 'Ethnic minorities and the eugenic promise: the Transylvanian Saxon experiment with national renewal in inter-war Romania',

European Review of History/Revue europeenne d'histoire, 17/6 (December 2010), 861–80.

13. T. Lorman, *Counter-Revolutionary Hungary, 1920–25: István Bethlen and the Politics of Consolidation* (Boulder, CO, 2006).
14. Lendvai, *The Hungarians*, 386. Also Lorman, who shows, in addition to voting limits, the government party resorted to electoral chicanery to thwart the democratic process. Lorman, *Counter-Revolutionary Hungary*, 134–6.
15. Prime Minister István Bethlen, quoted in C. Gati, 'The Populist Current in Hungarian Politics, 1935–44', (Indiana University: Unpublished Ph.D. Dissertation, 1965), 14.
16. Eric Hobsbawm, *Nations and Nationalism since 1780. Programme, Myth, Reality*, 2nd edition (Cambridge, UK, 1993), 131–62. István Bibó wrote of this phenomenon as endemic to Central Europe. '...in these countries, "culture" has been a factor of enormous political significance. However, this does not mean as much the flowering of culture, as its politicization. As these countries did not exist in the Western European sense of unbroken historical continuity, it became the task of the national intelligentsia to discover and nurse the distinctive and separate...new or reborn nation and to justify the actual truth, namely that these new popular frameworks...were deeply rooted and alive' István Bibó, 'A kelet-európai kisállamok nyomorúsága', in Bibó, *Valogatott tanulmányok*, vol. II (Budapest, 1986), 223. See also Keith Hitchins, 'Interwar Southeastern Europe Confronts the West', *Angelaki* 15/3 (December 2010), 9–26, esp. 20.
17. Victoria de Grazia, 'Mass Culture and Sovereignty: The American Challenge to European Cinemas, 1920–1960', *Journal of Modern History* 61/1 (March 1989), 61. De Grazia, writing generally, not of Hungary specifically, calls the national 'practically impossible to define'. Jim Leach warns, however: 'It is thus tempting to dismiss "national character" as a myth that obscures the diversity and contradictions....While this may [be] true, however, the myth can have powerful effects on filmmakers, on government policy, and on the response of spectators to the films'. Leach, *British Film* (Cambridge, UK, 2004), 5.
18. Miklós Kozma, 'A legélesebb magyar fegyverek egyike a magyar rádió', *Cikkek, nyilatkozatok, 1921–1939* (Budapest, 1939), 115.
19. Anders Linde-Laursen, 'Taking the National Family to the Movies: Changing Frameworks for the Formation of Danish Identity, 1930–90', *Anthropological Quarterly* 72/1 (January 1999), 18.
20. This Andrew Higson quote is the premise of nearly every text with the words 'national cinema' in the title produced from the late 1980s through the present. Higson, *Waving the Flag: Constructing a National Cinema in Britain* (Oxford, UK, 1995), 8.
21. *MOL* – K26, ME, 1919, XXXVII t., 6311 M.E. sz., 4 December 1919. This Trade Ministry memorandum to the prime minister endorsed the Republic of Councils' concept of a national film propaganda ministry, even after the fall

of the Republic. The Trade Minister argued a 'Hungarian national film factory' was necessary to preserve the nation's purity and honor. 'Purity', Andrew Lass tells us, is often an expression of desire for a 'successfully integrated population' in the absence of integration. Lass, 'Aesthetics of Distinction in Czechoslovakia', in I. Banac and K. Verdery (eds), *National Character and National Ideology in Interwar Eastern Europe* (New Haven, CT, 1995), 62.

22. Lee Grieveson, 'On governmentality and screens', *Screen* 50/1 (Spring 2009), 180–7, esp. 186–7.
23. The Hungarian terms are *nemzeti filmipar* and *keresztény nemzeti filmgyártás.*
24. Zsolt Kézdi-Kovács, 'Film als kultureller Integrationsfaktor in Mitteleuropa', in Helmut Pflügl (ed), *Mein 20. Jahrhundert: Der Traum vom Glück/Az Én XX. Századom: Visszaszámlálás* (Wien, 1999), 17. See also Daniel Biltereyst, Richard Maltby, & Philippe Meers, 'Cinema, Audiences, and Modernity: An Introduction', in *Cinema, Audiences and Modernity: New perspectives on European cinema history* (New York, 2012), 9; and Anna Manchin, Anna Manchin, 'Imagining modern Hungary through film: debates on national identity, modernity and cinema in early twentieth-century Hungary', in ibid., 64–80.
25. The authors of works on Central European film have taken longer to accept this judgment. Works such as Marek Haltof's *Polish National Cinema* continue to assert the existence of 'characteristic features and elements, recognized locally and internationally as distinctively Polish – what one might call a recognizable "national accent"'. Even the collection *Cinema of Central Europe,* despite its title and arguments about shared aspects of culture, insists that 'there is little case for regarding the four cinemas [Hungary, Poland, Czech, & Slovak] as constituting a 'regional cinema' or for identifying a common Central European or East Central European cinema. It is much more logical to regard them as 'national' cinemas which…enjoy a common geographical space and shared histories' See Haltof, *Polish National Cinema* (NY, 2002), vii; Peter Hames (ed), *The Cinema of Central Europe* (London, 2004), 6.
26. The works of Brubaker, Anderson, Duara, Gellner, Hroch, Bhabha, Billig, Brass, Gentile, and Hobsbawm & Ranger are among the essential texts.
27. Examples of this scholarship include Jeremy King's *Budweisers into Czechs and Germans* (Princeton, NJ, 2002) and 'The Nationalization of East Central Europe: Ethnicism, Ethnicity, and Beyond', in M. Bucur and N. Wingfield (eds), *Staging the Past: The Politics of Commemoration in Habsburg Central* (Purdue, IN, 2001), 112–52; Pamela Ballinger, *History in Exile: Memory and Identity at the Borders of the Balkans* (Princeton, NJ, 2003); James Bjork, *Neither German nor Pole: Catholicism and National Indifference in a Central European Borderland* (Ann Arbor, 2008); Paul Hanebrink's *In Defense of Christian Hungary* (Ithaca, NY, 2006); Pieter Judson's *Guardians of the Nation* (Cambridge, MA, 2006); Holly Case's *Between States* (Stanford, CA, 2009); Cynthia Paces' *Prague Panoramas* (Pittsburgh, PA, 2009), Tara

Zahra's *Kidnapped Souls* (Ithaca, NY, 2008) and 'Imagined Noncommunities: National Indifference as a Category of Analysis', *Slavic Review* 69/1 (Spring 2010), 93–119; Nancy Wingfield's *Flag Wars and Stone Saints* (Cambridge, MA, 2007); Chad Bryant's *Prague in Black* (Cambridge, MA, 2007); the 'Sites of Indifference' articles in *Austrian History Yearbook* 43 (2012), 21–137; and Randall Halle's *The Europeanization of Cinema. Interzones and Imaginative Communities* (Urbana, IL, 2014).

28. The scholarship on national cinema and theory behind it exploded in the 1990s, with regional, transnational and global studies moving to the fore in the 2000s. Comprehensive surveys of French, Australian, British, South African, Italian, Canadian, Mexican, Spanish, German, 'Nordic', 'Balkan', Chinese, Polish, Austrian, Indian, Iranian, and many other national cinemas, a boom attributable primarily to Routledge, Wallflower, Berghahn, I.B.Tauris, Texas, and a handful of other publishers. Work on national cinema now includes every continent and a real range of methodologies. The main divisions are between those in film studies who favor textual analyses and highly theoretical approaches tied to semiotics and to those who favor a more historical methodology and analyses of political and economic structures and institutions. Selections from the literature and theory of national cinema include Chris Berry, 'From National Cinema to Cinema and the National', in V. Vitali and P. Willemen (eds), *Theorising National Cinema* (London, 2006), 148–57; Stephen Crofts, 'Concepts of national cinema', in J. Hill and P. C. Gibson (eds), *The Oxford Guide to Film Studies* (Oxford, UK, 1998), 385–94; John Hill, 'The Issue of National Cinema and British Film Production', in D. Petrie (ed), *New Questions of British Cinema* (London, 1992), 10–21; Tom O'Regan, *Australian National Cinema* (London, 1996); Alan Williams (ed), *Film and Nationalism* (New Brunswick, NJ, 2002), 1–88.
29. E.g. Ian Jarvie, 'National Cinema: A Theoretical Assessment', in M Hjort & S. MacKenzie (eds), *Cinema & Nation* (London, 2000), 76; Philip Rosen, 'History, Textuality, Nation: Kracauer, Burch and Some Problems in the Study of National Cinemas', *Iris* 2/2 (1984), 71; Mette Hjort & Scott MacKenzie, 'Introduction', in Hjort & MacKenzie (eds), *Cinema & Nation*, 4; John Orr, 'The Art of National Identity: Peter Greenaway and Derek Jarman', in J. Ashby & A. Higson (eds), *British Cinema, Past and Present* (London, 2000), 327; Stephen Crofts, 'Concepts of national cinema', 385.
30. Brian Porter-Szűcs, 'Beyond the Study of Nationalism', in K. Jaskułowski & T. Kamusella (eds), *Nationalism Today*, (Oxford, UK, 2009), 4; Berry, 'From National Cinema to Cinema and the National', 153; Leach, *British Film*, 5–6.
31. I will use the concept of national character (a.k.a. essence, soul, spirit, or nature) as developed by Katherine Verdery and contemporary Central European thinkers. The notion of the national character is that it:
    (1) Provides the mechanism of linking the individual to the collective by providing specific characteristics that distinguish the national group.

(2) Makes the claim that its characteristics are either heritable/genetic, and if not, passed on experientially. In either case, advocates assume them to be natural.
(3) Produces a spirit or feeling, some internal sense, of attraction and belonging to a particular group.

See Verdery, 'Introduction' and 'National Ideology and National Character in Interwar Romania', in I. Banac and K. Verdery (eds), *National Character and National Ideology in Interwar Eastern Europe* (New Haven, CT, 1995), esp. xvii–xix, 84–98. Also Peter Mandler, *The English National Character: The History of an Idea from Edmund Burke to Tony Blair* (New Haven, CT, 2006).

32. Rogers Brubaker, *Ethnicity without Groups* (Cambridge, MA, 2004), 1–87, esp. 11–18; and Brubaker, *Nationalism Reframed. Nationhood and the National Question in the New Europe* (Cambridge, UK, 1996), esp. 21. See also V. Vitali and P. Willemen, 'Introduction', in Vitali and Willemen (eds), *Theorising National Cinema* (London, 2006), 6–7; and Rick Altman, *Cinema/Genre* (London, 1999), 198.
33. In this manner, my work is different from existing scholarship on the interwar and wartime Hungarian film industry, which focuses primarily on films, not the industry itself, and largely accepts the idea of a territorially-bounded Hungarian cinema industry. See endnotes 52–4.
34. Nationalists and antisemites insinuated 'cosmopolitanism' rejected nationalism. Jewish film elites virtually unanimously disagreed, arguing no contradiction existed between their Hungarian nationalism and a Central European cosmopolitanism that accepted toleration, assimilation, and a multi-ethnic national community.
35. Hungary's gross national product, for example, declined by 55 percent between 1929 and 1932. A. C. Janos, *East Central Europe in the Modern World. The Politics of the Borderlands from Pre- to Postcommunism* (Stanford, CA, 2000), 137.
36. 'Culture of étatism' from Janos, *The Politics of Backwardness*, 313ff.
37. *BFL*– Nb. docs re: trial of László Balogh, Nb. 1699/1945, 663. István Balogh, quoted in Jgykv. felvétetett az ONFB 1943. December 10-iki üléséről', 5. Demographically, Hungary was very top-heavy. It had only one metropolis, Budapest, which accounted for more than one-fifth of the entire Hungarian population.
38. Throughout the work, I will use the phrase 'Jewish question', a direct translation of the Hungarian word *zsidókérdés*. As it was a commonly used term, I will not place it in quotation marks. The definition of the term 'Jewish' was inconsistent, ascriptive and evolved over time, although it always presumed Jews act as Jews, not as Hungarians. Because most contemporary Jews rejected the idea that they belonged to a separate race/ nation, I will use a lower case 's' when referring to antisemitism.

39. During the Dual Monarchy period, less than half of the residents of the Hungarian portion of the Empire were ethnically Hungarian. The Hungarian ruling elite thus welcomed the relatively recently emancipated Jews as citizens. The Jews reciprocated, assimilating (which sometimes meant converting to Christianity) and contributing vast intellectual and financial resources to the Hungarian nation-building efforts of the late 1800s and early 1900s. When one includes the Jews, the ethnic Hungarian population of Greater Hungary increased to slightly over 50 percent, providing the ruling Hungarians a numerical justification for their authority.
40. Between 1920 and 1935, Jews averaged 9.9 percent of all university students. In 1920, there were approximately 450,000 Jews in Hungary. As of the 1930 census, Jewish Hungarians constituted 5.1 percent of the population. Mária M. Kovács, *Liberal Professions and Illiberal Politics* (Washington, DC, 1994), 64; Randolph L. Braham, *The Politics of Genocide. The Holocaust in Hungary,* vol. 1 (Boulder, CO, 1994), 80.
41. Even in the late-1930s, after years of attempting to squeeze Jewish members from the professions, Jewish Hungarians still constituted large numbers of Hungary's liberal professionals. For example, 38 percent of all architects working in building construction in 1937 were Jewish. In 1936, 73 percent of Hungary's professional journalists were Jewish. Nearly 44 percent of all doctors in 1940 Hungary were Jews or converts from Judaism. Over 50 percent of all lawyers were Jews or converts from Judaism in 1939. Jews also played a 'seminal role' in the Hungarian performing arts. Statistics from A. Kovács, *A zsidóság térfoglalása Magyarországon* (Budapest, 1922); R. Patai, *The Jews of Hungary. History, Culture, Psychology* (Detroit, MI, 1996), 519–30 and Kovács, *Liberal Professions and Illiberal Politics,* 63. For 1930 statistics on the percentages of Jews in specific professions, see Braham, *The Politics of Genocide,* 80–1.
42. The question of the centrality of Jews to Hungarian modernization and the development of the Hungarian middle-class has become a major theme of Hungarian historiography. See, e.g., the works of Viktor Karády and Gábor Gyáni and their debates in *Budapesti Könyvszemle* in the late 1990s.
43. Patai, *The Jews of Hungary,* 520, 439. In 1910, Jews owned no less than 85 percent of Hungary's financial institutions, and there is little reason to believe this number changed substantially before 1938.
44. I consider the problem of defining Jews in Chapter 4.
45. On disenfranchized minorities producing 'national' cultures, see Bhabha, *Nation & Narration,* 303–19.
46. The term 'transnational' refers to horizontal exchange, where sites of production transcend borders and co-financing or cross-border capital flows occur, or where other filmmaking processes are internationalized.

47. Gary B. Cohen, 'Preface', in N. Wingfield (ed), *Creating the Other: Ethnic Conflict and Nationalism in Habsburg Central Europe* (New York, 2004), vii.
48. 'Minor cinema' is Mette Hjort's descriptor for Danish film, but is fitting in the Hungarian context. Mette Hjort, *Small Nation, Global Cinema. The New Danish Cinema* (Minneapolis, MN, 2005), ix–x.
49. Excellent works on 'Europeanness' includes Hitchins; 'Interwar Southeastern Europe…', 9–26; Andrea Orzoff, *Battle for the Castle: The Myth of Czechoslovakia in Europe, 1914–1948* (Oxford, UK, 2009); and Case, *Between States*. In film studies, scholars have taken to the term 'glocalization' to describe this process of defining a distinct small-scale identity for consumption in international markets.
50. No less than Thomas Elsaesser, a doyen of film studies, is representative of this approach, claiming that in the 1920s, '…Hollywood became the not only the dominant force; it was very successful in dividing the Europeans among themselves. For a brief period in the in the late 1920s, it seemed the Russians might be Europe's inspiration. Instead, from 1935 onwards, it was Nazi cinema that dominated the continent until 1945'. Elsaesser, 'European Cinema. Conditions of Impossibility?' in *European Cinema. Face to Face with Hollywood* (Amsterdam, 2007), 13. For Hungary, see e.g. Károly Nemes, *Miért jók a magyar filmek?* (Budapest, 1968), 31; Beverly James, 'Hungary', in G. Kindem (ed), *The International Movie Industry* (Carbondale, IL, 2000), 172.
51. 'Extreme insignificance' is how Ulrich Gregor and Enno Patalas describe Hungarian wartime film in *Geschichte des Films* (Gütersloh, Germany, 1962), 287.
52. See studies of interwar or wartime British, Polish, Nazi, and Fascist cinemas. Higson, *Waving the Flag*; Jeffery Richards, *Films and British National Identity* (Manchester, UK, 1997) and his edited text *The Unknown 1930s: An Alternative History of the British Cinema* (London, 1998). For Poland, Sheila Skaff, *The Law of the Looking Glass: Cinema in Poland 1896–1939* (Athens, OH, 2008). For Germany and Italy, the best English language examples are Sabine Hake's *Popular Cinema of the Third Reich* (Austin, TX, 2002) and Steven Ricci, *Cinema and Fascism: Italian Film and Society, 1922–1943* (Berkeley, CA, 2008).
53. Exceptions to this rule include the works of Zsolt Kőháti on the silent period, including, *Tovamozduló ember, tovamozduló világban. A magyar némafilm 1896–1931 között* (Budapest, 1996) and his series 'Magyar film hangot keres', *Filmspirál* II/2-1 (1996), 67–131 and II/3-2 (1996), 1–52; the works of Tibor Sándor on the Jewish question, particularly *Örségváltás* (Budapest, 1994) and *Őrségváltás után: zsidókérdés és filmpolitika, 1938–44* (Budapest, 1997); the occasional paper or journal article produced through the Hungarian National Film Archive or the ELTE film journal *Metropolis*; Gy. Balogh and J. Király's, *'Czak egy nap a világ…' A magyar film műfaj- és stílustörténete, 1929–1936* (Budapest, 2000), and Anna Manchin's multiple works.

54. Andor Lajta, the longtime editor of the premier and official contemporary film trade magazine, *Filmkultúra,* and the annual *Filmművészeti Évkönyv,* published only one book, *A tizéves magyar hangosfilm 1931–1941* (Budapest, 1942). His manuscript on the sound period, is far more comprehensive, but was never published. 'A magyar film története. V. A magyar hangosfilm korszak első 16 éve, 1929–44', (Budapest: Theater and Film Studies Institute unpublished manuscript, 1958). Sadly, István Langer's 'Fejezetek a filmgyár történetéből, I–II.rész, 1919–48', (Budapest: Hungarian Film Institute unpublished manuscript, 1980) is also only available to scholars who visit the Hungarian Film Archive.
55. Nemeskürty, a prolific writer, editor, and former head of MAFILM, is the acknowledged authority on interwar Hungarian film. He sometimes is the only source referenced in English language texts on Hungary. This is understandable, as prior to Cunningham's work in 2004, Nemeskürty's *Word and Image: History of the Hungarian Cinema,* trans. Z. Horn and F. Macnicol, 2nd edn, (Budapest, 1974), was the only Hungarian text covering the 1929–44 period in English. Examples of the Nemeskürty-reliant literature include B. Burns, *World Cinema: Hungary* (Trowbridge, UK, 1996), 1–35; M. J. Stoil, *Cinema beyond the Danube: The Camera and Politics* (Metuchen, NJ, 1974), 1–58. Instead of merely recapitulating Nemeskürty's writings, some works actually include segments written by Nemeskürty himself. See J-L. Passek (ed), *Le cinéma hongrois* (Paris, 1979).
56. Those histories that do focus on culture tend either to be limited to newspapers and traditional high culture – such as music, literature, visual arts, and theater – or they concentrate on folk culture.
57. Susan Hayward, 'Framing National Cinemas', in Hjort & MacKenzie, (eds), *Cinema & Nation,* 89.
58. The phrase 'national identity of excuse' belongs to György Csepeli, in *National Identity in Contemporary Hungary,* trans. M. D. Fenyo (Boulder, CO, 1997), 91.
59. The films of this period have experienced a nostalgic renaissance. Many are available on DVD or viewable on one of several Hungarian cable channels. How to interpret this nostalgia goes beyond the reach of this text.
60. The term 'ideological dissonance' is Kevin Johnson's. See Johnson, 'Kulturelle (nicht-)Vermischung. Nation, Ort und Identität in tschechisch-deutschen Mehrsprachenversionen', in J. Roschlau (ed), *Zwischen Barrandov und Babelsberg. Deutsch-tschechische Filmbeziehungen im 20. Jahrhundert* (Hamburg, Germany, 2008), 78.

1

When Silence became Loud: The Silent Era and the Origins of Sound

Introduction

Economically enfeebled and still smarting from what its citizens universally regarded as the criminal truncation of their lands by the 1920 Treaty of Trianon, the traumatized and suddenly ethnically homogeneous state of Hungary had great incentives to redefine its culture. When sound came to the silver screen in the late 1920s, Hungary's political and film establishments quickly determined that a vibrant film culture – consisting of self-sufficient film production and consumption – could do just that. It could be the primary means of imagining, preserving, and highlighting the intrinsic value and national character of Hungary, its people, its 'civilization', its modernity, and its place in Europe. Not only would motion pictures advance Hungarian nationalism, they would also bring employment, cross-border trade, and the promotion of social stability, all important national objectives.

This chapter will explore how Hungary's political and cultural elites rose to the challenge of creating a viable sound movie culture in one of modern history's most unpromising epochs. Conditions in Hungary were particularly inauspicious. The private studios of the silent age had failed, and in an era of larger and costlier productions, Hungarian elites believed that only the state could marshal the capital necessary for international-quality sound filmmaking. The state was loath to accept this role, however, lacking the requisite funds. Government officials took a 'you-first' approach, hedging their bets until they were sure that private financiers, possibly foreign, possibly domestic, were willing to invest in sound film production. Putting aside this slow start and their residual fears about the mass nature of film, by the early 1930s, few cultural elites questioned whether sound film should be the pedestal upon which the national culture rested; neither did they doubt that the state must construct the broad scaffold supporting the industry.

The prominence of the state created space for tension, contradiction, and moderation that would both plague and protect Hungarian cinema through the mid-1940s. Partial state control of Hungary's only sound studios made the vertical integration of private producers, distributors, and theaters that came to be the cornerstone of Hollywood's success impossible in Hungary, except through complete nationalization. Hungary's governing oligarchy refused to consider this option, tarred as it was by association with Hungary's 1919 Bolshevik Republic of Councils. Instead, the Horthy-era film establishment opted for a complicated arrangement featuring sporadic state interventions in a relatively free-market system. The results of this compromise were mixed, fostering a system that, while resilient and answerable to public desires, also became inbred and corporatist, rife with potential and actual corruption. The main state institutions' officials who created the film industry's legal structures and enforced its regulatory schemes held major stakes in Hungary's only two film studios and dominated their boards of directors. As government agents, these directors were bound to enact state priorities which sometimes ranked profit below other goals, such as protection from market forces and nationalist or antisemitic imperatives. As stakeholders in the studios, their decisions clearly affected the potential dividend they would receive. In an oligopoly such as interwar Hungary, these conflicts of interest, this hubris of bureaucracy, was not unique to the film industry. However, they significantly influenced the ways members of Hungary's film establishment contemplated the idea of a national film industry, thus limiting the industry's development in the 1930s. These institutional and legal strictures, the personalities behind them, and the cleavages they created are some of the issues explored in this chapter.

Hungary's mixed measures did manage to provide the country with sound film production capability by late 1930, a capacity not widely shared by contemporary European states east of Germany and west of the Soviet Union.[1] But looking inward provides an incomplete picture. As the Depression-stoked nationalist winds blowing across Europe whistled through Hungary, they brought human and financial capital across its borders, capital that was critical in the initial establishment of Hungarian sound film production. International and transnational developments were vital to the promotion of Hungary's early national sound film culture. Both interest from American and French investors attempting to devise new strategies for surviving the transition to sound and later the Depression, and the expulsion of Hungarian filmmakers from Germany radically impacted Hungary's changeover to the new audio-visual medium.

The Silent Era

When the first moving picture arrived in Europe's fastest-growing metropolis in 1896, Budapest was consumed with the millennial celebration of Hungarian settlement in the Danubian basin. Thus, the debut of Edison's kinetoscope was a relatively small part of a larger triumphal event. Film exhibitions had begun just months prior, and the first Hungarian attempt to produce a motion picture coincided with the June 1896 festivities. Arnold and Zsigmond Szikaly, using newly acquired Lumière equipment, filmed Emperor Franz Joseph as he strolled through the *Városliget*, Budapest's renowned park.[2] The brothers' dream was grandiose: charge customers to see the Emperor if they did not witness his visit in person. Their technique, however, left much to be desired. Lacking familiarity with the new equipment, they frequently decapitated the Emperor. Audiences, particularly the *Kaisertreu* rising middle classes, found this gaffe appalling. The Szikaly experiment was a disaster.

In many ways, this anecdote is a wonderful metaphor for the early history of Hungarian silent and sound film. Grand expectations coupled with awful execution proved a minor obstacle in the long-term development of a durable film industry. Undeterred by the Szikaly failure, the population of Budapest embraced the motion picture with alacrity. Budapest's rapidly growing ranks of cosmopolitan middle and upper classes and their coffeehouse culture proved perfectly suited for film.[3] By the turn of the century, hundreds of Budapest's coffeehouses had already become occasional movie theaters.[4] By 1914 there were around 110 permanent cinemas in Budapest alone, giving it one of the highest per capita rates of any European capital.[5] At the outbreak of World War I, Budapest already had such a mature cinema-going culture that Greater Hungary[6] represented three to four percent of the world's film consumption.[7] France's Pathé, Denmark's Nordisk, and several German and Austrian film firms established offices in Budapest between 1907 and 1915. According to a 1920s account written by a proud Hungarian, only Denmark, the United States, France, Italy, and Germany had similarly advanced film cultures.[8]

In terms of silent film production, Hungary made incredible gains between the 1912 release of *Sisters* [*Nővérek*], the country's first silent narrative feature film, and the end of the Great War. 'I can see amazing things, possibilities in film which the narrow confines of the stage can never encompass', wrote Mihály Kertész, the filmmaker whom we know better as Michael Curtiz.[9] He was not the only one to express such sentiments. He and his compatriots were engaged in a significant number of film projects when hostilities broke out, and the diminishing supplies of foreign film resulting from the fighting propelled their work even further. During the first year of the war, a hodge-podge of photographers and journalists-turned-filmmakers teamed up with stage actors-turned-film-actors to make 18 feature length films or narrative 'film scenarios'.[10] Wartime annual production figures rose steadily, and by the end of 1918, Hungarian filmmakers working in over two dozen studios in Budapest and Kolozsvár, Transylvania, had released well over 250 motion pictures.[11] In the 1917–18 film year, only the United States and Denmark made more movies than Hungary. The director Alfréd Deésy alone made 31 films between 1917 and 1918.[12]

In 1918, during the period described by contemporaries and film historians alike as the first golden age of Hungarian film, Hungary counted Alexander [Sándor] Korda and Michael Curtiz among its premier directors, Vilma Bánky, Lia Putty, Lili Berky, Paul [Pál] Lukács, Béla Lugosi, and Victor Varconi [Mihály Várkonyi] among its lead actors, and Béla Balázs among its young theorists.[13] Korda himself became a star, editing the journal *Cinema Weekly* and adopting the persona of Hungary's first cinema mogul. A Jewish-born country boy who learned his craft in Paris, Korda married one of his studio's starlets, Mária Farkas, developed an affectation for cigars, and moved to the Hotel Royal on Budapest's Ring Road. Living well beyond his means, Korda projected the image of a great man, adding glamour to the new medium. Yet there was more to Hungary's burgeoning cinema culture than Korda's charisma. Budapest was home to more cinema seats per capita than London, Paris, or Rome. Hungarian movie-goers could choose between 17 journals exclusively devoted to film.[14] Hungary's top *literati,* including Zsigmond Móricz and Frigyes Karinthy, heartily endorsed the new art form, giving it a high-brow cachet it did not enjoy in other parts of Europe where debates over film's primary functions as entertainment, art, education, or propaganda were far more rancorous.[15] Film rapidly became one of the most important, if not the most influential, forms of mass culture in Hungary during the first two decades of the twentieth century. By 1917, Hungary's leading film studio, the Star Film Factory and Production Company, had opened branches in Berlin, Vienna, and Zurich, all part of a plan to aggressively market Hungarian fare throughout Central Europe.[16] Hungary had become 'one of the cradles of European film'.[17]

Figure 1: Michael Curtiz (Mihály Kertész). Public domain.

Even the destruction of Austria-Hungary in 1918 did not completely ruin the fortunes of Hungarian silent films. Film production continued during the short interlude of the Mihály Károlyi-led liberal republic. In fact, when Károlyi's government collapsed and Hungary experienced the world's second successful Communist revolution in March 1919, the film industry flourished. Nationalized by the Béla Kun-led Republic of Councils and coordinated nominally by the György Lukács-run Commissariat of Public Education, Hungarian studios produced an incredible 31 films in the four months the Republic existed.[18] This zenith occurred despite the fact that Hungary remained blockaded by the Allies and suffered invasions by its Czech, Serb, and Romanian neighbors. Among the made-in-Hungary features were foreign-directed works meant for distribution in Italy and Germany, as well as a number of Hungarian films rushed by plane to Bolshevik Russia.[19]

The significance of the relationship between state and cinema cannot be overstated, even though the promise of a Communist transformation of the film industry soon faded.[20] When the Red Army failed to rescue Hungary's Bolsheviks, the collapse of the Republic of Councils soon followed. In early August 1919, Kun and his comrades fled for Austria, leaving the army-less Social Democrat Gyula Peidl little chance against the counter-revolutionary forces of Admiral Miklós Horthy. The combination of Horthy's ascension to power as Regent in March 1920 and the signing of the Treaty of Trianon by his government in June had a devastating effect on the film industry. After churning out at least 220 films in the 30 months between 1917 and the collapse of the Republic of Councils, the industry began a free fall.[21] Some of Hungary's oldest and most successful studios, including the Star, Corvin, Astra, and Uher companies, were 'denationalized', reorganized, and then permitted to re-start production. But there was no sudden burst, no return to the levels of production seen in 1918 and half of 1919. Capital became scarce and the White Terror, Hungary's authoritarian and avowedly antisemitic counter-revolution carried out by Horthy's allies, wreaked havoc upon the motion picture business. Many of Hungary's top filmmakers, tainted in the eyes of the new conservative ruling elite by their connections with the political left or accident of birth, were ushered out of the country, if not arrested and tortured. Horthyites interned Alexander Korda, the Jewish, intellectual, cigar-chomping symbol of Hungary's cosmopolitan film industry, and threatened to 'beat the shit out of that Communist kike...'[22] Later released unharmed, he took this scare seriously and soon followed his compatriot Michael Curtiz, who had fled the country in 1919. Other prominent film professionals, Jews in particular, were not so lucky. The director Sándor Pallós, for example, was beaten to death by Horthy's minions.[23] Waves joined Curtiz and Korda, spilling out of Hungary and heading to London, Paris, Berlin, New York, and Hollywood. This direct linkage of culture and politics caused a bleeding of talent that mortally wounded the country's silent film production capacity. By 1923, the majority of Hungary's wartime film professionals had emigrated and the industry was populated by 'mediocrities, novices, or, at best, competent journeymen'.[24]

Fueled by an integral nationalist spirit and antisemitic zeal, the Horthy government reached further into the infrastructure of film production, distribution, and exhibition. It confiscated cinema licenses from Jewish owners and used them as bonuses for political comrades, completely destabilizing the business by shunting aside trained managers and sources of capital in exchange for cash-poor, greenhorn theater owners. Producers became more hesitant to invest, unsure not only about the filmmakers they hired, but also concerned that their products would be left to bungling theater owners who did not know how to market movies or manage crowds. Producers were further hampered by loss of access to hundreds of theaters which now resided in foreign rather than Hungarian lands as a result of the Great War border alterations. A new Hungarian censorship regime may have also contributed to the movie malaise, although the creation of the National Censorship Committee [*Országos Mozgóképvizsgáló Bizottság* – hereafter OMB], which simply signed off on scripts pre-production, was not a major factor in the downturn.[25] Populated by film neophytes, the OMB was more a reminder to self-censor than a heavy-handed apparatus. Still, when added to the lack of filmmaking and exhibition talent, and the loss of markets, the results were devastating.

As the 1920s unfolded, Hungarian film production rapidly withered. Hungary's last operating studio, the vaunted Star, closed its doors in 1929.[26] Distributors relied on imports from Germany, the United States, France, Denmark, England, and Italy as domestic production evaporated.[27] Even these choices narrowed, as much of Europe experienced the same production crisis as Hungary. Finland's only remaining filmmaking firm went bankrupt in 1929, while Norwegian feature production fizzled to one by 1931.[28] British movie production, which had nose-dived in the early 1920s, remained in distress in the late 1920s.[29] Italian output collapsed as international markets disintegrated, dropping from 38 films released in 1928 to two in 1931.[30]

'In the history of Hungarian cinematography 1929 will have been perhaps the darkest and saddest year', wrote Andor Lajta, the editor of Hungary's *Film Arts Almanac.*[31] On the eve of the Great Depression and the dawning of the sound era in Hungary, the ruination of the domestic film industry was complete, and the prospects for salvation were poor. Hungary was an impoverished, small, and weak Central European state. It was culturally constrained, or less generously, 'linguistically challenged', owing to the fact that its approximately 8.7 million inhabitants spoke a tongue understood by few outsiders.[32] The lifeblood of its film culture was entirely imported, and few beyond Hungary's urban middle classes even had the means to attend exhibitions regularly. Hungary had little to give the motion picture world beyond the talent it had long before expelled and the few films it had already offered. As the silent age came to a close, Hungarian film culture was virtually inert.

The Transition to Sound

After the premier of the world's first sound film, *The Jazz Singer,* Frances Goldwyn, spouse of MGM founder Sam Goldwyn, allegedly touted the opening as 'the most important event in cultural history since Martin Luther nailed his theses on the church door'.[33] Hyperbole or not, there is no denying that the advent of sound had a massive impact on the nature of the film medium, the film trade, and representations of culture and identity. Following its 1926 American debut, 'the talkie' was introduced throughout Europe between 1927 and 1929. Cultural critics the world over both decried the change and wrote of its enormous potential to revolutionize film as an art form.[34] Producers, audiences, and governments all were forced to rethink their relationships to the medium. In essence, sound created the modern culture industries, mixing new technologies, big business and high finance, and various components of the arts to a greater extent than silent pictures had required. Sound movies were more difficult to craft than their silent predecessors and significantly more costly. Cinematography had to be more precisely choreographed, actors and actresses could no longer freely *ad-lib*, and directors could not scream instructions during filming. Talkies required more equipment, substantial sets and studios, and larger, better trained crews. Filmmaking became a profession. Increased costs forced filmmakers to exercise more stringent self-censorship, as the financial risks associated with a film that did not win the censor's approval grew exponentially. Complexity and expense drastically reduced the number of films produced, increasing the market value of each movie.

Sound also revolutionized exhibition. Sound projection required a substantial capital investment in new equipment, and many small theaters all over Europe – nickelodeons, pubs, and even town or village cinemas – hastily folded. The average size of theaters rose, and new construction projects were large, not small. With fewer films available and distribution costs skyrocketing, regulation increased and business practices became more formalized. Running a theater could no longer be a mere hobby. As cinema's status evolved, so did movie audience demographics. These changes spurred exhibitors to organize in order that their concerns not be ignored by authorities accustomed to listening only to producers and distributors.

Even more importantly, sound raised expectations that film would become a national medium. Some, particularly Soviet theorists, viewed this negatively, blaming sound for depriving film of much of its unique artistic and universal character. Others welcomed sound's emphasis on cultural specificity and the new techniques it afforded for projecting difference.[35] Commentators across Europe rightly predicted that films would be produced in various national languages, and that this process would politicize them. Distributors throughout Europe unofficially embargoed non-majority languages. In the first years of sound, French films were rare in Germany, Hungarian films banned in Czechoslovakia, and German films absent from Poland. Furthermore, no one wanted films 'in American.'[36] This challenge to American supremacy, film nationalists hoped, would automatically attract distinct local audiences, leading European nations to 'reestablish sovereignty over their cinema establishments…'[37] Although silent films were not devoid of cultural symbols, language and music stripped film of much of the flexibility the speechless format had enjoyed. No longer could intertitles simply be translated and replaced. No longer could national censors merely cut scenes they deemed unacceptable and splice the movie back together. With sound and dialogue, film, over time, became much more narrative and much less ambiguous. Editing and censoring, like production, became more complicated and, potentially, subject to greater nationalist strait-jacketing.

The majority of contemporary commentators worldwide agreed that sound nationalized film. Unlike any other medium, film could represent the nation in space, sound, and narrative, offering an unprecedented means of imagining and constructing a national culture. Film would map the cultural terrain upon which a new national identity might be composed. By bringing together dance, music, theater and other visual arts, film professionals privileged their medium as *the tool* for synthesizing the national culture, fabricating a new national civic consciousness, and creating new concepts of group belonging and allegiance. In Hungary, this notion became axiomatic and film production and regulation assumed great symbolic and psychological importance in intellectual and governmental quarters.

Because sound film came of age as elites sought to naturalize national identity, to present it as an unchangeable reality, Hungary's cultural gatekeepers described the capacity to produce sound film in highly charged, nation-succoring terms. Some framed support of sound production as an existential struggle. Foreign language film besieged more than just language, the core of Hungarian culture which elites imbued with inherent moral force. It caused indifference, threatening Hungarian thought, Hungarian spirit, and even Hungarian political sovereignty.[38] It placed Hungary's rightful cultural standing in Europe at risk. Indeed, the lack of a vital film culture was demonstrative of Hungary's political, economic, and cultural impotence. For some in the film industry, foreign domination of distribution, purported foreign control of Hungarian theaters, and Hungary's inability to produce its own films served as painful reminders of the condition described as 'material Trianon', a condition of subservience based on ever increasing obligations to the West.[39]

Other Hungarian commentators envisioned a developed film industry and a lively film culture as positive signifiers that would prove Hungary's modernity to the West and distinguish it from the cultural 'wastelands' of the Balkans.[40] The imperative to make sound film was thus a national mandate, a means of catalyzing a broader cultural renaissance.[41] Because their country's small size and difficult language meant foreign productions were unlikely to be made in Hungarian, they argued, domestically produced sound film was crucial to inculcating and promoting a more widespread habit of movie-going. They presented this construction of a national film culture as necessary not only to preserve the economic well-being of those engaged in the motion picture business, but for 'national-educational' reasons as well.[42] In other words, Hungary needed its film culture in order to nationalize Hungary, both by informing the people about who they were and teaching the world about the Hungarian nation.

The Founding Fathers: Hungary's Film Elite

'A nation is born when a few people decide that it should be', wrote the Hungarian historian Paul Ignotus.[43] His description seems especially apt for Hungary at the beginning of the age of sound. In a land as bound by tradition, stratified by education, and divided by class as Hungary, the tasks of teaching the nation about itself and creating the nation's culture could not be left to chance. In the late 1920s and early 1930s, an unstable union of creative personnel and 'technicians of power' decided that their industry must lead the resuscitation of a down-and-out Hungarian nation.[44] The collected interests and opinions of this influential group of several hundred men and a handful of women formed the aesthetic, commercial, and ideological matrix for cinema. It was due to, and often in spite of, this film elite – this self-appointed concatenation of film professionals, cultural critics, and government officials involved in the regulation and supervision of the movie industry – that Hungary became the force it did during the early age of sound. Their actions revived a flatlining business, and their discussions and debates outlined the rhetorical visions of the national motion picture industry through the 1930s and 40s.

In some ways, these progenitors of sound emerged in the same manner as cinematic movers and shakers in any other country. Theater critics became film critics, composers and stage actors began working in film, and innkeepers became theater owners. However, some of the fixtures of the new Hungarian film establishment were perfect examples of an intermixing of government, business, and culture unique to the smaller and less democratic European states, a condition that Andrew János described as 'backwardness' and George Mosse saw as a dangerous politicization of art.[45] Case in point was János Bingert, a police captain with a Ph.D. whose interest in movies catapulted him to mogul status. His love of films led the Interior Minister, in 1924, to make him the Ministry's secretary in charge of film matters. In this capacity he traveled to North America to study the film business, later making similar trips throughout Europe. When the state took possession of the bankrupt Corvin studio in 1928, the Interior Minister appointed Bingert as the atelier's Managing Director, a position he held until 1944.[46]

Another example was Miklós Kozma, a cavalry officer who had distinguished himself in the service of Admiral Miklós Horthy's counterrevolutionary movement. This enigmatic figure served as a Habsburg officer during World War I and joined the Hungarian National Defense Association [*Magyar Országos Véderő Egylet* or MOVE], heading its news and propaganda department, after the breakup of Austria-Hungary. By mid-1919 he rose to become MOVE's chief, elevating himself to the center of the powerful anti-Communist counterrevolutionary group based in Szeged. In 1923 he took charge of the private Hungarian Telegraph Office and its newsreels and foreign propaganda subsidiary, the Hungarian Film Office [*Magyar Film Iroda*, hereafter MFI]. Kozma eventually became, in the words of Mária Ormos, a 'captain of media' who played a major role in the shaping of the Hungarian film industry.[47] Hungary's Censorship Committee was another example of this amateur-state goulash. Its staff consisted mainly of policemen and Interior Ministry administrators, and it had only one member who was an acknowledged film expert at the time of its origin. The relationship between state and cinema, and its substantial impact on the evolution of the industry, was largely mediated through this quasi-government, quasi-private faction of Hungary's sound fathers.

Jewish Hungarians were also prominent in the film arts during the interwar period, both as creators and as providers of capital. Béla Gaál, one of the two dominant directors of the 1930s and a film school founder, was certainly a member of the film elite. The same was true of István Gerő, the brash stockbroker who purchased theaters, equipped them for sound, and became one of the Hungarian film industry giants. They were joined by many other Jewish actors, actresses, theater owners, distributors, and film journalists as powerful voices in the world of motion pictures. Their presence in the power structure made the film elite an odd mix during the 1930s. Jewish stockbrokers who owned theaters, censors and studio heads who had trained as policemen, film journalists who espoused national socialism – all convened to form the flammable cast of characters who determined the film industry's future.

The Film Profession and the State: Searching for Sounds of Life

Given the makeup of this 'imaginative community', it should be no surprise that its members rarely agreed on anything, let alone how to construct an industry they ardently hoped would project a coherent vision of Hungary.[48] The general economic languor, moribund indigenous film production, and sharply declining ticket sales made the task of sustaining a vital film culture truly daunting. When the first sound film shown in Hungary, Al Jolson's *The Singing Fool*, premiered in Budapest's Forum Cinema in September 1929, Hungary's film culture was critically ill. Hungarian film production had ceased, a condition both caused and made intractable by the continued flood of filmmaking talent out of Hungary. The *Hungarian Film Courier* [*Magyar Filmkurír*], an independent trade paper, lamented that 'all of Hungary's talent is elsewhere'. In Hollywood, 'the Hungarian colony has significant weight'.[49] In Berlin, 'Hungarians are found in every studio'. In Vienna, Hungarian was 'one of the languages of the industry'. Even England had Hungarian film professionals. 'The empty, silent Hungarian studios' were the only places the Hungarian language was not heard.[50] This exodus of talent led some Hungarian film experts to doubt whether Hungary had the human capacity for quality filmmaking, even if it somehow discovered the financial resources. While our professionals might have a lot of energy and desire, wrote the director Dezső Németh, they 'don't have the practice and the knowledge that allow foreign films to reach such a high standard'.[51]

For exhibitors and distributors, the situation was just as dire. The official number of licensed Hungarian film theaters totaled 507 in 1932, but the film community estimated that of those, maybe 200 actually operated.[52] Closed theaters increased unemployment, bedeviled distributors, and threatened Hungary's film culture.[53] The directorate of the Stylus Film Company summarized the industry consensus starkly: 'The 1930 business year was a catastrophe for the film profession'.[54] Distributors, unable to pass on to theater owners expensive licensing fees and import surcharges, which were rising with concomitant increases in sound production costs, now found themselves in serious jeopardy. Rather than crush struggling theaters with excessive exhibition costs and thus destroy their own client base, distributors resorted to importing untranslated, cheap, and frequently lower-quality films. Theater owners claimed this strategy all but assured the disaffection of Hungarian audiences, especially in the rural areas where most spectators spoke only Hungarian.[55]

Statistics appear to support this claim. Budapest ticket sales, declining even before the advent of sound, dropped precipitously by 1930.[56] Thereafter, attendance figures for rural Hungary continued to spiral downward.[57] Representatives of the National Association of Movie Theater Licensees [*Magyar Mozgóképüzemengedélyesek Országos Egyesülete*, hereafter MMOE] estimated that countryside attendance decreased a whopping 40 percent from 1929 to 1932 and that 65 percent of the population lacked access to sound film.[58] These statistics also revealed significant class and geographic disparities. While Budapest and a few other large cities were well-provisioned with sound theaters, rural towns and villages were not.

These developments led Hungarian film elites to warn of a mortal crisis for the profession and its 'cultural mission'. The Depression-induced drop in disposable incomes, deprivation due to theater closure or distance, limited programming, lower-quality films, and dearth of Hungarian-language filmmaking all combined to produce alienation from film, especially among the Hungarian working classes and rural masses. Such an erosion of Hungary's film culture would ruin the industry's prospects. Even if Hungary gained the capacity to make its own films, distributors and theater owners feared that there would be no audiences in the few remaining theaters to patronize them, apart from urbanites with cosmopolitan sensibilities.[59]

Conditions were so grave that nearly all of Hungary's cultural and political leaders agreed that only the state possessed the means and the power to rescue Hungary's flailing film culture. Whether the state could bring stability and economic salvation was a questionable proposition. The Horthy government's record of intervention in and subvention for Hungarian film in the silent era was checkered, a mix of misconceptions and golden handshakes, offset by the occasional good judgment. It had failed to prevent the collapse of Hungarian silent film production and stood by as movie theaters closed by the hundreds. Limited by a lack of funds and imagination, and guided by ideologies tinged by backwardness, antisemitism, and discomfort with international commercial capitalism, Horthyites preferred to add to existing regulation and avoid addressing underlying problems. This practice had lasting consequences for the developing sound industry.

Rather than create a new apparatus which would take advantage of the cathartic potential of sound, the Hungarian government left major elements of the silent film world largely intact. Specifically, regulators made mere tweaks to structures from the 1920s which governed the funding of film production, censorship, and taxation. Most egregiously, they did not alter the 1921 cinema license revision. This regulation stripped theater licenses from their rightful owners without compensation and awarded them to war veterans, especially disabled war veterans, and war widows. Signed as Hungary's anti-Jewish terror was winding down, the decree rewarded those who had served and suffered in the Great War at the expense of 'untrustworthy' Jews.[60] The regulation allowed the Interior Minister to redistribute licenses to the National Association of Retired Officers, the League of Awakening Hungarians, and other rightist, often antisemitic organizations.

By bringing 'dilettantes' who were 'professionally uninformed' and bereft of cash into the film industry, this change magnified the crisis the Hungarian cinemas experienced during the 1920s and especially during the early years of sound.[61] License holders found themselves forced either to close their doors, as many did, or to circumvent the law in search of cash and expertise. Recognizing the catastrophic impact of this legislation, the Interior Minister authorized the 'official' return of Jewish theater owners to the profession by legalizing partnership arrangements in 1923.[62] This gave the struggling theater managers an ironic choice: go bankrupt, or lease their licenses, often back to the original Jewish license holders whose theaters they indirectly stole. These outstanding legislative maneuvers did little to promote good will or industry stability. Nor did further official disbursement of theater licenses. The government sometimes awarded the most lucrative licenses to high-ranking army officers, widows of high-ranking officers (many of whose husbands died after, not during, the war), and government bureaucrats as part of their retirement packages. These 'big pensioners' frequently leased these posh perks or sold them outright after receiving them, collecting a substantial premium.[63]

Theater licensing re-emerged as a hot-button issue with the rise of sound film and the resulting fall in the number of cinemas. Hungary's two most powerful professional film organizations, the MMOE, which represented the interests of theater owners, and the National Association of Film Producers [*Országos Magyar Mozgóképipari Egyesület* – hereafter OMME], which represented producers and distributors, both petitioned the Interior Minister to revise licensing procedures. The MMOE contended that corrupt practices awarded the licenses for Hungary's top theaters to 'banks, high-ranking government officials, and trusts' rather than those who 'deserved' them.[64] Similarly, the OMME claimed in 1932 that there were many financiers ready to invest, but would only do so if the government overhauled the flawed licensing system.[65] These requests came to naught, as state authorities preferred their system of patronage and antisemitism to profitability and meritocracy. This legacy would burden the industry through the 1930s and early 1940s.

Theater licensing was just one government incursion into the film industry tracing from the early years of Horthy's reign. In response to the monetary instability of the early 1920s, stagnating film production, and a growing concern for the cinema's influence on the moral and cultural well-being of the nation, the Hungarian government enacted sweeping film legislation in 1925. These dictates created expansive regulatory structures which governed all segments of the film industry, with some alterations, through 1944.[66] The single most important feature of the 1925 decrees was the creation of a Film Industry Fund [*Filmipari Alap*, hereafter FIF] to encourage domestic production. A twelve-person board, which included representatives from the most powerful government ministries and professional film bodies, supervised the day-to-day operations of the Fund.[67] The heads of the two pre-existing organs charged with regulating Hungarian film, the Film Council and the Censorship Committee, assumed roles as president and vice-president of the Fund's supervisory board. The Fund immediately became one of the most important forces in the business, enjoying broad jurisdiction. It could authorize or annul production contracts, arbitrate disagreements between film producers and studios, and regulate the release of government-backed credit. It also oversaw production of films it deemed mandatory for all cinemas to screen. To pay for these responsibilities and initiatives, the Hungarian government assessed per meter fees on all films, which by the end of 1932, totaled nearly HP 2.2 million.[68]

In late 1928, the Fund made its most critical transaction. With domestic production dwindling and no private investor interest, the FIF purchased the former jewel of Hungarian silent production, the Corvin studio.[69] The Fund then renamed the studio Hunnia and began planning its rehabilitation. It also reconstituted Hunnia as an independently-run company, albeit one intricately linked to the government. A range of government ministries cooperated with the Fund to select Hunnia's corporate leadership, placing their own representatives on its board of directors and in 1930 appointing János Bingert as Hunnia's Managing Director. Many of Hunnia's board members were the same men who held high-ranking positions on regulatory agencies, such as the Film Council, the Film Industry Fund and the Censorship Committee, and in key state ministries, such as the Prime Minister's, Interior Minister's, Commerce Minister's and Religion and Education Minister's offices.[70]

The complicated and incestuous relationship between those who supervised the Hungarian film industry and those who operated an ostensibly for-profit studio was a conflict of interest that did not go unnoticed by the more vocal figures in the film community and film trade press. József Pakots, a screenwriter in the silent era and a Social Democratic parliamentarian in the era of sound, persistently called for the abandonment of the FIF, arguing that government interference was handicapping the film world and preventing, rather than facilitating, private investment.[71] *Film News'* [*Filmújság*] editor called for the 'foul-smelling links' between Hunnia, the FIF, and the OMB to be severed.[72] Distributors complained not about who ran the Fund, rather the interests they represented, denouncing a perceived unhealthy favoritism toward the industry's production sector. Instead of calling for greater privatization, some distributors called upon the government to earmark monies for them. A distribution fund, they asserted, would guarantee that importers could buy quality films and sell them to exhibitors at reasonable prices, improving film culture in general.[73]

Cinema owners, through the MMOE, critiqued the FIF as well. They bemoaned the fact that as of October 1930, the government had collected over HP 1.5 million in fees, leisure taxes, and import duties yet had nothing to show for it save a non-operating studio [Hunnia] whose modernization was incomplete. MMOE members had borne these costs with 'patriotic feeling', dutifully honoring their obligation to take all possible measures to help spread Hungarian culture. The government, they felt, impeded their 'natural inclinations' to fulfill this lofty goal by refusing to effectively subsidize and incentivize Hungarian movie making.[74] Debates over the role of government in the movie business in the early 1930s had yielded a clear consensus: most producers, distributors, and exhibitors concurred that the government was not promoting their wellbeing, the Film Fund was not meeting expectations, and national interests were not being served.

Outside of Hunnia, there was widespread resentment that the Film Industry Fund did not support the production of newsreels by the Hungarian Film Office [MFI]. The MFI was an exemplar of the patronage-based, public-private amalgam Hungary's oligarchs created in the interwar era. Founded in November 1923 as a subsidiary of the Hungarian Telegraphic Office, the MFI was, in theory, a private company. In practice, however, similar to Hunnia, the Hungarian Film Office was a quasi-governmental operation, perhaps an even more confusing web of private and public ties than the state-owned Hunnia facility. Like Hunnia and the FIF, the MFI's structure reflected Hungary's ambivalent relationship with capitalism and its propensity for government intrusion. The Foreign Ministry owned 40 percent of the firm's original shares.[75] This interest, augmented by critical seats on the company's Directorate, gave the Foreign Ministry control over Hungary's 'most influential and powerful propaganda tool–film', and insured that the MFI remained an 'official instrument'.[76] The MFI Directorate included representatives of various government departments and film oversight agencies, and even several erstwhile Hunnia board members.[77] While it did not receive FIF support, its production costs were partially underwritten by the City of Budapest.

The MFI enjoyed a degree of state favor and preference that made true private companies envious. Until late 1935, the MFI did not even possess its own studio. Instead, it rented filmmaking space from the Pedagogical Film Factory, an educational film company owned and operated by the City of Budapest. This compact meant that the city's subventions for MFI newsreel and educational film production sometimes returned to it in the form of rent. An additional example of how the Film Office occupied a nether world between independence and state control was that the Ministry of Religion and Education planned and bought nearly all of the MFI's culture, science, and educational films, but provided no additional grants for regular operating costs. Further, these state-created films still had to pay censorship fees and follow the normal Censorship Committee approval process.

Its profusion of state ties, however, did win the MFI one gem: a newsreel monopoly. Although numerous companies produced newsreels in the 1920s, the government made the exhibition of MFI products obligatory in 1926. Compulsion to buy the newsreels placed a heavy burden on smaller, struggling theaters, who lobbied for FIF support as the means for making newsreels affordable. Their cries reached a crescendo by 1932, but produced no results.[78] The authorities' early 1930s denial of direct subsidies to the MFI, which in 1932 was Hungary's largest film producer, illustrates that Hungary's leaders did not have the stomach for unrestricted capitalist competition nor for complete state-led centralization of film production. Their quest for a third way led authorities to grant the MFI some advantages and deprive it of others, leaving it in a nebulous condition that pleased few and befuddled many.

Even before it became clear that the FIF would not be the panacea for Hungary's film ailments, motion picture constituencies searched for ways to protect the existing film culture and to encourage domestic production. Proposals ranged from tax exemptions and utility rate discounts to legislated limits on film imports, schemes popular in the press and Parliament. Advocates cited the quota and contingency systems adopted by Britain and Germany as examples of the methods Hungary could use to resuscitate its film culture. Others suggested mandates that foreign films screened in Hungary be dubbed or subtitled in Hungary by Hungarians, foreshadowing legislation later passed in Italy, France, and Spain that all films be made or subtitled in the mother tongue.

Import regulations and distribution fees carried the day, adopted in a June 1929 Prime Ministerial decree. This ruling required any company which imported 20 feature films into Hungary to fund the production of a Hungarian feature at least 1,500 meters in length. The film had to be made in Hungary and cost more than HP 60,000.[79] Government officials, deliberately following the example of Great Britain's 1927 Films Acts, hoped that their legislation would result in foreign funding of movie making in Hungary, just as it had in Great Britain. They believed, in turn, that this would 'assure the more effective development of the Hungarian film industry or [at least] would secure consumer markets for Hungarian films'.[80] Whether the decree would have accomplished its goals is unclear, given Hungary's lack of existing production capacity. In place barely seven months, undermined by the government's and film establishment's ambivalence toward the regulation, and under severe pressure from American and German film companies and governments, Hungary's rulers diluted the legislation.[81] As of February 1930, distributors could earn import exemptions simply by agreeing to fund 400 meter shorts rather than full-length features. By June 1930, a scant four months after its first concessions, the Hungarian government completely succumbed, abandoning a system that it never had the will to enforce. In its stead, the government instituted a series of fees and taxes as an alternative to the production requirement, levies that filled FIF coffers.[82] While the choice of making film in Hungary remained, it was far cheaper for distributors simply to pay the surcharges.

The failure of Hungary's quota experiment is one of the first illustrations of how Hungarian sound film policy was shaped by global influences. Pressure from domestic sources certainly influenced the Hungarian government's quota roll back. Hungarian theater owners and film distributors spoke out vociferously against protectionism, contending that any laws discouraging imports would worsen the crisis for theaters, already failing at unprecedented rates. As customers disappeared, a cascade of bankrupt distribution companies would follow, they warned. More importantly, however, Hungarian authorities were cowed by memories of recent boycotts of European markets led by American distributors. American majors had humbled Hungary when it first adopted import legislation in 1925, cutting off the country until Hungary cleansed its legislation of all compulsory production requirements in April 1926.[83] If Hungary's bureaucrats had short memories, a more recent case reminded them of Hollywood's persuasiveness. In 1929, France organized the *Chambre Syndicale de la Cinématographie Française* and upped the ratio of indigenous film to foreign imports from one to seven to one to three. American companies objected, halted exports between March and September 1929, and strong-armed the French into restoring the old ratio. When America and Germany began rattling their sabers at the mere hint of quotas in the summer of 1929, Hungary found itself facing dual embargoes. Hungarian authorities immediately caved.[84] The quota program never gained traction because all sides realized compensatory domestic production, meaning Hungarian work capable of replacing foreign film in significant numbers, was years away. Surcharges quickly proved preferable to quotas as a means of preserving the remnants of the national film culture.

Ill-conceived initiatives such as these damaged industry confidence. Thus, while Hungary remained mired in economic crisis in 1931–2, state promises of support for domestic film production stimulated neither enthusiasm among native creditors nor faith in the Film Industry Fund.[85] This timidity meant that prior to 1933 the FIF generally stockpiled its monies. The Fund received few proposals between 1930 and 1932 which met the basic qualifications needed to release state support: a complete screenplay, a cast, and a set of backers. As a result, Hungary's film elites forlornly awaited the return of those 'world famous Hungarians…who play across the ocean', whom they fantasized would triumphantly land with buckets of cash, ready to lavish their largesse on Hungarian film production because 'in their hearts they remained good Hungarians…[and] good businessmen'.[86] This listless self-deception and the paralysis it inspired frustrated many in the Hungarian film community. Not until 1933 did changing domestic and geopolitical conditions, combined with the state's willingness to grease the wheels of production, stimulate Hungarian capitalists to take the first steps necessary to advance native film production.

Help from Abroad: The Transnational Origins of the Early Hungarian Sound Film Industry

As the government backed away from import regulations and domestic capital markets remained airtight, a consensus emerged among film elites. To reinvigorate the national film culture they must embrace the fact that film remained an international medium. Unlike the United States, France, Germany, or the Soviet Union, Hungary was far too small to rely on its domestic market. It could not expect to put out more than a handful of profitable sound features each year aimed at its home market alone. Strained relations with most of its neighbors, exacerbated by linguistic isolation, precluded the export of Hungarian film to much of central and Southeastern Europe.[87] For Hungary to become a viable sound film producer, the industry needed to access foreign markets and attract investment from abroad. Without this boost, there was little chance that Hungary's indigenous investors would risk their monies on an unproven commodity such as Hungarian film. Therefore, instead of protectionism, Hungary moved toward policies and practices premised upon open, not closed, borders; promotion of a multiplicity of languages rather than Hungarian exclusively; and the belief that Budapest must become an international film center in order for Hungary's national film culture to flourish.

Film elites thus advocated for partnerships, including co-productions aimed at the three dominant language markets – English, German, and French – and multiple language version movies.[88] One of the early proponents of this position was Hunnia director János Bingert. Foreign capital, he suggested in early 1930, was the solution to the national film industry's shortfalls. If foreign companies could be convinced either to invest in joint productions or the construction of sound studios in Hungary, rewards would be mutual and multipliers exponential. Hungarian studio rental costs would be 30 to 40 percent cheaper than in Germany, Bingert estimated. External investment would persuade Hungarian banks to extend credit. The banks would then be more open to long-term cooperative arrangements with the government. Loan guarantees and even additional government subsidies would pave the way for rapid domestic growth.[89]

Bingert's colleague Gyula Pekár, also one of the most important members of Hungary's film establishment, concurred. A writer involved in film since 1900, Pekár was the head of the Film Council and the president of the Film Industry Fund. He too believed that Hungary must look beyond its borders, disagreeing with Bingert only where he cast his gaze. Pekár was initially partial to France, whereas Bingert favored American and German partners. Thus, the two most important figures in Hungarian film production believed foreign financing was an absolute prerequisite to initiate and sustain domestic sound film production. Only after attracting capital from abroad did these men, government functionaries themselves, believe they could convince other Hungarian officials to authorize expenditures.[90]

In 1930, Hungarians made overtures to Germany, Austria, Hollywood, and France, highlighting Hungary's comparative worth to potential suitors. Officials cited mushrooming regulation of foreign language films in Italy, Spain, and even France as reasons American and German film companies should move operations to Hungary. The *Magyar Filmkurír* courted foreign firms by touting the low labor expenses and plentiful language and acting skills that made Hungary the 'most suitable place for the creation of a center for European sound film'.[91] The April 1930 visit of Paramount's Adolph Zucker (Zukor) and Jesse Lasky demonstrates the resonance of these arguments. Made credible by the opening of Warner Brothers/First National studios in Britain and Germany and the expansion of the European operations of several other American majors, rumors were rife, though exaggerated, that Paramount planned to open a film studio in Budapest. A report filed by the American Minister to Hungary on the Zucker/Lasky European fact-finding trip noted that the Paramount bosses told Hungarian authorities that the 'talkie had become essentially a national problem' to which Paramount would respond by producing in multiple languages. Zucker informed his counterparts this could be done in Hungary. A Hungarian émigré, Zucker cited not only his 'sentimental attachment' to Hungary but '…the histrionic and musical ability of the Hungarians and the colorful settings that are characteristic of this country'.[92] These comments, some of which the *New York Times* repeated, raised the expectations of Hungarian authorities.[93] A Paramount sound studio would make Budapest the commercial hub of middle European film production and, Hungary's film elites believed, inspire the return of world-class Hungarian talent.[94] International capital, in other words, would trigger the re-nationalization of Hungary's prodigal film professionals.[95]

Although Paramount made several Hungarian-language films in 1930, employing some of Hungary's best personnel, the films were not made in Hungary. Rather, they were part of a larger multiple language version [MLV] experiment conducted at the famous Joinville studio outside Paris. Attempting to amalgamate Fordist standardized production and casts from a variety of European countries, Paramount's MLVs proved disastrous. Both of Paramount's Hungarian-language version movies, *The Doctor's Secret* [*Az orvos titka*] and *The Laughing Lady* [*A kacagó asszony*], flopped, even though they featured some of Hungary's most beloved stars.[96] The failure of the MLVs reinforced what Hungarian elites believed to be a truism: different national audiences required authentic products. Simply translating a Hollywood script would not insure the success of a film.[97] The American majors, despite currency restrictions which made it difficult for them to extract funds from Hungary, soon reached similar conclusions.[98] Universal, which had mulled the possibility of using its Hungarian income for Hungarian-language production, determined that it was not worth the risk. With dubbing too expensive and technically underdeveloped, a financially unstable Paramount decided in early 1932 that neither MLVs nor single version films for a Hungarian-language market of less than 15 million were sound investments. In 1933, Paramount's bankruptcy quashed all remaining hopes that the firm might build a Central European Joinville-like studio in Budapest.

Even as Hungary's courtship with American investors evanesced, and film professionals warned of impending doom, the arrival of a French white knight revived Hungary's film fantasy. In late 1929, Gyula Pekár had initiated talks with a group of Francophone professionals who hoped to expand French cinematographic influence in Central Europe. The two sides agreed to form a Franco-Hungarian production group that would draw on the Film Industry Fund resources. The Hungarian Foreign Ministry initially greeted this proposal with skepticism, citing anti-Hungarian prejudices on the part of the French press and the Fund's tendency to involve itself in failed enterprises.[99] While his superiors registered reservations, Ministerial Advisor Zoltán Gerevich bubbled with enthusiasm. He believed the cooperative deal would jolt Hungarian film production back to life and might have important implications in terms of cultural diplomacy.[100]

Gerevich's optimism was not misplaced. The Parisian banker of Hungarian origin Lajos Mannheim and the French filmmaker Adolphe Osso (a.k.a. Alfred Savar, formerly the head of Paramount's French branch) became convinced that there was potential in the Hungarian market just as Paramount's interest in Hungary was waning. Operating under the assumption that inexpensive Hungarian talent could create quality films for foreign markets, Osso combined forces with Hungarian associates with whom he had worked at Joinville. He used Mannheim's money to form or recapitalize a number of Budapest firms between late 1931 and early 1932. The Mannheim/Osso companies – Osso Film, Minerva Film, and later City Film – were all to play an outsized role in the early evolution of the Hungarian sound motion picture business.[101]

The husband of Ica Lenkeffy, a Hungarian star of French silent cinema, Lajos Mannheim was the perfect Franco-Hungarian intermediary.[102] His cash infusion was the most significant commitment made to early Hungarian sound film, larger than any concurrent private or public Hungarian contribution. It was also a diplomatic initiative. During the Osso Film board's initial gathering, members spoke of shouldering the task of 'addressing a long-recognized need to improve both cultural and business relations between France and Hungary'.[103] Affiliated with this effort was a virtual who's who of Hungary's early screenwriting vanguard. Menyhért Lengyel, later to arrive in Hollywood and have a long and outstanding film-writing career; Zsolt Harsányi, a veteran playwright, novelist, screenwriter, film journalist, and cultural theorist; Lajos Zilahy, one of Hungary's leading playwrights and screenwriters in the 1920s and 1930s; and Jenő Heltai, the poet, novelist, playwright, screenwriter, cabaret-writer, and nephew of Theodore Herzl, were all among the original stockholders of Osso and Minerva. Heltai, who is reputed to have coined the Hungarian word *mozi,* meaning movie, was also the leading figure in the Hungarian management of Minerva.[104]

Although these endeavors included some of Hungary's intellectual vanguard, the enterprises, by-and-large, were French. Osso Film, 88 percent owned by Adolphe Osso, and Minerva, 90 percent owned by Mannheim, were precisely the heroes Hungary's film establishment fancied.[105] The mere announcement of the formation of these production companies inspired a spike in confidence in the Hungarian national sound film industry. The editors of the journal *Magyar Filmkurir* boasted in February 1932, before a single Franco-Hungarian project had commenced, that the French should be offering Hungary the following toast: 'Long live Hungary! The Paris film profession is saved!'[106] Zealous film magazine editors claimed the French had guaranteed production of up to 16 Hungarian films.[107] Rhetoric aside, Osso and Minerva literally did become the driving forces behind Hungarian film production in 1932. That year, in the recently finished, state-of-the-art Hunnia studio, three companies produced eight features. Osso was responsible for five, the Hungarian company Phőbus made two, and Germany's Ufa one.

Of the total HP 2.58 million invested in filmmaking in 1932, 2.1 million came from foreign sources, at least 1.63 million of which, or over US $400,000, was French.[108] Funds from Osso and Minerva supported seven of the fifteen Hungarian-language features completed between 1932 and 1933, not to mention several more foreign language pictures. The journal *The Technology of Film* [*Filmtechnika*] correctly declared that Minerva, Hunnia, and Osso *were* Hungarian film production, revitalizing 'an otherwise more-or-less humdrum Hungarian film world'.[109] Utilizing large Hungarian crews, the French-backed projects gave Hungarian film production a legitimate foundation, providing both income and vital experience to a cadre of Hungarian film professionals. In addition to bankrolling production, Osso Film built and provisioned a second sound studio at the Hunnia settlement in exchange for free rental of the studio. Construction alone employed over 150 people and amounted to an additional investment of over HP 250,000. Including the sound cameras, lamps, editing boards and other equipment furnished by Osso and Minerva, the total French outlay was nearly one-half million *pengő*.[110] This investment was the keystone, the start-up money that would establish Budapest 'as the filmmaking capital of Central and Eastern Europe'.[111] Without a doubt, the French connection proved to be an essential impetus – psychologically, experientially and materially – for the maturation of the Hungarian sound film industry.

A closer examination of the rise and fall of the French-funded film forays, however, shows how the competing strains of Hungarian nationalism complicated and eventually destroyed this international film venture. Osso and Mannheim founded their Budapest companies with the intent to produce not only Hungarian films, but French films for export. They hired the Italian director Carmine Gallone to make their first two features, *Fils d'Amerique* and *Le roi des palaces*. Both films were produced in French only, and neither was distributed in Hungary. Osso Film also made three additional features in 1932, *Flying Gold* [*Repülő arany/Rouletabille aviateur*], *The Waters Decide* [*Ítél a Balaton/Tempête/Menschen in Sturm*], and *Spring Shower* [*Tavaszi zápor/Marie, légende hongroise*]. All three were large multi-language productions, the first directed by István Székely, the latter two by Pál Fejős. *Flying Gold,* a film international in appeal yet reflective of its Hungarian milieu, epitomized the Osso episode. Hungarian film historians describe *Flying Gold,* a mystery based on a Gaston Leroux novel, as director István Székely's attempt to find a competitive, distinctly Central European form of the criminal action film.[112] The plot of *Flying Gold* took audiences from Paris, through Vienna, to Budapest, and did so through the most modern of means: the airplane and automobile. In this manner, the picture was a proxy for the economic, cultural, and political connections Osso, Mannheim, and their Hungarian counterparts sought to create through their joint ventures. *Flying Gold* was an early attempt to leverage international backing to forge a Hungarian cosmopolitanism which put Budapest on par with other European capital cities. This cosmopolitanism thus functioned as national, helping to reconnect an isolated Hungary to Europe.

All five of Osso's 1932 Budapest-made films won widespread critical acclaim. Pál Fejős' *The Waters Decide,* a multi-language adaptation of *Romeo and Juliet* which took place in a fishing village on Hungary's Lake Balaton, was Hungary's top grossing film in 1932.[113] In *The Waters Decide,* Mária, the daughter of a Hungarian fisherman, is unable to marry her true love because of long-lasting family animosities. Yet, she is also unable

Figure 2: The aerial camera used in *Flying Gold.* Source: MaNDA, Budapest.

to resist her heart. Caught violating her marriage vows, she is sentenced to the supposedly traditional penalty for unfaithful women: cast onto the stormy waters of the Balaton in a rowboat, she is provided neither oars nor protection from the rain. When both her husband and her lover swim out to save her and her lover begins to struggle, Mária begs her husband to help save his drowning competitor. Her husband recognizes the judgment of the fates, obliges his wife, and succumbs in the process. The film concludes with the lovers locked in an embrace, surrounded by the Balaton's angry waves.[114] In *The Waters Decide,* Hungary had its first patently Hungarian tragedy made for the sound screen.

The stridency of Hungarian hopes for these works and the extent to which Europeans perceived almost everything through nationalist lenses was palpable in the buzz created by the production of these films. Hungarian newspapers and magazines started referring to the Hunnia studios as the 'Franco-Hungarian Hollywood'.[115] Osso Film tried to capitalize on its special relationship by placing a prominent graphic near the bottom of its advertisements for films it imported into Hungary that read 'Don't forget the symbol for great Hungarian films!'[116] By mid-1932, Osso, Mannheim, and their Hungarian partners were planning the production of eight films in Budapest in 1933 and the distribution of numerous others. Not only did the future of Hungarian film, thanks to its French saviors, look promising, but some Hungarian film elites believed they had won great international clout by contributing to the survival of French-made European film, just as *Magyar Filmkurír* had bragged months earlier. The prospect that Budapest might become a commercial center for European film production buoyed Hungary's film establishment. This auspicious rebirth would surely convince Hungarian émigré motion picture luminaries to return home. Such was international film nationalism. International finance, particularly monies from France, would affect the creation of a Hungarian national film industry.

Unexpectedly, in October 1932, before two of the three Osso/Minerva films even premiered in Hungary, the companies indefinitely suspended all plans for further production in Hungary beyond mid-1933.[117] Jenő Heltai, the Hungarian writer and principal partner in Osso's Minerva branch, offered two reasons for the abrupt change: the Hungarian press' mistreatment of the French, and the failure of the Hungarian government and population to extend the French film professionals the 'proper courtesies'.[118] While Heltai believed Hungary's film establishment recognized and accepted the necessity of foreign language film production in Hungary, in his opinion the broader press was far more nationalistic and less tolerant of 'foreign imperialism', especially on the part of France. The anti-French opinion Osso/Minerva's founders had hoped their goods would dissolve, had in fact grown more acerbic. Inflamed by France's diplomatic alliances with all of Hungary's 'enemies', especially Romania, Czechoslovakia, and Yugoslavia, this sentiment represented a persistent obstacle for French-owned companies attempting to profit from Hungarian cultural nationalism. The average Hungarian patriot would not deign to see a film that presented potentially unflattering portraits of Hungarian life, particularly if produced by a company whose country was the guarantor of the hated Treaty of Trianon.

Included in what Heltai referred to as 'proper courtesies' were concerns about government treatment of Minerva and Osso Film. Several of Osso's imports, in addition to its aforementioned Hungarian-made films, had run afoul of the Hungarian Censorship Committee. In the name of national purity and morality, the authorities either required Osso and Minerva to do additional editing or denied requested exhibition licenses.[119] Additionally, Minerva's board may have interpreted the Budapest Municipal Business Committee's (*Cégbizottság*) attempt to force the company to change its name as Francophobia.[120] While not unduly harsh, these actions indicated how different conceptions of Hungarian national interests conflicted and undermined efforts to foster Franco-Hungarian cooperation.

Osso itself was also responsible for its failures. While the company's board claimed that it had not made miscalculations, economic uncertainty obviously impacted the gates collected by Osso movies and thus influenced Osso's decision to stop making films in Hungary.[121] The Great Depression had finally arrived in France, far later than in much of Europe, and by the end of 1932 the company was hemorrhaging *pengő*.[122] Osso's investors found themselves far less willing to stomach deficits than they had in months prior, given the increasingly bleak international economic outlook and the continued stagnation of Hungary's domestic movie market. The curtain thus fell on the Osso/Minerva Hungarian drama, one that by the companies' own accounts brought Hungary more than HP 2 million of capital and resulted in the creation of 15 Hungarian, French, German, and English language features and culture films in Hungarian studios.[123]

If the demise of Paramount's multi-language version films was symptomatic of a broader but temporary decline for American films and American cultural influence in Central Europe, and the failure of Osso's ventures symbolic of a more general and permanent French political and economic retreat from Central Europe, then Hungary's pursuit of German assistance represented the most logical strategy for building the nation's filmmaking capacity. With the July 1930 settlement of the Western Electric versus Tobis-Klangfilm dispute over sound standards and patent rights, international agreements now placed Hungary in the German sound zone.[124] This cartel arrangement clarified which sound equipment Hungary should purchase, finally removing an obstacle which had impeded the outfitting of Hungary's Hunnia studio. At the studio's April 1931 opening, Hunnia's directors offered the same grandiose claims that accompanied Hungary's other international film dalliances. Once again they prophesied that '...with time the success of the Hunnia film factory will make it the center of film production of the Eastern European nations'.[125]

German initiatives nearly validated this prediction in the early 1930s, as funding from and political developments in Germany had a massive influence on the evolution of Hungarian filmmaking. To start, German sources financed the filming of the first smash Hungarian sound feature, *Hyppolit the Butler* [*Hyppolit, a lakáj,/Er und sein Diener*] in 1931. *Hyppolit* was the archetype of the international layering characteristic of nascent Hungarian sound film. The movie was co-produced by Sonor Film, a German-Czech company headed by the Czech producer Albert Samek, and the Hungarian firm of Emil Kovács and Associates. Samek hired two Hungarians then working in Berlin, the director István Székely and the screenwriter Károly Nóti, to write indentical German and Hungarian scripts, which the pair completed in Paris. The resulting comedy, filmed in Budapest, spotlighted the clash between a *nouveau riche* transport company owner and the traditional aristocratic butler his wife hires in an attempt to speed the family's assimilation into the upper classes. The movie was a huge success due primarily to the inspired acting of its stars and the fact that it was the first Hungarian-specific sound film. Familiar parts of Budapest, familiar Hungarian slang, and familiar middle-class family settings attracted

Figure 3: The cast and crew of *Hyppolit*. Seated at the center and wearing a bowtie is Hunnia Chief János Bingert. Source: MaNDA, Budapest.

domestic audiences in droves. The film was highly profitable and catapulted its director, Székely, and several of its stars, including Gyula Kabos, Gyula Csortos and Pál Jávor, to prolific and acclaimed sound careers.[126]

A blend of German and Austrian capital and production companies played nearly as significant a role as Osso in the gestation of early Hungarian sound filmmaking.[127] Two of the three feature films produced in Hungary in 1931 were made in German and with German money. In addition to *Hyppolit,* Sonor Film also made *A Car, No Money* [*Ein Auto, kein Geld*] in the Hunnia studios. Produced only in German by the Austrian Jacob Fleck, *A Car* was never released in Hungary. While *Hyppolit* and *A Car* did not portend a flood of German capital into Hungary, they did mark a beginning. Ufa commissioned its first film to be made entirely in Hungary in 1932 not because its leadership believed, as Paramount claimed to, in Hungarian histrionics, or out of some broader cultural-political vision, as Adolphe Osso and Lajos Mannheim professed, but for much more mundane reasons. *The Old Villain* [*A vén gazember, Der alte Gauner*] was designed primarily to bypass Hungarian currency restrictions and win Ufa import fee waivers. Directed by Heinrich Hille and based on a Hungarian novel, *The Old Villain* pitted a spendthrift aristocrat against his well-intentioned but embezzling estate manager. Like the Sonor films and some Paramount MLVs, Ufa's work brought together different casts to make separate versions filmed by the same crew. The effort resulted in a HP 470,000 windfall for the Hunnia studio, the largest take from a single film to date. Prospects for greater cooperation with German film companies looked bright indeed. Géza von Bolváry, a Hungarian director working in Germany, sang Hungary's praises in a February 1932 article, trumpeting: 'The biggest surprise…is that Hungarian film is sensational business….It is worthwhile for every German producer to work in Budapest'.[128]

While von Bolváry urged German producers to head to Hungary, German distributors already had. Between late 1930 and mid-1932, German movies made unprecedented gains. Before sound, Hollywood had dominated the Hungarian market, as it had dominated nearly every European market. However, Hungary's movie-going public, predominantly urban and far more fluent in German than English, naturally gravitated toward German-language film with the initial spread of sound. At the outset of 1930, three American sound features premiered in Hungary for every German one. Yet by October 1930, 12 of Budapest's 13 premier theaters were playing German-made, German-language films.[129] The *Deutsche Tages-Zeitung* reported that Hungarian theaters were overflowing, and that Budapest did not have large enough theaters to hold the audiences for German films.[130] By late 1932, the Hungarian market turned decidedly in Germany's favor. Three German pictures opened for every two American, and Germany had captured nearly 50 percent of the entire Hungarian feature film market.[131] From 1930 through early 1933, German film in Hungary experienced a short-lived zenith and in the process prevented a broader deterioration of Hungary's film culture.

Hungary's relatively tussle-free acceptance of foreign language film made it exceptional among its European counterparts. In Czechoslovakia the public rioted in protest against German-language film; in Poland audiences sometimes became violent when viewing German films or American films subtitled in German; and in France raucous crowds destroyed theaters playing English-language films.[132] Certain Hungarian constituencies, on both the left and right, did, of course, oppose the Babel-like profusion. The right railed against German film in particular, denigrating the Budapest movie-going public which 'lustily suckles at the breast of German-Jewish film culture'.[133] Even centrist film journalists worried that the dominance of German film might damage Hungarian culture. '...German-language film[s] are Germanizing Hungary', wrote László Zsolnai, editor of the independent film trade journal *Filmújság*. 'Little by little', he continued, 'foreign pronunciations are picked up and distance [us] from Hungarian culture'.[134] But the government, desiring friendly relations with Germany, parried these attacks during the first years of the 1930s. Interior Minister Béla Scitovszky resisted pressure to block imports of German sound film out of the prescient fear that Germany might retaliate by dismissing some of the hundreds of Hungarians making films in Berlin. Clearly, the movie-going public sided with Scitovszky by voting with its feet. Audiences and officials seemed united in support of long-term German-Hungarian cinema industry cooperation.

Just as Germany seemed poised to consolidate its export gains and to expand investment in Hungarian film production, German economic and political developments prevented these initiatives from taking root. In its haste to construct its own 'national' film industry and to protect itself from the economic trauma of the Depression, Germany passed legislation between 1930 and 1933 that deeply affected Hungary's entire film culture. These laws, known as the *Verordnungen über die Vorführung ausländischer Bildstreifen* defined then re-defined 'German' and 'foreign' films. The third variant, passed on 28 June 1932, required that for a picture to be considered German, it meet all of the following stipulations:

> It must be owned and copyrighted by a German company; produced by a studio in Germany; its script, sound, and music must be written and performed by German citizens; its producer and director must be German citizens; and finally 75 per cent of production personnel must be German citizens.[135]

As a result of this decree, Hungarian-origin film professionals faced an agonizing choice: keep their citizenship and risk unemployment and expulsion from Germany, or forfeit their Hungarian citizenship and hope to keep their jobs. Several renowned Hungarian-origin stars and directors, some of whom later became the most successful participants in cinema under the Nazis, possessed German citizenship or were Austrian citizens who years later were absorbed as Reich citizens.[136] Most, however, did not have this option and thus hundreds of Hungarian film professionals residing in Berlin soon found themselves out of work.[137] The resulting 1932 deportations of these jobless Hungarians caused great consternation in Budapest, producing vengeful calls for boycotts of German films and expulsions of German stage players in response to this 'pogrom against Hungarians'.[138] According to German accounts, the new law forced nearly 300 Hungarians from Berlin.[139]

Cast out, Hungarian émigrés flooded the studios of Budapest, Hollywood, Vienna, London, and Paris.[140] They were followed by a second wave soon after the Nazi seizure of power. The Nazis' June 1933 Fourth Film Order made it virtually impossible for Hungarians, particularly Hungarian Jews, to choose German citizenship. Internationally known Hungarian film directors István Székely, Béla Gaál, Paul Czinner, László Vajda, and Viktor Gertler; actors Szőke Szakáll, Ernő Szenes, Gitta Alpár, Rózsi Bársony, and Franciska Gaál; composers Pál Ábrahám and Miklós Brodszky; screenwriter Károly Nóti; editor Endre Marton; and producer József Paszternák (Joe Pasternak), all of whom had flitted among European and American studios, now found themselves barred from Berlin, where they had been responsible for upwards of 120 films.[141]

Germany's transformation presented both new challenges for Hungary and unforeseen opportunities to rethink Hungarian film production and film culture. Hungary's cultural and political leaders did not initially recognize the potential represented by the sudden influx of talent from Germany, instead viewing the expellees as a burden. A mid-1932 article in the premier Hungarian film trade magazine *Film Culture* [*Filmkultúra*] reflected this sentiment. Rather than welcome the returning film professionals, *Filmkultúra's* editors worried that these men and women would merely add to Hungarian unemployment lines because the domestic industry was too underdeveloped to provide jobs.[142] This fear, however, was misplaced. As the émigrés returned, Prime Minister Károlyi's government began to publicize its willingness to subsidize any domestically produced Hungarian-language movies. Following this, banks began to offer more favorable terms and new domestic Hungarian investors agreed to underwrite film projects.[143] Several French and American film companies moved their Central European branches out of Berlin and into Budapest, and Universal even agreed to fund one Hungarian film in 1933.[144] Hungarian producers soon realized that German chauvinism offered them not only the chance to re-establish the allegiances of home audiences, but an opportunity to attract new spectators from among 'the 150 million viewers of German film' sure to be unsatisfied by the 'dross' likely to come from a nazifying Germany.[145]

The backlash German actions provoked spread beyond production to every facet of Hungarian film culture, including viewership. Thus, the heyday of German film in Hungary proved short-lived. Near the end of 1932, when the Third German Film Order sparked the expulsion process Minister Scitovszky feared, the rise in German imports leveled off, and over the next year declined sharply, falling by 40 percent. In the 18 months from 1933 to mid-1934, Germany's share of the Hungarian market plummeted from 60 percent to less than 19 percent.[146] Repulsed by political developments in Germany and a *Der Stürmer*-led hate campaign targeting Budapest and Hungarian stars recently booted from Berlin, segments of the moviegoing public, particularly the Jews of Budapest, conducted quiet boycotts of German film.[147] Critics were offended by what they saw as a qualitative change in German production. Hungarian distributors responded to their audiences, backing away from German imports. To maintain its film culture and to continue to provide its theaters with material, Hungary turned back to America. Hollywood was all too happy to reclaim its position as the preeminent supplier to the Hungarian market, and did so with impressive speed. Still, the precipitous drop in German imports left a void, one that Hungarian production, while still minimal in early 1934, now had the opportunity to fill.

Conclusion

Reconstructing the structures of Hungary's film business in the early 1930s, industry elites found themselves both hamstrung by the country's cinema history and liberated from it. Hungary's leaders dreamt of regaining the status they had in 1918, when the motion picture business thrived, Greater

Hungary was intact, and the value of Hungarian culture was unquestioned. But at the start of the sound age, the problems which had destroyed silent cinema culture remained unresolved. Where to find capital, both monetary and human, for an industry even more reliant on both due to the sound transformation, remained a mystery. The problem of the incestuous oligarchic bureaucracy, born of the silent era, worsened. State bureaucrats acted haphazardly to mediate the harms of capitalism through targeted, yet underfunded interventions, all the while creating rules to enrich themselves. They rationalized their actions and the decrees they issued by waving a flag of nationalism sometimes tainted with antisemitism. But in some respects, the early 1930s were different from the preceding period. The severity of the economic crisis and rapidly changing geopolitics inspired political elites to back off or rethink both their nationalism and antisemitism. As a result, they hailed the return of expatriate Jewish-Hungarian professionals, retreated from restrictions on foreign films, and welcomed international initiatives which contributed to the preservation of Hungarian film culture. They embraced the transnational nature of film. These haphazard compromises ironically gave the industry some resilience, established infrastructure, and enabled Hungary's film elite to begin thinking seriously about the national cinema culture they wished to foster.

Notes

1. Among the states in this region in 1931, only Sweden, Poland, Czechoslovakia, Hungary, and Austria could make sound film.
2. John Cunningham, *Hungarian Cinema: From Coffeehouse to Multiplex* (London, 2004) 6, 203n.5–6.
3. See John Lukacs, *Budapest 1900: A Historical Portrait of a City & Its Culture* (New York, 1988), 139, 146–54.
4. Ibid., 178–9; Kőháti, *Tovamozduló ember*, 15–16. Coffeehouses and inns, often owned by Jewish proprietors, became the first theaters and help explain the disproportionate numbers of Jews in Hungarian film exhibition.
5. Henrik Castiglione, 'Budapest mozgóképüzemei', *Magyar Statistikai Szemle* VII/2 (February 1929), 210–17. Reprinted in *Filmspirál* VII/26-2 (2001), http://epa.oszk.hu/00300/00336/00010/budapesti.htm. Castiglione claims 108, John Cunningham claims 110, Ignács Romsics claims 92 in 1912. See Cunningham, *Hungarian Cinema*, 7; Romsics, *Hungary in the Twentieth Century*, trans. T. Wilkinson, (Budapest, 1999), 77.

6. Greater Hungary, which in 1915 had 270 cinemas, is the term that refers to the geographical Kingdom of Hungary that existed within the Habsburg Empire until 1918.
7. Kőháti, *Tovamozduló ember*, 46.
8. József Patkos, quoted in Kőháti, *Tovamoduló ember*, 46.
9. Kertész, quoted in Marton, *The Great Escape*, 21.
10. Nemeskürty, *Word and Image*, 21.
11. Ibid. *A képpé varázsolt idő*, 116. According to Nemeskürty, the Hungarian film industry produced 18 features in 1914; 26 in 1915; 47 in 1916; 75 in 1917; and 109 in 1918.
12. Nemeskürty, *Word and Image*, 33. Deésy held the wartime production record, making a total of 34 films between 1915 and 1918. Mihály Kertész, the most prolific of the early silent film directors, made somewhere between 38 and 47 films 1912–18, while Alexander Korda produced some 19 films in the same period. Other less well-known directors made scores of films during the war. For film production statistics, see Nemeskürty, *Word and Image*, 29–36 and *A képpé varázsolt idő*, 116–17; Kőháti, *Tovamoduló ember*, 154, and Graham Petrie, 'Hungarian Silent Cinema Rediscovered', *The Hungarian Quarterly*, XXXVIII/147 (Autumn 1997), 152–60.
13. On the 'golden age', see Kőháti, *Tovamoduló ember*, 91.
14. Castiglione, 'Budapest mozgóképüzemei', 210–17; Nemeskürty, *Word and Image*, 37.
15. Victoria de Grazia, *Irresistible Empire. America's Advance through Twentieth-Century Europe* (Cambridge, MA, 2005), 304.
16. Kőháti, *Tovamoduló ember*, 95.
17. Mira Leihm & Antonín Leihm, *The Most Important Art. Eastern European Film after 1945* (Berkeley, CA, 1977), 11.
18. Nemeskürty, *Word and Image*, 45.
19. Nemeskürty, *A képpé varázsolt idő*, 186.
20. Júlia Kormját, 'A mozi a kommunista társadalom', *Vörös Film* (3 May 1919), reprinted in Zsolt Kőháti (ed), *A Magyar Film Olvasókönyve (1908–1943)* (Budapest, 2001), 67–9. Komját was Lukács' deputy.
21. Kőháti, *Tovamoduló ember*, 154.
22. Marton, *The Great Escape*, 30.
23. Nemeskürty, *Word and Image*, 55.
24. Cunningham, *Hungarian Cinema*, 25.
25. In its first ten years (1920–9) the OMB prohibited, on average, four to five percent of the films it reviewed. Virtually all forbidden films, however, were foreign productions. Márk Záhonyi-Ábel, 'Filmcenzúra Magyarországon a Horthy-korszak első évtizedében', *Médiakutató* (Summer 2012), http://www.mediakutato.hu/cikk/2012_02_nyar/12_filmcenzura_magyarorszagon/

26. Nemeskürty reports Hungarian production numbers as: four films made in 1922; 14 in 1923; seven in 1924; two in 1925; three in 1926; three in 1927; and one in 1928. Nemeskürty, *A képpé varázsolt idő*, 224.
27. Kőháti, *Tovamoduló ember*, 307–8. The two most popular films in Hungary in 1928, *Hungarian Rhapsody* [*Magyar rapszódia*] and *The Csárdás Queen* [*Csárdáskirálynő*], were Hungarian-themed products of Austrian and German companies. On Hungarian silent cinema's collapse, see Kőháti, *Tovamozduló ember*, 286–331.
28. Tytti Soila, 'Finland', and Gunnar Iverson, 'Norway', in T. Soila, A. Söderbergh Widding, and G. Iverson, *Nordic National Cinemas* (London, 1998), 40–1, 110.
29. Kenton Bamford, *Distorted Images: British National Identity and Film in the 1920s* (London, 1999).
30. Ricci, *Cinema and Fascism*, 69.
31. Andor Lajta, 'Az év tőrténete', *Filmművészeti Évkönyv* (1930), 193.
32. Janos, *East Central Europe in the Modern World*, 5. Janos indicates the 1930 population was just under 8.7 million.
33. Frances Goldwyn, quoted in Alan Williams, 'Historical and Theoretical Issues in the Coming of Recorded Sound to the Cinema', in R. Altman (ed), *Sound Theory, Sound Practice* (New York, 1992), 131.
34. Béla Balázs, for example, wrote: 'The sound film will teach us to analyze even chaotic noise…and read the score of life's symphony. Our ear will hear the different voices in the general babble and distinguish their character as manifestations of individual life. It is an old maxim that art saves us from chaos…. The vocation of the sound film is to redeem us from the chaos of shapeless noise by accepting it as expression, as significance, as meaning…' Quoted in 'Theory of the Film: Sound', in E. Weis and J. Belton (eds), *Film Sound: Theory and Practice* (New York, 1985), 116.
35. Richard Maltby & Ruth Vasey, 'The International Language Problem: European Reactions to Hollywood's Conversion to Sound', in D. Ellwood & R. Kroes (eds), *Hollywood in Europe: Experiences of Cultural Hegemony* (Amsterdam, 1994), 68; J. Forbes & S. Street (eds), *European Cinema* (London, 2000), 30.
36. Leon Brun, quoted in Skaff, *The Law of the Looking Glass*, 118.
37. de Grazia, *Irresistible Empire*, 315.
38. Gyula Gyárfás, 'Magyar film-feltámadás!' *FK* IV/4 (1 April 1931), 1–2. Gyárfás claimed that the danger to the Hungarian language placed Hungarian political sovereignty at risk. A note on *Filmkultúra*. Edited by Andor Lajta, it was the film industry's official, 'most valuable' and most widely circulated trade magazine in the 1930s. It was funded jointly by government sources and every sector of the film business. Mária Kovács, 'A Filmkultúra', in I. Nemeskürty, I. Karcsai Kulcsár & M. Kovács (eds), *A magyar hangosfilm története a kezdetektől 1939-ig* (Budapest, 1975), 342–3.

39. Gyula Gyárfás, 'Magyar filmszinházak eltrösztösödése!' *FK* III/6 (1 June 1930), 1–2. Gyárfás used the term *anyagi-Trianon* [material Trianon] to describe foreign control of Hungarian film interests. Any reference to Trianon, which Hungarians universally regarded as unfair, vindictive, and illegitimate, evoked a visceral response.
40. In interwar Great Power parlance, Hungary was generally included in the geographical designation Southeastern Europe but excluded from the pejorative category of the Balkans. The Hungarians themselves seemed to believe that Western Europe constantly needed reminding that Hungary was somehow distinct from the lands Western Europeans depicted as their 'uncivilized other'. For the seminal discussion of the cultural construction of the term 'Balkan' see Maria Todorova, *Imagining the Balkans* (New York, 1997). On the importance of film cultural supremacy, see: 'És most jöjjön az állam!' *Magyar Filmkurír* III/ 7 (20 February 1929), 1.
41. In an article titled 'The Americans and Hungarian film production', the editors of *Moziviläg* claimed sound film production would bring about a flowering of not only Hungarian film but literature, acting, and culture in general. 'Az amerikaiak és a magyar filmgyártás', *Mozivilág* II (XIX)/19 (11 May 1930), 1–2.
42. The frequently used Hungarian phrase is *nemzeti-nevelő.*
43. Paul Ignotus, *Hungary* (London, 1972), 44.
44. My definition of the film elite combines ideas from Thomas Saunders and George Mosse. Saunders' full definition of the film establishment is the 'creative personnel (producers, directors, performers and screen authors), entrepreneurs (board directors, distributors and theater owners), and an army of journalists, critics, advertisers and miscellaneous camp followers'. Thomas Saunders, *Hollywood in Berlin: American Cinema and Weimar Germany* (Berkeley, CA, 1994), 18. For Mosse's 'technicians of power' quote, see Mosse, *The Culture of Western Europe. The Nineteenth and Twentieth Centuries,* 3rd edition (Boulder, CO, 1988), 297. Mosse determined that by the 1920s, the elite could be understood as a 'class of leaders' who 'constituted the center of thought...determined...its own ideology...[and] directed the course of politics and society'.
45. Janos, *The Politics of Backwardness,* esp. 201–312. Mosse, *The Culture of Western Europe,* 297ff.
46. Nemeskürty, *A képpé varázsolt idő,* 233. Nemeskürty describes Bingert as 'excellently prepared from a professional standpoint'.
47. Mária Ormos, *Egy magyar médiavezér: Kozma Miklós* (Budapest, 2000), 2 volumes. Kozma selected Zoltán Taubinger-Tőrey to manage the MFI. Taubinger-Tőrey was a lawyer whose previous government work involved supervision of the state tobacco monopoly. Ibid., 232.
48. On imaginative communities, see Aviel Roshwald, *The Endurance of Nationalism* (Cambridge, UK, 2006).

49. The apocryphal stories of the sign over the entrance to Paramount's studios reading 'Just because you are Hungarian does not mean you are a genius' and Otto Preminger's later lament that to make it in Hollywood, you should be required to do more than just speak Hungarian, lend credibility to this assertion.
50. 'És mit exportálunk mi?' *Magyar Filmkurír* III/3 (20 January 1929), 6.
51. Dezső Németh, 'Magyar hangosfilmgyártás', *Mozivilág II* (XIX)/13 (30 March 1930), 1.
52. Gyula Gyárfás outlined the worst case scenario, claiming that 200–300 theaters ceased operating or folded between 1929 and 1932 and that in 1932, only 178 Hungarian theaters regularly screened films. Gyárfás, 'Végső pusztulás fenyegeti a magyar filmszinházakat', *FK* V/4 (1 April 1932), 8. The catastrophic reduction in operating theaters occurred in early 1930 as theaters tried to equip themselves for sound and the international credit crunch hit full force. Gyula Décsi, an official of the MMOE, reported that as of April 1930, only 120 of Hungary's theaters operated daily and approximately 80 more operated inconsistently. Décsi's speech is summarized in 'Szomorú számok a mozi súlyos válságáról', *Mozivilág II* (XIX)/15 (13 April 1930), 1. See also Lajta, 'A magyar film története. V. A magyar hangosfilm korszak első 16 éve, 1929–44', 60, 64; Langer, 'Fejezetek', 91.
53. Ruth Vasey incorrectly claims that although large numbers of theaters closed, there was no decline in total theater capacity. Vasey, *The World According to Hollywood, 1918–1939* (Madison, WI, 1997), 88.
54. *BFL* – Stylus Filmipari Rt. – Cg 11047. 'Jelentése a Stylus filmipari részvéntársaság igazgatóságának az 1930 Január 1-től 1930 December 31.-ig terjedő üzletévre vonatkozóan', 15 February 1931.
55. Károly Lajthay, 'Számok és valóság a magyar filmgyártás kérdésében', *FK* IV/11 (1 November 1931), 9. The general stagnation and lack of quality films, suggested the director Lajthay, precluded the development of 'film favorites' in Hungary. Without such favorite actors, actresses and directors, audiences lost their desire to see films.
56. According to Henrik Castiglione, from 1927–30, Budapest ticket sales plummeted nearly 39 percent. In 1927, theaters averaged 108 percent of capacity; in 1930 they were only 65 percent full. Annual ticket sales in Budapest in 1927 were 12.496 million (37,000 per day); in 1930, they declined to 7.664 million annually (23,000 daily). Castiglione, 'Három év a statisztika tükrében', *FK* III/12 (1 December 1930), 3–5; and Castiglione, 'A mozgóképszinházak viszonyai az új népszámlálás tükrében', *FK* IV/5 (1 May 1931), 2–4.
57. Castiglione estimated that in 1933, 30 percent of rural theaters were financially untenable. Castiglione, 'A vidéki filmszinházak az uj népszámlálás tükrében', *FK* VI/5 (1 May 1933), 1–4.

58. 'Negyven százalékkal visszaesett a vidéki filmszinházak látogatottsága', *FK* V/4 (1 April 1932), 8–9, quoting Sándor Fenyő. See also Pál Morvay, 'Magyarország filmszinházákultúrája a statisztika bonckése alatt', *FK* VI/7–8 (1 July 1933), 3; Lajta, 'A magyar film története', 64–6; Gyula Gyárfás, quoted in 'Conditions in Hungary', *New York Times* (18 October 1931), 112.
59. This was the mantra repeated in the film trade journals, particularly by representatives of the MMOE such as Gyula Gyárfás, Gyula Décsi, Henrik Castiglione, and others. See notes 52, 55–8.
60. 'Mi van a gseftelő közintézmények moziengedélyével?' *Filmújság. Zsolnai László Hétilapja* II/9 (4 March 1933), 1. Filtering through often unreliable statistics, Zsolt Kőháti reports that the Theater Revision of 1921 affected 52 of the 88 cinema licenses in Budapest. Kőháti, *Tovamozduló ember,* 183.
61. Langer, 'Fejezetek', 26. See also 'A moziengedélyek és a rokkantkérdés', *Mozivilág* XXI/46–7 (27 November 1932), 2.
62. Decree 6900/1923 B.M. sz. See also Sándor, *Őrségváltás*, 35–6.
63. 'Meg kell szüntetni az álláshalmozást a moziengedélyesek körében', *Filmújság. Zsolnai László Hétilapja* I/1 (10 December 1932), 3.
64. 'Rövidesen elkészül a belügyminiszterium mozireviziós javaslata', *Filmújság. Zsolnai László Hétilapja* I/1 (10 December 1932), 6.
65. Rezső Vári, 'A szűnetelő mozgóképszinházak jogi helyzetének rendezése', *FK* V/9 (1 September 1930), 2–3.
66. The 1925 legislation regularized censorship fees, fire and police protection fees, and film luxury taxes.
67. Created by Prime Minister Bethlen's order 6292/1925, 6 November 1925, the Fund's governing board included representatives of the Prime Ministry, Ministry of Commerce, Ministry of Religion and Education, and Interior Ministry. By 1928, the OMME, MMOE, and the Musicians' and Writers' Union had also joined.
68. Langer, 'Fejezetek', 91–2. Langer cites an MFI 'Memorandum' 58/1105/1932.
69. Ibid., 40. The Fund put HP 810,000 toward purchase and modernization.
70. From 1929–32, for example, in addition to Bingert, Hunnia's Directorate included Elek Horváth, Zoltán Bencs, Frigyes Pogány, and Ernő Fehér. Horváth was, in December 1930, a deputy state secretary in the Interior Ministry, President of the Censorship Committee, and a member of the board of the Hungarian Film Office, Hungary's other quasi-governmental film company. By 1931 a Ministerial Advisor in the Prime Minister's Office, Bencs was a longtime ranking member of the OMB, the Film Council, and the FIF. Pogány, in the early 1930s the president of the OMME, was a top official in the Ministry of Religion and Education, and the Vice-President of both the Film Council and the FIF. Fehér was a mechanical engineering specialist from the Commerce Ministry who, like most of his comrades, also served on the Film Council and FIF.

71. 'A rosszul értesültek', *Mozivilág* II (XIX)/7 (16 February 1930), 1; 'Miért van szükség a Filmípari Alap megszüntetésére', *Magyar Filmkurír* III/7 (16 February 1930); "A Magyar filmgyártás legfőbb akadálya: a Filmalap', *Magyar Filmkurír* III/26 (29 June 1930), 1–2.
72. 'Kozma Miklós lesz az átszervezett magyar filmgyártás vezetője', *Filmújság. Zsolnai László Hétilapja* II/1 (7 January 1933), 1, 6.
73. 'Tőkehiány miatt visszafejlődik az európai filmgyártás', *FK* V/12 (1 December 1932), 3–4. One of the leading advocates of this idea was the Hungarian producer/distributor of international repute, Joe Pasternak.
74. Sándor Fenyő & Gyula Gyárfás, 'A magyar filmgyártás ügye', *FK* III/10 (1 October 1930), 9–10.
75. Foreign Ministry share ownership is confirmed by numerous sources, such as the Foreign Ministry document cited in the following footnote and a Foreign Ministry letter 3 September 1929. *MOL* – KM, K66-cs474, 1940, III-6/c: Film, szinházügyek; Közérdekeltségek Felügyelő Hatósága to the MFI Directorate, 4.310/1937, 8 November 1937. *MOL* – MKI, K429-cs59-t2. Vegyes MTI levelezés 1938–9; and numerous documents in the MFI document collection, *MOL* – MFI, K675-cs1-t2, 1928–44. See also Ormos, *Egy magyar médiavezér*, 103.
76. *MOL* – KM, K66-cs335, 1937, III-6/c: Film, szinházügyek. KÜM Kulturális Osztály to Foreign Minister, 4 December 1937. From 1928 on, the Directorate included at least one Foreign Ministry delegate, with Zoltán Gerevich and Baron Lajos Villáni [Villany] serving together from the mid-1930s into the 1940s.
77. Elek Horváth, Dr. Zoltán Bencs, and Dr. Román Felicides, all government officials and Hunnia board members, also sat on the MFI Directorate in the early 1930s. Additionally, Felicides served on the OMB.
78. Zsolt Kőháti, 'Magyar film hangot keres (1931–1938)', *Filmspirál* II/2-1 (1996), 97; 'A magyar mozgószinházak végső vergődése és sulyos válsága 1932 első félévében', *FK* V/5 (1 May 1932), 2.
79. Decree 2.900/1929 M.E. sz. A 1,500 meter film ran approximately one hour; HP 60,000 were worth around US $10,500 in 1930. The 5.71 *pengő* to the dollar rate is from Gyula Borbándi, *Der ungarische Populismus* (Mainz, 1976), 54, footnote 55.
80. *MOL* – BM, K158–cs3, OMB, 1932. Alapsz. 11, 1929 – 'A m.kir.minisztériumnak 2.900/1929 M.E.sz. rendelete, a magyar filmgyártás fejlesztésről szóló 4.963/1925 M.E.sz. rendelete módositása'.
81. Andor Lajta, the editor of *Filmkultúra* and post-WWII authority on interwar Hungarian film, indicated that the government itself refused to enforce the film production requirement because producing silent film was a waste of resources. See Lajta, 'A magyar film története', 25.
82. Importers paid censorship, import, and a per meter surcharges on films exhibited in Hungary.

83. *NARA* – RG59 [State], M-973, Roll 211. 864.4016, letters concerning Hungarian film. See also Edward G. Lowry, 'Certain Factors and Considerations Affecting the European Market', Internal MPPDA memorandum, 25 October 1928 (MPAA Archive, 1928 Foreign Relations File), reproduced in A. Higson & R. Maltby (eds), *'Film Europe' and 'Film America': Cinema, Commerce and Cultural Exchange 1920–39* (Exeter, G.B, 1999), 364–6.
84. Vasey, *The World According to Hollywood,* 8.
85. 3080/1931 M.E. sz., issued by Prime Minister István Bethlen in May 1931, authorized subventions. This act, unfortunately, had had little effect, considering there were few films eligible for the subsidies and the FIF did little to facilitate disbursement. Nor did Gyula Károly's re-authorization of the act, 2670/ 1932 M.E. sz., in June 1932, have much effect. See 'Megjelent a külfoldi filmek újabb pótdijairól szóló kormányrendelet', *Mozivilág* XX/24–5 (21 June 1931), 1; and Nemeskürty, 'A magyar hangosfilm első evei 1931–38', 31.
86. 'Erdélyi Mór m.kir. kormányfőtanácsos nyilatkozik a filmszakma fontos kérdéséiről', *Mozivilág* XX/38 (20 September 1931), 4–5.
87. I use Central Europe geographically to refer to Germany, Poland, Czechoslovakia, and Austria. Southeastern Europe refers to Yugoslavia, Romania, and Bulgaria. Hungary is included in both terms.
88. One of Hungary's well known émigré actors, Mihály Várkonyi [Victor Varconi], believed that co-productions were Hungary's only choice, as independent Hungarian sound filmmaking was impossible. 'Várkonyi Mihály Budapesten', *Mozivilág* II (XIX)/17 (27 April 1930), 3.
89. János Bingert, 'Magyarország és a hangosfilmgyártás', *FK* III/2 (1 February 1930), 1–3. Czech authorities, incidentally, adopted the same strategy. Petr Szczepanik, 'Undoing the National: Representing National Space in 1930s Czechoslovak Multiple-Language Versions', *Cinema & Cie* 4 (Spring 2004), 56.
90. See Bingert's comments in a speech before the Hungarian Film Club reported in 'A magyar hangosfilmgyártás lehetőségei', *Magyar Filmkurír* III/5 (2 February 1930).
91. 'Budapestre az európai hangosfilmcentrumot', *Magyar Filmkurír* III/10 (9 March 1930).
92. *NARA* – RG59 [State], M-1206, Roll 6, 864.4061/Motion Pictures 38. J. Butler Wright to the Sec State, Despatch No. 572, 30 April 1930.
93. 'Tells Foreign Film Plans', *New York Times* (21 May 1930), 35.
94. Gyula Pekár, 'Lesz magyar film!' *Magyar Filmkurír* III/23 (8 June 1930), 3.
95. Among the world-renowned professionals of Hungarian ancestry or citizenship working abroad were a contingent in London including the director Sándor Korda (Alexander Korda); his brothers Vince and Zoltán, a set-designer and director, respectively; his wife Mária; the writer Laurence Pearson (Lajos Biró); the costume designer Marcell Vértes; and the composer Miklós Rózsa (who also worked regularly in Hollywood). A commuter

between Hollywood and London was the actor Leslie Howard, star of *The Scarlet Pimpernel, Pygmalion,* and *Romeo and Juliet.* In Hollywood one found a good number of Hungarians in powerful positions. Adolph Zukor (Adolf Zucker) founded the Paramount Studio, and Vilmos Friedmann (William Fox) did the same for Twentieth Century Fox studios. The actress Vilma Bánky and the actors Mihály Várkonyi (Victor Varconi), László Lőwenstein (Peter Lorre), and Béla Blaskó/Arisztid Olt (Béla Lugosi) worked in various European capitals before settling in the United States in the late 1920s or early 1930s. Among the other Hungarians in Hollywood in the 1920s and 30s were three future Oscar winners, the directors Mihály Kertész (Michael Curtiz) and György Cukor (George Cukor), and the actor Pál Lukács (Paul Lukas). In late twenties Germany resided a massive group of influential Hungarians. Included were screenwriters László Vajda the elder, writer of *Westfront 1918,* Károly Nóti (Karl Noti), and Menyhért Lengyel (Menyhért Lebovics); the directors Géza Bolváry, Pál Czinner, Arzen Cserépy (of *Fredericus Rex* fame), and József Baky (later the director of the German color smash *Münchhausen*); the composers Pál Ábrahám and Miklós Brodszky; and the players Károly (Puffy) Huszár, Paul and Attila Hörbiger, Szvetiszlav Petrovics (Ivan Petrovich), Szőke Szakáll (S.Z. Sakall/Jenő Gerő), Franciska Gaál, Gitta Alpár, Rószi (Rose) Bársonyi and Kató Nagy (Käthe von Nagy). Pál Fejős could be found in Hollywood, Paris, and Vienna in the late 1920s and early 1930s. In Paris was the cinematographer Rudolf Máté (of *La Passion de Jeanne d'Arc* renown). Incidentally, the majority of the aforementioned individuals were Jewish, and many permanently relocated to the US in the 1930s.

96. Nemeskürty, 'Előzmények', in Nemeskürty, et al. (eds), *A magyar hangosfilm története a kezdetektől 1939-ig,* 15; Maltby & Vasey, 'The International Language Problem', 87.
97. Ginette Vincendeau, 'Hollywood Babel: The Coming of Sound and the Multiple Language Version', *Screen* 29/2 (Spring 1988), 37. Kristin Thompson estimates the cost of each foreign version at $30,000–40,000 beyond the cost of the original language production, versus the approximately $2,500 it cost to subtitle the average silent film. See Thompson, *Exporting Entertainment,* 160.
98. Romsics, *Hungary in the Twentieth Century,* 139–40; David S. Frey, 'National Cinema, World Stage: A History of Hungary's Sound Film Industry, 1929–44' (Columbia University: Unpublished Ph.D. Dissertation, 2003), 61–2.
99. *MOL* – KM, K66-cs164, 1930, III-6/c, Filmügyek. István Csáky to KM Sándor Khuen-Hédervári, 27 December 1929.
100. *MOL* – KM, K66-cs164, 1930, III-6/c, Filmügyek. Zoltán Gerevich to KM Khuen-Hédervári, 31 December 1929.
101. On 22 January 1932, the Societé Anonyme des Films, a French front firm for Mannheim and Osso's money, contributed 50,000 P to the 'Magus' Irók és Szinészek Filmgyártó Rt. The company was reorganized and renamed Osso

Film. Its objectives were to make film in Hungary, make film for export to France, and import French film into Hungary. *BFL* - City Film Rt, Cg26059. 'Jgykv'. of 'Magus' Irók és Szinészek Filmgyártó, 14 January 1932.

102. Péter Ábel, *Magyar filmgyártó, kölcsönző vállalatok filmlaboratóriumok törzslapjai /1928–38/*' (Budapest: Unpublished MFI project, 1 July 1972), entry 101.
103. *BFL* - City Film Rt, Cg26059. 'Jgykv.' of 'Magus' Irók és Szinészek Filmgyártó Rt, 14 January 1932.
104. G. István Reményi, *Ismerjük őket? Zsidó származású nevezetes magyarok arcképcsarnoka* (Budapest, 1997), 54.
105. *BFL* - City Film Rt, Cg26059 and Minerva Filmgyártó KFT, Cg33470.
106. 'Lenkeffy Ica lesz a párisi Osso-filmgyár budapesti-sztárja', *Magyar Filmkurír* VI/4–5 (3 February 1932), 7.
107. Lajta, 'A magyar film története', 32. Lajta claims the original investment in the second Hunnia studio was premised on free studio rental for up to 16 films.
108. Langer, 'Fejezetek', 90. *BFL* - Hunnia Filmgyár Rt. - Cg29830 - 3949. Közgyülés, 2 June 1933.
109. 'Minerva, Hunnia, Osso,' *Filmtechnika* III/9 (June 1932), 9–10. Italics mine.
110. Lajta, 'A magyar film története', 32.
111. 'Francia lobogó', *FK* V/3 (1 March 1932), 1.
112. Balogh & Király, *'Csak egy nap a világ...'*, 110.
113. *MOL-Ó* - Hunnia, Z1123-r2-d5, Igazg. iratok, 1940–3. Undated [1944] report 'Hunnia Filmgyár Rt'.
114. Balogh & Király, *'Csak egy nap a világ...'*, 178–89; and Balázs Varga (ed), *Játékfilmek 1931–1997* (Budapest, 1998), 17. The Hungarian version of the film no longer exists.
115. This assertion appears in the March-April 1932 issues of *Magyar Filmkurír.*
116. Osso advertisement in *Moziviláq* XXI/27–8 (17 July 1932), 3.
117. Osso partnered with the Hungarian banker turned part-time producer Sándor Winter, and several existing projects went forward under the Osso name well into 1933, including *Miss Iza* [*Iza Néni*], *Ghost Train* [*Kísertelek vonata*], *Everything for the Woman!* [*Minden a nőért!*] and *Rákóczi March* [*Rákóczi induló*]. Several were produced in German, English, and French. Winter eventually took over all Osso assets and folded them into City Film.
118. Jenő Heltai, 'A francia filmprodukció mérlege', *FK* V/10 (1 October 1932), 1–2.
119. The import of the French film *Paris Béguin* [*Pipascok és apascok*], for example, was forbidden, edited, and still rejected by the OMB. See *MOL* - BM, K158-cs4, OMB, 454/1932 eln.
120. The *Cégbizottság* attempted to force Minerva to change its name for the arcane reason that a mythological name did not properly identify the type of service the company provided. *BFL* - Minerva Filmgyártó KFT, Cg33470.

121. *BFL* – City Film Rt, Cg26059. 'Letter to Shareholders from the Board of Directors', 31 December 1932.
122. 'Osso Film Rt Mérlegszámla 1932. December 31-én', *A Budapesti Közlöny Hivatalos Értesitője* (13 April 1933), 15.
123. Osso monies funneled through the Minerva, Osso, and City film companies funded five Hungarian features. The larger total includes several French and Hungarian cultural films. *BFL* – Minerva Filmgyártó KFT, Cg33470. Dr. Róbert Palágyi [Minerva's lawyer] to Budapesti Kir. Törvényszék, 30 November 1932, 3.
124. On the sound patents/protocols dispute, see Thompson, *Exporting Entertainment,* 148–58 and Douglas Gomery, 'Europe converts to sound', in Weis and Belton (eds), *Film Sound: Theory and Practice,* 28–30.
125. 'Hunnia Filmgyár', *FK* IV/4 (1 April 1931), 10–11.
126. Hunnia documents list *Hyppolit* as one of Hungary's greatest box office successes, 1931–44, and contemporary press coverage confirms this. *MOL-Ó* – Hunnia, Z1123-r2-d5, Igazg. iratok, 1940–3. Undated [1944] report 'Hunnia Filmgyár Rt'.
127. Even before Hungary had its own sound studios, Germany demonstrated a willingness to engage in co-productions with Hungary. The German film giant Ufa and its production head Erich Pommer were actually responsible for the first entirely Hungarian-language sound feature film, *Sunday Afternoon* [*Vasárnap délután*], directed by Hans Schwarz. Much of this dual German-Hungarian language picture, which premiered in Hungary in March 1930, was filmed in Hungary, but it most of its personnel were German and the film was prepared in Ufa's Babelsberg studios. That Hungarian martial marching songs were sung even in the German version became a point of pride for Hungary's elites.
128. Géza Bolváry, 'Ma mar nem lehet kételkedni a magyar filmgyártás lehetőségében', *Magyar Filmkurír* VI/4–5 (6 February 1932), 6.
129. *AAA* – GsB, Fach 218, Akt. VII, 16; Filmwesen, Bd. 1 – 1935–8. GsB (Schoen) to AA, 1 October 1930, B.Nr.659 XVI, 4.
130. 'Für und gegen die deutschen Filme in Ungarn', *Deutsche Tages-Zeitung* (8 October 1930).
131. *MOL* – BM, K158-cs7, OMB 1936, 188–9. 'Nyilvántartás az OMB által megvizsgált filmek évi vizsgálatának részletes statisztikai adatairól 1920'. évtől-1932. év dec. 31-ig. Statistics for the full 1932/3 film year show Germany holding a 3:2 edge, much of which resulted from sales agreements made prior to Germany's June 1932 enactment of the third film law and Hitler's appointment as Chancellor of Germany in January 1933. Depending upon the source, the ratio was either 114:72 (Langer) or 105:64 (Castiglione). Langer, 'Fejezetek', 120; Henrik Castiglione, '1933/34 Grafikonja – Adatok a szézon mérlegéhez', *FK* VII/6 (1 June 1934), 5–7.

132. For Czechoslovakia, Nancy M. Wingfield, 'When Film Became National: "Talkies" and the Anti-German Demonstrations in Prague', *Austrian History Yearbook* XXIX/1 (1998), 113–38; for Poland, *Deutsche Tages-Zeitung* [Karlsbad], no. 270, (21 November 1930), 5; Skaff, *Through the Looking Glass,* 74–5; for France, Martine Danan, 'Hollywood's Hegemonic Strategies: Overcoming French Nationalism with the Advent of Sound', in Higson & Maltby, (eds), *'Film Europe' and 'Film America',* 230. For evidence that the Hungarian response to German-language film differed from its neighbors, many of whom saw the German sound film threat as analogous to that posed by Hollywood and English, see Nataša Ďurovičová, 'The Hollywood Multilinguals, 1929–1933' in Altman (ed), *Sound Theory, Sound Practice,* 150. Although Czech nationalists opposed German-language film, Czech audiences, more than any others in Central Europe, welcomed German-language film, at least throughout 1934. Petr Szczepanik, '"Tief in einem deutschen Einflussbereich." Die Aufführung und Rezeption deutschsprachiger Filme in der Tschechoslowakei in den frühen 1930er Jahren', in Jan Distelmeyer (ed), *Babylon in FilmEuropa. Mehrspachen-Versionen der 1930er Jahre* (Hamburg, 2006), 89–102; Gernot Heiss and Ivan Klimeš, 'Kulturindustrie und Politik. Die Filmwirtschaft der Tschechoslowakei und Österreichs in der politischen Krise der dreißiger Jahre', in Heiss & Klimeš (eds), *Obrazy času/Bilder der Zeit* (Prague, 2003), 402.
133. *AAA* - GsB, Fach 218, Akt. VII, 16; Filmwesen, Bd. 1 - 1935–8. Pressebericht Nr. 222, 2 October, 1930. Report on lead article from *Magyarság.*
134. 'Rövidesen megjelenik az uj filmrendelet, amely kötelezővé teszi a magyar filmek játszását', *Filmújság. Zsolnai László Hétilapja* II/6 (11 February 1933), 1.
135. Saunders, *Hollywood in Berlin,* 234; *BA* – Reichskanzlei R43 1/2500, 67–72; Henning von Boehmer and Helmut Reitz, *Der Film in Wirtschaft und Recht* (Berlin, 1933), 64–6.
136. Among those able to continue working in Germany or who worked for the Reich after the *Anschluß* were Käthe von Nagy, Márta Eggerth, Marika Rökk, Hilde von Stolz, Mária von Tasnády, Joseph von Baky, Géza von Bolváry, Paul and Attila Hörbiger, and Géza Cziffra.
137. German authorities specifically targeted the Hungarians and Austrians who participated in German filmmaking when they adopted the new film laws. Wolfgang Mühl-Benninghaus, *Das Ringen um den Tonfilm. Strategien der Elektro- und der Filmindustrie in den 20er und 30er Jahren* (Düsseldorf, 1999), 351–2.
138. *AAA* - GsB, Fach 218, Aktenzeichen VII, 16; Filmwesen, Bd. 1 - 1935–8. Pressebericht, Nr.171, 8 August 1932 – Report on *Újság* article 'Magyarischer Pogrom auf deutschen Filmen'. Also 'A német filmkontigens-rendelet', *FK* V/7–8 (July-August 1932), 1.

139. 'Ausweisung ungarischer Filmleute aus Berlin', *Neues Politisches Volksblatt* (17 July 1932). István Nemeskürty, drawing from Péter Ábel, claims that 'until 1933, Berlin was a Hungarian film city'. Nemeskürty, 'A magyar hangosfilm első évei 1931–1938', in Nemeskürty, et al. (eds), *A magyar hangosfilm története a kezdetektöl 1939-ig*, 38; and Ábel, 'A berlini Magyar filmkolónia krónikjából', *Filmkultúra* 5 (1966), 93–8. Overall, some 900 film professionals 'were made homeless'. Günter Peter Straschek, *Filmemigration aus Nazideutschland* (Westdeutsche Rundfunk, 1975), documentary.
140. Tobias Hochscherf, 'Kennen Wir Uns nicht aus Wien? Émigré Film-makers from Austria in London, 1928–1945', *The Yearbook of the Research Centre for German and Austrian Exile Studies* (2006), 137–9. Hochscherf misidentifies several émigrés as Austrian when in fact they were Hungarian citizens.
141. Rene Geoffroy, *Ungarn als Zufluchtsort und Wirkungsstätte deutschsprachiger Emigranten (1933–1938/39)* (Frankfurt-am-main, Germany, 2001), 268–71, esp. 268; and Ábel, 'A berlini Magyar filmkolónia krónikjából', 93–8.
142. 'A német filmkontigens-rendelet', *FK* V/7–8 (July-August 1932), 1.
143. By 1934, the total number of films produced in Hungary was 12, far exceeding the seven produced in 1933. Of that total, eight were solely funded by domestic Hungarian investors, doubling the previous year.
144. Its German operations experiencing difficulties, Universal agreed to make one film in Hungary: the successful German-Hungarian version film *Skandal in Budapest* [*Pardon, tévedtem*], directed by István Székely. The film's success, however, did not result in additional joint projects.
145. Gyula Gyárfás, 'A Berlini tűzcsóva a film szempontjából', *FK* VI/3 (1 March 1933), 1–2.
146. Henrik Castiglione, '1933/34 Grafikonja – Adatok a szézon mérlegéhez', *FK* VII/6 (1 June 1934), 5–7; 'A magyar filmpiacon 1934. augusztus 1.-től 1935. julius 31.-ig 237 magyar és külföldi film jelent meg', *FK* VIII/11 (1 November 1935), 10–11; *MOL* – BM, K158-cs6, OMB, 1934–5, Alapsz. 365/1934.
147. E.g. 'So sieht er aus', *Der Stürmer* 13 (1933), reprinted in Loacker & Prucha, *Unerwünschtes Kino,* 15, is a vicious attack on the Hungarian Jewish actor Szőke Szakáll that mocks 'Judapest' as the 'final destination for the fat trimmed from the second rate film world'.

2

Constructing the Fantasy of Hungary: Elite Concepts of the National

Introduction

Similar to its neighbors, Hungary struggled through a post-Great War identity crisis, its elites frantically searching for durable, usable notions of the national 'essence', 'soul', or 'character'.[1] Nearly all agreed that culture and aesthetics were expressions of this difference, whether that difference originated in ethno-cultural conceptions of a national ontology, geographical determinism, or biological notions of race.[2] Sound film's very transnational and cosmopolitan origins offered Hungary a wealth of nationalizing opportunities, particularly as domestic production began to surge. Hungarian bureaucrats, from censors to diplomats, were tantalized and confused by these possibilities. Their psychoses about Hungary's past and future place in Europe emerged as they plumbed film's potential. They alternately manipulated and grumbled about stereotypes as they worked to shape and protect the nation's audiovisual imagery or to promote culture as a means for Hungary to regain regional ascendance. Industry participants wrapped themselves in the flag, conflating self-interests and patriotic/national interests. Critics and politicians used film symbolically, decrying and promoting imagery according to their nationalist ideologies.

Buried among these competing interests were important areas of agreement. Critics, bureaucrats, and industry players on all sides concurred that the sound film medium contained unparalleled potential for displaying Hungary's unique 'ethnological colors and external forms'.[3] Even those with disdain for popular culture concurred that sound catapulted film to the top of the artistic pantheon, beyond painting, sculpture, theater, radio, and the industrial arts. No other medium had comparable potential for 'documenting and defining' Hungarian national culture, claimed the director of populist films, István György.[4] Cinema, echoed a top ranking censor, was 'an excellent tool…for bringing into existence a uniform world of feelings and thought, a singular national taste [able to] penetrate the entire society'.[5] This could best be accomplished by fostering a film industry from a 'national point of view', suggested Interior Minister József Széll. In an exchange with *Reichsminister* Joseph Goebbels, Széll agreed that 'Film must suit each *Volk*, and each *Volk's* film must assume a characteristic place in the market'.[6] National film production was, in the words of OMB member Sándor Jeszenszky, 'one of the life and death questions for Hungarian culture'.[7]

The belief that film had a mission beyond entertainment and matters of money, that it could and should aid the country's search for greater social and cultural cohesion and provide a stronger sense of national identity, became an industry-wide rhetorical consensus. Henrik Castiglione, the film industry's statistical guru, succinctly summarized this position. 'Hungarian film', he declaimed, 'is an instrument of the nation, for which there is not and cannot be a program; rather it must be a religion, a spiritual integration into the future'.[8] The majority of Hungary's film community agreed with Castiglione's extravagantly stated position, although most realized that spiritual integration and national unity did, in fact, require a program. This meant that the film community had to determine what it meant for a film, and a broader film culture, to be Hungarian. Not surprisingly, this is where the consensus broke down and the fissures that emerged during the industry's founding became cavernous. Whether cinema was nation-effacing or a site of national indifference, as some contemporaries saw it,[9] or a place where alternative visions of Hungarian modernity were articulated,[10] or even a combination of both, rhetoric linking 'national', 'Hungarian', and motion pictures became pervasive in the 1930s, propelled by the explosion of sound. In this way, Hungary's experience mirrored that of every nation whose state and industry cooperated and competed as they designed so-called national film industries or national cinemas. Rather rapidly, discussions of 'national film' and a healthy 'national film culture' became surrogates for the basic question of 'What is Hungarian?' These discourses and battles over authenticity became enmeshed in and helped shape all of the potent public debates of the day. Populist philosophies, racial nationalism, antisemitism, liberalism, democracy, and competing visions of modernity permeated considerations of Hungarian movie production, distribution, and exhibition.

Within the film community, 'nation' and 'Hungarian' became potent discursive vessels, filled and re-filled according to whatever was politically, economically, religiously, diplomatically, culturally, or aesthetically expedient.[11] Often interpretations were facile, expressions of self-interest, race-based frustration, boasts of cultural superiority, or denials of inferiority. But as the industry rapidly expanded and the discussion became substantive, it inevitably spawned a range of sometimes incompatible conceptions of what constituted the 'national essence' or the national interest. This chapter will provide an overview of the rhetoric and practice of film's perceived national, nationalizing, and nationalist missions through the later-1930s. It will explain some of the ways groups within Hungary's film establishment – Hungary's official government image makers and regulators, its private film producers, its distributors, exhibitors, and critics – attempted to influence representations of Hungarian national identity and define their actions as national.[12] Through the 1930s, I suggest, concepts of nation, the national, and acceptable nationalism were increasingly mediated, constituted, and conflated through film and discourse about cinema. The Hungarian film industry became a focal point for communal contests over identity and provided a stage upon which a surprising variety of visions could arise, compete, and coexist.

The Cultural Politics of the 'Official Nation'

Their state deprived of much of its geopolitical significance, military might, and economic integrity after the conclusion of World War I, Hungary's interwar rulers had little choice but to read tremendous power into culture. Hungary's oligarchs believed the Entente had misunderstood the Hungarian nation, resulting in its unfair treatment at the Paris peace talks in 1919–20. Count Kunó Klebelsberg, the Minister of Religion and Education for much of the 1920s, convinced his colleagues to begin a worldwide campaign promoting Hungarian culture and education in order to correct Western misapprehensions. Once convinced that Hungary belonged among them, the Great Powers would restore Hungary's rightful place in Europe.[13] Along with Klebelsberg, Hungary's leaders soon came to see culture as the most powerful arrow in their diplomatic quiver. They determined 'that the demonstration of "cultural superiority" was the only path' towards renewed international clout and ultimately to overturning the despised Treaty of Trianon.[14] In this vein, Hungary's elites not only grasped but amplified the West's clichéd imperialist rhetoric of the civilizing imperative and dominance as a form of birthright. Hungary's language and the 'civilization' it spawned were not only of special value, but when combined with Hungary's unique position in Europe as the natural authority in the Carpathian and Danubian basins and the Christian bulwark against the spillover of Asia, the cultural diplomatic discourse of nation produced the following logic: the extension of Hungary was the extension of Europe.[15]

While Klebelsberg was just beginning to think about the project of defining Hungarian culture and forging an attractive, uniform national identity in the post-Trianon era, he enlisted Hungary's censors in this effort, empowering them to draw sharper lines demarcating Hungarian national identity. In 1920, Klebelsberg brought together government officials and industry representatives to form the Film Council. He tasked them with appraising moving pictures and supervising the industry according to 'moral, national-defense, national-educational, literary, and artistic points of view'.[16] Nearly simultaneously, the Interior Minister created the Censorship Committee [OMB], authorizing it to evaluate all film products.[17] Among the OMB's guiding principles were requirements that films be forbidden or sanitized if they 'ran counter to the ideas of the nation-state'; 'injured or endangered state security or relations with other states'; or 'brought harm to the prestige of the state, army, or state authorities'.[18] As a result, censors became consciously and unconsciously engaged in state and nation building. To define what ideas ran counter to the nation-state, they needed to discern for themselves what ideas were at the nation-state's core. To determine what might harm the public order, they had to formulate who or what constituted the public.

By means of inclusion and exclusion, bestowing or denying permission for screening, Hungarian censors recognized that they held enormous power to determine what images would be part of the national imaginary. OMB president, Elek Horváth, welcomed new members in 1928, noting that in the new sound motion picture age, 'when morals and ideals originating in other nationalities' spirits are ambitiously distributed through film…it is these that are the scales upon which Hungarian national thought and national feeling are measured, and it is through them the goals of the national state, and the moral opinion of Hungarian society are filtered'.[19] At this same meeting, Béla Scitovszky, the Interior Minister to whom jurisdiction over the Censorship Committee fell, offered what he called 'nation-building directives in the censorship of film'. The censors must be reactive, protecting the public, public morals, foreign relations, the purity of the Hungarian language, and the quality of art. They must also be proactive, ordered Scitovszky, creating the frameworks for national feeling and thought, formulating and preserving the state's most holy values and ethics, and combining social morality with good taste.[20] Because film had a role in legitimating a particular cultural vision of what was Hungarian, Hungary's censors played a critical role in establishing the boundaries of national identity by defining what stood outside that vision, outlining the 'preferred expressions of [the] national culture'.[21]

It is difficult to know how the censors interpreted this task through the 1930s, as most of the documents explaining their decisions have been lost or destroyed.[22] However, archival remnants from 1930–1 indicate that the OMB initially felt compelled to re-evaluate and re-configure its role as the epidemic of 'sound film fever' infected Hungary. In 1930, the majority of Hungary's operational theaters converted to the new technology. Sound film's market share rocketed from around three percent in 1929 to over 50 percent one year later.[23] With sound film came language specificity, which in turn resulted in cultural specificity, meaning intrinsic differences in forms of dialogue, difficult-to-translate idioms, and various other cultural signifiers. Sound created new categories for censorship, such as poor audio quality and objectionable lines of dialogue, as well as less obvious ones, including behaviors made problematic when accompanied by speech and music the censors could interpret as anti-Hungarian. During a period of confusion and adjustment, the number of sound feature films the OMB prohibited climbed to well over ten percent by 1931.[24] The spotty records indicate the Censorship Committee rejected sound features at nearly double the rate of silent ones, demonstrative of new concerns about the transformed medium's power.[25]

As the decade of the 1930s began, the principles and ideals the OMB voiced in the silent era continued to structure their designations of what was Hungarian. OMB decisions stressed the need to uphold moral purity, social stability, monarchical legitimacy, anti-Bolshevism, and reverence for Christianity.[26] Censors refused permits to films endorsing pacifism, revolution, or democracy, such as the *Captain of the Guard* and *All Quiet on the Western Front*. They also rejected domestically made films endorsing Jewish-Christian intermarriage or portraying Hungarian characters as excessively boorish, such as *Country of Promise* [*Igéret országa*] and *What Time Is It, Zsuzsi?* [*Hány óra, Zsuzsi?*]. These efforts to restrict images during the first years of sound amounted to a strategic campaign, a mobilization of acceptable stereotypes in order to limit the spectrum of political and social thought in Hungary and to demarcate Hungarian identity.[27] However, as will become clear, the censors did not sustain a consistent campaign through the 1930s, as the domestically produced films they permitted offered a very different sense of nation than that detailed in OMB guidelines. After an early 1930s adjustment period, Hungary's censors seem to have adopted a more hands-off approach, at least for films made in Hungary. They preferred the light touch: to eliminate problem elements in pictures by guiding production rather than prohibiting films.[28]

While some of the censors' initial visions may have coincided with those stereotypes of Hungarians in circulation in the West, their notions of nation contained an inherent double standard. Within the borders of Hungary, a few less than positive characterizations were not only permissible but necessary for the art form to have legitimacy. Thus, generalizations about a 'Hungarian character' portraying Hungarians as sincere but haughty; intense but lazy; well-mannered but quick-tempered and passionate; courageous but sometimes barbarous; good-natured but prone to drink or gamble; European but foreign; distinct from both Jew and German – might be fine in Hungarian film, as they were in theater and literature.[29] However, in the global pool of audiovisuals where representations of Hungarians and Hungary were rare and the range of identifiers far more limited, these same images were potentially damaging.[30] When Western films mobilized Hungarian stereotypes, portraying Hungary as ruled by a backward aristocracy of disconnected 'inferior conservatives' concerned more with maintenance of ancient constitutional prerogatives, oligarchic privileges, and revision of the Treaty of Trianon than with alleviating the woes of the general Hungarian populace, these generalizations, though not incorrect, infuriated official Hungary, even as censors reluctantly permitted these same characterizations in indigenous works.[31] This seemed a paradoxical stance for political elites convinced of the existence of uniform national cultures.[32] Seeking to square the circle, some of Hungary's state image makers became obsessed with countering Western assumptions about Hungary. They saw sound film as the means of rescuing Hungary's honor, correcting misperceptions, and improving Hungary's international reputation.[33]

In May 1930, Zoltán Gerevich, the Foreign Ministry's representative on the National Censorship Committee, informed his superiors about the new cultural and political implications of language-specific films and how they complicated bureaucratic concerns about the images of Hungary and 'Hungarianness' [*magyarság*].[34] He urged the Foreign Minister to persuade his colleagues in other government departments to respond to the portrayals of the 'so-called Hungarian', especially the Austro-Hungarian and the Hungarian army, in talkies made outside of Hungary. When a Hungarian is spoken of or in a movie, claimed Gerevich, he is either a drunk, purposely excluded, or demeaned. These characterizations, he felt, were 'not just annoying, but totally insulting to the Hungarian nation' and demonstrative of the West's mistaken perceptions.[35] He reiterated the refrain that these misapprehensions had led to the imposition of an unfair peace upon Hungary in 1920, and their persistence was responsible for Hungary's continued diplomatic and economic woes.[36]

As Gerevich voiced his concerns, a contingent of government officials and representatives from various professional film associations and organizations had already begun planning a response to Western mischaracterizations of Hungary. They devised a rough outline for state involvement in startup Hungarian sound film production. They recommended that the government bankroll 50–60 short films each year for educational and 'propaganda' purposes both at home and abroad. While many in the film business supported the use of newsreels and non-fiction shorts, known as *Kulturfilme*, both for cultural diplomacy and for keeping the 'flames of revision' burning inside Hungary, the preference for *Kulturfilme* provoked disapprobation in some quarters.[37] *Cinema World* [*Mozivilág*] editor Zsigmond Lenkei objected that 50–60 short films would earn less than four to five feature films, be aesthetically sterile, and be ignored by target audiences. Lacking the propaganda power of features, short films, Lenkei concluded, would mean truncated incomes and truncated influence for a truncated Hungary.[38]

Trade Minister János Bud appeared to agree with Lenkei when he celebrated the opening of the Hunnia studio in April 1931. Bud affirmed that with the capacity to make sound features the foundation could be laid for the creation of a true Hungarian product, whether for domestic or international distribution. According to Bud: 'Should Hungarian capital work in the studio, Hungarian workers and intellectuals earn their bread there, Hungarian art triumph there, and Hungarian public opinion support its creative results, Hungarian film production will come into being'.[39] This unsophisticated truism established the early parameters for what most of Hungary's officials meant when they invoked the terms Hungarian and national: any product in the Hungarian language, made by Hungarian citizens, watched by Hungarian audiences, would magically, *sui generis*, have 'Hungarian' qualities.

Eventually persuaded by the arguments made by Gerevich, Bud, and others, the Foreign Ministry jumped into the breach. Aiming to fashion more positive national images, but aware of its limits in terms of resources and clout, the Foreign Ministry took a tiered approach. First, it instructed Hungarian diplomats to work to restrict 'objectionable' images of Hungary and Hungarians in foreign films. Second, the Foreign Ministry commissioned a variety of sound *Kulturfilme*. Third, in 1932, it ordered the remaking of the feature-length propaganda film *Hungária*. Since its debut in 1928, *Hungária* had been the primary foreign service vehicle for dissemination of robust images of Hungary. Made with Regent Horthy's blessing at the Hungarian Film Office under the supervision of the Council of Ministers and the Foreign Ministry, *Hungária* enjoyed unprecedented high-level government support.[40] Designed to create a more 'balanced' picture of Hungary and to demonstrate the vitality of its national culture to foreign audiences, the silent version was enormously successful.[41] It highlighted Hungary's geography, demography, and ethnography, concentrating on Hungary's developed intellectual society (ranging from its schools, libraries, institutes, and universities to its hospitals and technologically-advanced industrial practices), its ancient and lively culture (defined predominantly through language and music), and even its religious diversity. When introducing the film at a major Stockholm showing, Hungary's *Chargé d'Affaires* in Sweden declared *Hungária* to be an explication of Hungary's national cultural uniqueness, its place in Europe, its *raison d'être:*

> ...the reason why the Hungarians did not get lost, melted into the surrounding comparatively enormous nations is that we brought culture, original, national culture with us. I won't argue whether this culture was a higher or a lower one compared to that of the new neighbors – probably in some respects it was higher and in other respects less developed – but it was a culture, and an old one too...[that] is quite a different one [compared] to that which you find in other European and American states.[42]

But by 1932, a silent propaganda piece, even if visually stirring, was proof more of a country's obsolescence than its strength. The Foreign Ministry recognized this and, embracing the value of film as a diplomatic tool, ordered *Hungária* to be re-worked as a true sound film. The MFI project began in 1933 under the direction of László Kandó. The Foreign Ministry enlisted Ernő Dohnányi and Sándor László to write new music and text for an entirely reformulated work. In addition to being a cultural corrective, the Foreign Ministry hoped *Hungária* would attract tourism, cement foreign business contacts, and encourage cultural exchange. This inclusive endeavor brought together government organs and private companies, including several large Jewish-owned manufacturing works, who helped fund the project. The Film Industry Fund, however, was conspicuously absent from the list of supporters, preferring to remain on the sideline as Hungary's first sound propaganda experiment took place.[43]

When *Hungária* opened in Budapest's Royal Apolló in March 1934, all of Hungary's political and cultural elite attended, including Regent

Miklós Horthy, Prime Minister Gyula Gömbös, Foreign Minister Kálmán Kánya, future prime ministers Béla Imrédy and Kálmán Darányi, as well as Budapest's Lord Mayor Aladár Huszár. According to *Filmkultúra,* the 18 month process of revising and filming the approximately hour-long sound picture resulted in a final product that depicted Hungary's 'heroic resoluteness and strength of will in its struggle for existence and survival'.[44] An extant script of the film explains why *Filmkultúra's* editors voiced such inflated conclusions and how Hungary's film and diplomatic elites hoped to use the motion picture for the dual purposes of advancing foreign policy initiatives and molding national identities. Certainly, the primary motivation behind Hungary's interwar diplomacy was the desire to right the wrongs of Trianon, to regain the ancient Hungarian lands snatched away in 1920. Accordingly, the script's introductory sequence made immediate and savage reference to Hungary's interwar revisionist fixation. Dramatic music reached a crescendo as

> ...out of a plasticine of the old Kingdom of Hungary [was] sculpted a smaller, stylized embossed map, from which wonderful geographic unity [arose]. Above the map there [appeared] a terrible, vulgar fist which [slashed] the country five times with a bayonet. The smaller pieces [fell] off as the bayonet [engraved] the name 'Hungária' into the unmoving remaining piece.[45]

The crimes done to Hungary firmly established, the film then provided a panoramic view of Hungary, its cities, countryside, and historical landmarks. It demonstrated Hungary's virtue through its high level of technical achievement and the cultivated nature of its people. Whether the provocative opening remained in the final versions distributed abroad is unclear. Foreign Ministry documents indicate that its ambassadors and *chargés* had discretion over what they felt should be edited out, and it appears the gashing of the Hungarian map remained in many screenings.

Critics across the political spectrum and throughout Hungary's cultural elite agreed that the film was an immense propaganda and aesthetic success. *Budapesti Hirlap* lauded *Hungária* as evidence that Hungary had arrived as a filmmaking country, capable of creating pictures comparable to the best foreign wares.[46] *Pesti Hirlap's* review of *Hungária* found the cinematography exceptional, remarking that the film was constructed as if 'the California sun shined here', and adding that every person and every scene 'had the beauty of a wonderful painting'.[47] *Filmkultúra* pronounced *Hungária* one of the industry's most gorgeous productions.[48] Weeks after the initial showing, *Hungária* continued to receive 'the best possible critical reviews' domestically.[49] Distributed abroad, sometimes as a stand-alone feature but more often as a *Beiprogramm* attached to other features, the film enjoyed wide exposure and an overwhelmingly positive reception. By mid-1934, German, French, English, and Italian subtitled versions had already appeared in theaters in Warsaw, Washington, the Hague, Rotterdam, Amsterdam, Sofia, Brussels, Paris, Rome, Milan, Venice, London, Vienna, Berlin, Munich, Buenos-Aires, Alexandria, and numerous other cities. Hungary's diplomats were particularly pleased when reviews of *Hungária* designated Hungary as 'truncated' or 'wronged' and expressed sympathy for revision. These were achievements of inestimable value.[50]

Without a doubt, *Hungária* achieved three important goals. First, its artistic quality brought respect to the Hungarian film industry. Second, as a propaganda tool, it made Hungary appear attractive and instilled sympathy for Hungary's diplomatic positions, particularly lending credence to a symbolic geography of Hungary which included pre-Trianon lands. Third, with the success of *Hungária,* Hungarian film elites convinced themselves that they had finally manufactured an explicitly national film, one that truly represented Hungary as they wished it to be, with a contoured national culture, identifiable through costumes, behavioral norms, music, landscapes, and diverse intellectual and spiritual achievements.[51] They had begun to 'resurrect' the nation.

The global good fortune of *Hungária* was opportune, and the concurrent box office success of a handful of Hungarian features in France and Germany certainly helped persuade Hungarian authorities that Hungarian-made film possessed significant cultural, economic, and diplomatic value. Film industry figures now lobbied the government, contending that the Foreign Ministry had the obligation to actively promote the export of motion pictures. The movie producer and film journalist István Radó suggested that the government 'should think of the export of film [just as it thinks of the] export of beef and carrots' and it should actively engage in negotiations to open foreign markets.[52] In addition to filling the National Bank's vaults with precious foreign currency, wrote Tivadar

Csáky, the making and export of Hungarian film would have 'inestimable propaganda value' by showing the world 'Hungarian skills, Hungarian knowledge, [and] Hungarian creative abilities'.[53] Other film powerbrokers drew similar conclusions. Making film for export, wrote János Bingert, would show the world that Hungarians '...believe in and trust Hungarian culture; [have faith] in the strength of our own art, knowledge, and desires; and want to show that a truncated country is able to create new sources of strength despite all of its injuries and shortcomings'.[54] According to OMB member Géza Ágotai, assuring the widest distribution of 'representative [Hungarian] images', made by Hungarians, was absolutely essential to this effort.[55]

These comments illustrate that by mid-decade, Hungary's leadership had accepted film as part of the state's cultural-political arsenal, a weapon for shaping international opinion about Hungary's image and its future. To promote the wider circulation of Hungarian-manufactured images throughout Europe, the Foreign Ministry began incorporating reciprocal newsreel and *Kulturfilme* exchanges into the bilateral cultural agreements it negotiated.[56] Through the 1930s and into the 1940s, the Hungarian Foreign Ministry and the Royal Hungarian Tourist Office [*Magyar Királyi Idegenforgalmi Hivatal*] utilized the moving picture to promote a Hungary that contained a fantastic capital city, replete with all the history, culture, and urban entertainment that any international traveler would desire. Since cultural films about Budapest targeted the upper middle classes of Western Europe, they emphasized Budapest's spas, its night life, its fashion, and its notable architecture. Newsreels likewise tended to concentrate on events in Budapest, a natural consequence of the fact that almost everything newsworthy happened there. Other cultural films advertised Lake Balaton as a pan-European vacation spot. Still others portrayed Hungary as a technologically advanced country with top-notch manufacturing facilities, successful research scientists, and brave and debonair pilots. These protochronistic culture films were typical of the times; in some form or another all of Europe's filmmaking countries punched them out as they jostled over civilizational superiority.[57] Their existence in Hungary suggests that, depending upon the context, the official circles of Hungary's usually hidebound, gentry-led oligarchy could endorse modern, bourgeois visions of progress.

The Film Industry's 'Nations', 1933–8

The mid-1930s official imagery of Hungary as a territory and people primarily defined by Budapest and the Balaton, by progress and technological modernity, by spas and shopping, by scientists and pilots, largely coincided with the predominant imagery Hungary's private film producers began to offer in greater numbers, particularly after 1933. This was neither a natural nor an uncontested occurrence. Between 1931 and the end of 1933 movie companies made 18 features in Hunnia's studios, eight of which were domestically funded Hungarian productions. The remaining ten were co-productions or MLVs filmed in Hungary and intended primarily for foreign distribution. This meant that Hungarian private film enterprise remained chiefly transnational, utilizing cash and professionals from abroad. Cultural commentators of all stripes decried this situation as depriving Hungary of its ability to define itself even as filmmakers offered a different spin. Producers materialistically justified their activities and made claims for state assistance during these early years by suggesting that any film which projected the amorphous 'Hungarian spirit' [*magyar szellem*] or utilized Hungarian language and music was Hungarian. In light of the Joinville experiments in Paris, film elites even rationalized the making of Hungarian film outside of Hungary with non-Hungarian craftsmen. Although they saw the value of utilizing Hungarian casts and crews, not to mention mobilizing Hungarian capital, early film producers did not see these elements as necessary to produce film they considered 'national'.

Contentious debates about the nature of the Hungarian spirit and how to inculcate it into film dominated the early sound age. Worries about the quality of Hungarian film, its lack of profitability, its 'failure' to win new audiences and its non-national content were also common through the early 1930s. Important cultural commentators and government bureaucrats questioned whether Hungarian feature film production was sensible, and a minority remained unconvinced of the domestic motion picture industry's viability, promise, and intrinsic national character. Elemér Boross, writing in the well-read popular culture weekly *Délibáb*, issued this pessimistic critique in early 1933:

> To make Hungarian films, rather Hungarian language films, is not worthwhile…The Hungarian language area hardly offers suitable profit [or] secure numbers of paying viewers. We've

> already lost enthusiasm for made in Hungary...Hungarian films are bad, primitive, unenjoyable....The banana would also be cheaper, if we didn't import it – but on the pavement of the main boulevard neither is it possible to plant the banana....As of now, the good film and the good banana are only purchased, boldly, from foreign countries.[58]

Most filmmakers and critics, however, thought the stakes too high for Hungary to surrender to foreign suppliers. Domestic production, they preached, was imperative, particularly since by the mid-30s the Hungarian public was spending more on movie tickets, per annum, than on all other printed media combined.[59] Critics on the right, such as Béla Kempelen, demeaned all foreign film as 'distant from the spirit of the Hungarian people'. His critique was typical of European nativist and religious reactions to Hollywood goods in the 30s. Foreign movies induced 'depravity... moral destruction... crime...[and] addiction; ... [led] to fantasies, [incited] risky decisions, and [distracted from] the processes of thought....'[60] Worst of all, they could 'turn a pubescent girl into a prostitute'.[61]

Rightist critics and filmmakers thus warned that domestic production was vital, even if it could not be justified financially. 'An impossible venture' from a purely materialist perspective, the medium needed to seek 'spiritual profit', wrote the director István György. Domestic movies shouldered a 'sacred task' which included helping the nation overcome its Depression-induced ontological crisis, imparting a sense of pride and value, and preventing the further degradation of Hungarian culture.[62] Numerous voices across the spectrum of the film establishment, from censors and Foreign Ministry officials to centrist film journalists, chimed in with similar opinions. They beseeched Hungarian filmmakers to find their inspirations in Hungarian literature and history.[63] Pictures must feature Hungarian contributions to humanity, 'our homeland's history, the pearls of Hungarian literature...legends of medieval Hungary, Hungarian folk traditions, etc.'. They must be 'reflective of the Hungarian past', and suffused with lessons about virtue, the fulfillment of obligations, honor, and kindness.[64] This uniqueness, this creation of film based *on the strength of its national spirit* and Hungarian mentality, was the best didactic path and the surest route to wider acclaim, proposed OMB member Géza Jeszenszky. Deprived of exposure to this mentality, Hungarian audiences could become enthralled with the 'adventurous lives of American gangsters and the sugary, operatic existence of Berlin sales girls', losing touch with their distinctive brand of 'Hungarian happiness' in the process. To prevent this, Hungarian spectators required film that was not only national, but nationalist and nationalizing. In other words films should not alienate Hungary's masses from their own thoughts, 'lives…, and problems', but should inform audiences about who they were, what their culture entailed, and why that culture was worth defending.[65]

That film could contribute to a nationalist and nationalizing mission incorporating average Hungarians was welcomed by segments of the Hungarian populist movement [*népi mozgalom*]. Prior to the late 1920s, populism existed as a socio-cultural movement that occasionally became political. Its origins traced to the rare land reform initiatives of the late nineteenth century, and its 1920s variant grew out of a flourishing academic and literary interest in the peasantry.[66] Hungarian Boy Scouts, ethnographers, sociologists, and authors all went 'back to the land', either to document the hardships suffered by Hungary's non-urban masses or to locate some unchanging national essence in a peasant culture unsoiled by cosmopolitan modernity.[67] The umbrella of populism sheltered an enormous spectrum of thought, from liberal democrats championing moderate land redistribution, to communists campaigning for far-reaching reform, to religious reformers seeking social justice, to anti-modern, anti-liberal thinkers who saw renewal in a racist, organic form of nationalism.[68] All forms of populism extolled the peasantry, but as populism became more politicized in the 1930s, the voices of those who inveighed against modernization and liberal individualism or supported radical nationalism, if not autarky, became dominant.

Knowing that direct attacks on the *latifundia* or the ineffective bureaucracy would not be permitted by the conservative ruling elite whose very power was threatened by the democratic expansion and social reform they advocated, Hungarian populists generally cleaved to one of two paths. Either they focused on folk customs or they chose targets acceptable to the political oligarchs: the city, the capitalist, the foreigner (often the German), or the conflation of all three, the Jew.[69] The first option, the ethno-cultural path, included those who searched for folk customs by starting from the premise that modernity and the Great War had fragmented the cultural and spiritual unity of the Hungarian nation. If one could strip away the veneer of foreign customs and artificial political ideas, they believed, one would find the true social reality of Hungary.[70] The latter, ethno-racial path, included varieties of populism which targeted specific groups as alien. It would be a gross misstatement to claim that all populists in this group were antisemites, but accurate to say that antisemitism and *völkisch* racism found increasing acceptance, especially among these populist constituencies, during the 1930s. Variants of populism overlapped with Christian and integral nationalism, and the political rhetoric and cultural coding which emerged from this discourse eventually resulted in the demonization of Jews and Judaism in the 1930s and early 1940s. Shared phrases, such as 'Christian Hungary' and 'Hungarian Christian middle class', found new resonance and helped create an environment in which anti-Jewish political language became pervasive. At best, populism meant reform of the peasantry. At worst it created the necessary preconditions for the unraveling of the Hungarian Jewish assimilation project and the success of the Hungarian Holocaust.[71] As 'national characterology', either in the guise of populism and the focus on the rural or as a component of racist nationalism, became part of Hungarian mass politics in the late 1920s, the trend created disruptive ripples in Hungary's rapidly evolving film culture, particularly as the close of the 1930s neared.[72]

A small group of academics and writers, known collectively as the urbanists [*urbánusok*], stood in opposition to the populists, rightist nationalists, and racists.[73] By-and-large the urbanists were, as their moniker implied, from large cities, Western in their cultural tastes, and partial to foreign styles over nativist ones. They were disproportionately Jewish and often belonged to border-crossing intellectual circles.[74] Modernist and capitalist in their thinking, they sympathized with liberal or left-of-center democratic politics. They generally rejected the idea that simply living among the rural people, 'surrounded by the Turanian breath,…[inhaling] its ancient pagan fragrance into [one's] soul', would allow them to create an art that was 'instinctively Hungarian'.[75] Many of Hungary's top filmmakers, screenwriters, and distributors were partial to these beliefs or were themselves members of the urbanist camp.[76] Beginning in late 1933, as independent Hungarian production took off, they began to fashion a body of film which offered a vision of Hungary, its nation, and its way forward closely aligned with middle-class modernist and urbanist sensibilities.[77]

Unlike populists and others on the right who tended to dominate public discourse on questions of national identity, those with urbanist proclivities tended not to define their ideas as ideological. Rather, they saw themselves as practical nation builders, offering deeds, not words. Because, according to director István Székely, Hungarian films equaled Vienna and surpassed Prague in artistic quality, were on the verge of becoming a legitimate export item, and were '...beginning to count as part of European film production', these films were intrinsically nationalist.[78] They did a great service to the nation, even if their content was cosmopolitan in character. This concept of nationalism failed to sway the populists and their compatriots on the right. On the contrary, the right saw this as a disingenuous distortion of film's spiritual mission, leading to cultural decay and social alienation.

Given the positions staked out by official Hungary, one might expect state censors and FIF members to side with the populists and reject the works of Székely and others as lacking a Hungarian 'essence'. But they did not. Whatever their motives, they came to believe that bananas could grow on the boulevard, which allowed Hungarian filmmaking to bloom.[79] After years of limited use, the Hunnia studios started bustling. Between 1933 and 1938 the quantity of pictures made in Hungarian studios rocketed from 13 in 1934 to 37 in 1937, a startling number for such a small country. Additionally, Hungary's broader film culture showed signs of life. Nearly all of Hungary's licensed cinemas, as of December 1933, offered daily programming.[80] Beginning in 1934, the number of operating theaters grew steadily. By 1935, the average daily gate was between 42,000 and 51,000 movie tickets sold, meaning that over one-half of one percent of the population might be in cinemas on any given day.[81] Responding to audience growth, the Hungarian Film Office quite successfully ventured into feature film production in 1935. Hunnia and Hungary's banks began to offer better incentives for film production and easier access to loans. In 1936, a Film Bank appeared, run by Richard Geiger, a Jewish businessman and formerly one of Hungary's most important silent film producers and distributors. Geiger's bank became a new, albeit limited, creditor for domestic film production. Geopolitical change also promoted the expansion of Hungarian film. A decline in

German production due to the Nazi ascension to power meant fewer exports to Hungary, a situation made worse by strained Hungarian-German film relations. American films and augmented Hungarian production together compensated for this deficit.

The top grossing Hungarian-made films of early to mid-1930s were a form of 'popular cinema' which generally prioritized aesthetics and entertainment above avowedly nationalist content.[82] Romantic comedies based on Hollywood models such as *Hyppolit*, *The Dream Car* [*Meseautó*], *Salary, 200 a Month* [*Havi 200 fix*], *120 Kilometers an Hour* [*120-as Tempó*], *Miss President* [*Elnökkisasszony*], *Anniversary* [*Évforduló*], and *Hotel Sunrise* [*Hotel Kikelet*] presented an idealization of Hungary which rarely engaged prickly political questions. These films, however, did offer a clear concept of nation, envisaging the liberal cosmopolitan, consumption-oriented middle-class of Budapest as the symbol of modern Hungary.[83] Progress, liberation, and technology, particularly cars, were often mutually reinforcing *leitmotif*. The heroine of Béla Gaál's 1934 film, *The Dream Car*, for example, was a middle-class clerk in a Budapest company who becomes enamored with the most technologically advanced car available in the world while her boss becomes enamored with her. Endre Marton's screwball comedy *Miss President*, the third highest grossing film of 1935, featured a female company president who finds love with an unemployed engineer who invents a new weaving technology for textile manufacture. *Salary, 200 a Month*, a 1936 smash directed by Béla Balogh, embraced cars, Budapest, and upward mobility. It tells of how a job at a car factory in Pest makes marriage possible for a poor country couple. In László Kardos' 1937 hit *120 Kilometers an Hour*, a thoroughly modern banker's daughter moves through life with speed and daring, courtesy of her automobile. In the end, of course, she is lassoed and domesticated, marrying an up-and-comer whom she happened to hit with her car.

The automobile was more than a symbol of status, glamour, and bourgeois consumption. The car gave filmmakers a means of nationalizing Hungarian audiences. Cars transported audiences around national spaces, familiarizing them with the landscape of the nation's territory, bringing them into contact with the various peoples who constituted the broader national community. Automobiles also moved viewers through experiences more quickly, providing new mechanisms for apprehending time in

Figure 4: Cosmopolitan modernism at its height. Cars, consumption, and love are all brought together in the advertisement for Béla Gaál's 1934 blockbuster, *The Dream Car*. Source: MKT CD 1.

Figure 5: Hungarian modernism. The ambitious Hungarian reads the *New York Herald Tribune* with his friend and driver. Source: MaNDA, Budapest.

a new age.[84] Cars carried the progressive middle classes, including partially liberated women, allowing audiences to experience vicariously unprecedented freedoms. Perhaps more than any other technique, the car was the vehicle allowing Hungary's urbanist/cosmopolitan filmmakers to make their interpretations of the nation concrete.

Other popular films of the period featured the lives of doctors, lawyers, office workers, hotel owners, and bankers; the dancing and singing of urbane sophisticates in formal gowns and tuxedos; and the morality of those with the flexibility to marry either above or below their class. The success of these 'commerce', 'glamour' and romantic comedies buttressed the clichéd notion that escapism trumped all other motives when Hungarian moviegoers purchased their tickets.[85] An alternative, and perhaps more compelling assessment, however, is that this liberal, democratic vision of modernity, this valorization of a self-confident bourgeoisie's values and luxury, of technological and economic advance, and of class and gender liberation, had real popular appeal.[86]

Largely the creative efforts of liberal Hungarian Jews trained abroad, this body of film was fashioned after Hollywood and Weimar German models, and based on scripts that frequently were re-workings of plays or novels written by turn-of-the-century modernist Hungarian authors. While attractive to audiences, cultural critics of all stripes, excluding a set of bourgeois Budapest film magazine editors, pilloried these representations. It was not possible for cinema owners in the countryside, contended one owner János Füki, 'to manage [on a diet] of 'hypermodern' films' because true Hungarians would ultimately realize they could not relate to 'foreign stories about the lives, loves, and adventures of the urban middle class'. The middle classes, by this reading, stood outside the 'Hungarian spirit'.[87] These sentiments were echoed in parliament, where representatives suggested that providing rural audiences with ever more glimpses of unattainable middle-class Budapest life was divisive. By mid-1935, parliamentarians instead clamored for the reverse: more features about rural Hungarian lives, customs and dances, so that Budapest audiences could become better acquainted with 'real' Hungary.[88] Filmmakers should do a better job teaching the masses who they were supposed to be.

Over time, the film industry began taking these critiques more seriously. Columnists began to note Hungarian cinema's predictability and staleness. The film journalist András Komor wrote scathingly of his compatriots' use of a handful of staples in its lazy quest for 'moderate, uniform success':

> …in every film we find an elegant jobless engineer; there is always a stupid peasant; always gypsified gentry; and Gyula Kabos always has to fall into water. This is why in every film the storyline is forced…and there is always one scene where we hear jokes we have heard repeated hundreds of times. This is why there are no good Hungarian films.[89]

Director István György concurred. Hungarian film, he complained, was simply 'lacking Hungarianness'.[90] In direct response to this sort of sentiment and the political elite's desire to combat the growing popularity of fascism, populism, and socialism, the film industry expanded the national imagery. Rather than merely abandon the hypermodern, cosmopolitan middle-class and rely solely on traditional images and themes drawn from the history of Hungary, filmmakers mobilized more potent and inclusive national motifs. Through village idylls, 'senior' films, comedies about the unemployed, crime movies, or love stories, filmmakers sought to reach new audiences and, even if obliquely, confront the social and political problems of the day.[91] For example, the most oft-addressed theme of the decade was the loss of status. In their comprehensive study of the styles and genres of early Hungarian sound film, Gyöngyi Balogh and Jenő Király conclude the most common film hero was the déclassé landowner or the gentleman/professional degraded in social standing.[92] The unemployed doctor, the underemployed engineer, the aristocrat demoted to middle-class status, the peasant forced from the farm, the worker who becomes a hobo – these figures appeared on Hungarian screens with increasing regularity through the 1930s. The 1936 films *Man under the Bridge* [*Ember a híd alatt*] by László Vajda, and *It was me* [*Én voltám*] by Artur Bárdos, both contain prototypical examples of these protagonists.

Perhaps even more critical to the effort to broaden the national imaginary were representations of rural Hungary. Beginning in the middle 1930s, more diverse portraits of the lower middle-class and working masses, especially the peasants and the rural world in which they worked, reached Hungarian screens. A handful of films depicted the masses not only as part of the Hungarian nation but as its essence. They appropriated folk customs and images with abandon, and they presented rural locales as the seat of the nation. Life on the Hungarian plain, best represented by Georg Höllering's 1936 film *Hortobágy*, and films about the everyday travails of fishermen on the river Tisza, such as Géza von Bolváry's 1938 film *Flower of the Tisza* [*Tiszavirág*], characterized this trend. Yet reception is complex, and spectators did not always agree on the message these films purportedly conveyed. In fact, rural subject films often presented inherently contradictory messages or were interpreted as such. Critics from left and right, for example, took issue with Höllering's *Hortobágy*, a film based on a Zsigmond Móricz short story. The left disliked this docu-drama because of its idealization of the countryside, its glossing over the harsh reality of peasant life. The right objected to the fact that the film concluded with the farmer's son abandoning his traditional profession to become a mechanic, essentially adapting to the modern world rather than resisting it. Both sides took umbrage at the portrait of Hungary as backward.

Just as Hungarian feature producers began to change the content of their movies, those agencies engaged in culture film and newsreel production, the more 'official' producers of Hungary's international image, began a similar reconsideration of their material.[93] In a March 1936 letter to the Hungarian Foreign Minister, Ernő Lits, an official in the Hungarian embassy in Munich who himself made *Kulturfilme*, charged that foreign publics had been misled by big screen images of Hungary. Hungarian screenwriters and producers bought in to this deception, encouraged by foreign production companies to make romantic films in which 'the laughable figure of the carousing [Hungarian] *huszar* officer whiling away his fortune is the only image seen of Hungary abroad'. If Hungary wished to correct its image and project its reality, it must make culture films that show the true 'spiritual world of the Hungarian people, their unadulterated mentality, their actual character, their ethnography, and their folk culture'. Only in this way would Hungarians disabuse foreign audiences of the myths and misunderstandings about their nation.[94]

In the mid- and late 1930s, Hungary's state tourist agency, its diplomatic service, and its ministries charged with the oversight of education and film, listened to these suggestions and took action. The result was the proliferation of culture and education films painting Hungary as a pastoral country of small, happy villages of orchards and farms, wild boar and bear, folk dances and costumes. Hungary's Foreign Ministry partially funded shorts titled *Hungarian Village* [*Magyar falu*], *Hungarian Dances* [*Magyar táncok*], *Village in Celebration* [*Ünneplő falu*], *The Serenity of the Hungarian People* [*A magyar nép derüje*], *Hungarian Horse* [*Magyar ló*], and *The Hungarian Village Smiles* [*Magyar falu mosolya*], and actively distributed them globally. In addition, the Foreign Ministry selected several of these films for entry in the annual Venice Film Festival and other international exhibitions, thus insuring that the films would be seen by Europe's motion picture, political, and intellectual upper crusts. Hungarian diplomats even used culture films as gifts, providing Germany's *Reichsmarschall* and Master of the Hunt Hermann Göring *The Hunt in Hungary* [*Vadászat Magyarországon*] in 1938 in an obvious attempt to curry favor.[95]

By no means should one overstate the significance of a limited number of cultural shorts. They do illustrate, however, evolving thought about visual representations of Hungary. In response to perceived foreign desires and domestic political concerns, the pendulum of state imagery swung. Hungary's state advertisers began to market Hungary less as a land of spas and the birthplace of scientists and more as a nation of horseman and contented sharecroppers. János Pelényi, the Hungarian Ambassador to the United States, summed up the new approach: 'What interests foreigners…is that which they cannot see elsewhere…that which is characteristically Hungarian. First and foremost, this means national traditions, the lives of the people, their favorite means of entertainment, the world of the Hungarian plain, etc'. Even if stereotypical, the cultivation of these images served important touristic and diplomatic purposes. They produced, according to Pelényi, greater 'sympathetic attraction for our homeland' than images of modern life.[96]

While Hungarian culture films may have moved towards 'the people' more rapidly than features, the later 1930s was a transition period for all cinematic forms. Even if the mass of déclassé characters, rural-themed features, and culture films did not represent a paradigm shift in notions of Hungarian identity or a coherent social critique, such films did offer 'new and radically different way[s] of imagining the place of the rural in Hungarian modernity'.[97] According to historian Anna Manchin, Hungary's urbanist feature filmmakers sought a compromise. Certainly, they could not abide presenting the countryside in opposition to Budapest and as the place in which Hungary's authentic, untainted national culture resided. Instead, their most successful films of the later 1930s displayed rural and urban bourgeois cultures as coexisting and co-reliant. These films sometimes presented the urban bourgeoisie as the engine of regeneration for the rural countryside, bringing ideas, values, and even capitalism with them. Films such as László Vajda's 1936 *Three Dragons* [*Három sárkány*] and István Székely's 1937 films *Help, I'm an Heiress!* [*Segítség, Örököltem!*] and *A Girl Sets Out* [*Egy lány elindul*] all feature women with middle-class values of discipline, hard work, and frugality returning to the countryside and restoring foundering agricultural estates to prosperity. This body of films proposed a patriotic vision in which 'modern commercial culture was not harmful to rural Hungary, but rather could do much to democratize it and to facilitate its progress'.[98] As thinkers and politicians throughout interwar Europe sought a 'third way', elements of the film profession also prescribed a middle path to national unity. By promoting a filmic modernity which could enrich traditionalism, they would help the nation avoid the harms of both unrestrained capitalism and its communist opposite.[99]

During the later 1930s, while unabashedly modernist sentiment and 'traditional' representations of Hungary may have been fading, both continued to influence Hungarian cinema. Some of the most successful films of the later 1930s featured the stereotypical 'traditional' characters and themes to which audiences had become accustomed during the silent age. Examples of these smash hits included *Sister Maria* [*Mária nővér*], *Young Noszty and Mari Toth* [*A Noszty-fiú esete Tóth Marival*], and *Billeting* [*Beszállásolás*]. *Sister Maria,* a 1937 German-Hungarian co-production, was a sterilized saga featuring a gentry woman who marries for convenience rather than love and eventually ends up in a convent. A 1937 drama about a dashing, free-spirited, gambling-indebted Hungarian cavalry officer played by the 'Hungarian Errol Flynn' Pál Jávor, *Young Noszty* recounts the gentlemanly pursuit and ultimate conquest of a beautiful, rich woman who saves Noszty's honor by paying his gambling debts. *Billeting,* one of Hungary's three top grossing films in 1938, also featured officers, gentry values, and romance.

Yet even these examples were traditional only in the sense that they harkened back to Hungary's nineteenth-century golden age, which they did with an urbanist's eye. *Young Noszty* and other movies either aped the gentry or modernized it, often through the use of bourgeois females who rescued gentry men from their obsolete values. The 'traditional' world of the aristocrat in some cases took on a peculiar middle-class hue. In István Székely's 1936 *Danube Rendezvous* [*Dunaparti randevú*], aristocracy is reconfigured not as a matter of birth, but as a set of behaviors and status to which one could ascend through intellectual achievement.[100] Rather than mount a direct attack on the values and customs of aristocratic old Hungary, which they occasionally did through cabaret-inspired comedies, filmmakers preferred the oblique assault.[101] They frequently created Hungarian-American characters as foils and offered the United States as an ideal. Characters might go to America, enjoy life there, and even earn riches, but in most cases, the cinema émigrés come home. They might find Budapest 'preferable to Broadway', the *puszta* more appealing than the prairie, and the allure of the motherland, the patriotic pull,

Figure 6: You've got to love a man in uniform. One of the most successful Hungarian films of 1937, *Young Noszty,* drew from the deep well of Hungarian stereotypes. It featured the handsome officer who rides his horse with grace but plays cards without dexterity. His rich, beautiful love whips him into shape and preserves his honor by erasing his debts. Poster courtesy of the Ernst Galéria, Budapest.

irresistible.[102] Those who visit America often return with technical skills, bourgeois values, or great wealth. In 1937 alone, critical characters or plot lines constructed upon some form of American success were featured in Ákos Ráthonyi's *Pay, Madame!* [*Fizessen Nagysád*], Kardos' previously mentioned *120 Kilometers an Hour,* and László Vajda's *Borrowed Castle* [*A kölcsönkért kastély*].

Hungary in the 1930s was in transition, searching for identity. The corpus of the decade's entertainment films, and to some extent its *Kulturfilme,* represents a range of answers, a variety of coexisting and competing versions of the national synthesis. In many pictures, visions of the national were superficial, a matter of recycling the stock imagery that had filled directors' cabinets since the silent era: recognizable landscapes, folk costumes, and regal aristocratic characters. Others, however, had more depth. Some provided viewers an unvarnished endorsement of a vision of a cosmopolitan and technologically savvy Hungary modernized mainly by its recognizably Jewish Budapest middle-class. Others heroicized the downtrodden, celebrating those who had suffered losses but remained morally righteous. Some filmmakers presented spectators with an amalgamation, an attempt to incorporate rural populism into the urban-led march toward modernity. These concepts disputed idealizations of the rural countryside as pure and essential and shunted aside the notion of Hungary as the possession of the rural nobility. They presented a mixed and malleable national fabric that included a world outside Budapest. That these alternatives continued to be projected in the glory days of 'national characterology' in the later 1930s, as extreme nationalism, populism, and antisemitism flourished, is significant.[103] Liberal interpretations of the nation were largely drowned out by the right and center when raised in overt political discourse.[104] The majority of films made before 1939, however, are evidence that liberal members of the filmmaking elite did not cower in the face of growing right-wing pressure. Resisting populist parochialism, nostalgia for gentry values, and racial nationalism, they continued to produce visions of progress and unity premised on middle-class values and capitalism. The mass market success of their products indicates that these visions had appeal. This prospect did not sit well with conservative, populist, and far-right thinkers, who became more aggressive in their search for a truer national identity. They too redoubled their efforts to create positive, empowering symbols and narratives of Hungarianness, and over the course of the 1930s, became convinced that change must occur through radical reconstruction of the industry itself.

Figure 7: Off to the ball we go. The 1934 film *Ball in the Savoy* was representative of the operetta style whose internationalist nature made it indistinguishable from that produced by any number of other countries in the early 1930s. Except that they speak Hungarian, this couple could be found in Berlin, New York, Paris, Vienna, or London. Source: MaNDA, Budapest.

The National Interests: Film Industry Structure and the Intensification of the Jewish Question, 1933–8

As the film realm exhibited signs of life in 1933, the debate over how to best cultivate it entered a new phase. The issues were wide-ranging. From the extent of state involvement in the embryonic industry, to best practices for establishing industry order and rationality, dissension abounded. Concerns over Jewish and urban cosmopolitan elements of the business and the nature of film production also caused consternation. Between 1933 and 1938, all segments of the film trade addressed these concerns by cloaking their interests in the verbiage of nation, and increasingly, Christian nationalism. Analyzing the pressures which channeled the evolution of the motion picture business will explain how Hungary's 'national culture' transformed and why the products the movie world created varied as they did prior to the promulgation of antisemitic legislation in 1938 and 1939.

Questions of how the industry should develop became especially prominent in debates over rationalization. A term popular in the 1920s and 1930s, particularly in Germany where Hungary's émigré film workers likely first heard it mentioned, rationalization became a coded, symbolic concept in the Hungarian film industry's nascent stage.[105] Industry commentators used the words 'rationalize' and 'rationalization' to describe cost cutting and economic organization and as a means of associating the industry with particular visions of progress. Film was on the vanguard of Hungary's modernization process, and as a result, debates over rationalization, like those over film content and style, became proxies for debates over Hungary's brand of modernity.

Depending upon the constituency employing it and the concerns being addressed, the word rationalization was typically defined by self-interest and conflated with vanquishing more general motion picture ills. Producers saw rationalization as an excuse to cut the salaries of actors. Studio heads at Hunnia and the MFI hoped rationalization would mean more consistent provision of state subsidies, as well as a more efficient system for producers to obtain private financing and pay their bills. Government bureaucrats interpreted rationalization to mean a scheme of efficient and continuous film production, as well as sanctioned training of film professionals. Theater owners desired the laws affecting them be regularized and the burdens upon them reduced. Writers and screenwriters wanted legal rights protected. Every constituency framed rationalization as a means of promoting its own aims. As a result, the debates over rationalization often pitted film professionals and associations against each other. When they proved unable to settle their disputes cooperatively, they appealed to the state, enabling the state to gradually shift from facilitator to activist coordinator of the film industry.

Early state attempts to bring order to the messy movie world advanced with Gyula Gömbös' assumption of power. In October 1932, Prime Minister Gömbös unveiled his 95-point National Work Plan, a fascist-like design for the reinvigoration of Hungarian society. Noting the 'extraordinary importance of film in terms of its educational, cultural, propaganda, and entertainment influences', the 95-point plan promised the appointment of an official charged with reconciling the work of filmmakers and broader national goals.[106] Rumors abounded in early 1933 as Gömbös indicated he was ready to 'end the chaotic conditions' plaguing the film industry.[107] Betting money expected him to name his longtime friend, Miklós Kozma, one of the most influential personalities in the cinema world, as the commissar for film. 'Gömbölini', as he was nicknamed, did not. Instead he overhauled the government's role by mimicking Italian corporatist models. He formed an Inter-Ministerial Committee [IMB] consisting of representatives from the ministries of Interior, Commerce, and Religion and Education, with vague but wide regulatory powers over all film matters. It was to be assisted by a mixed private-state consultative body.[108] Film circles and American diplomats saw this legislation for what it was: Gömbös' effort to 'completely reorganize the operation and management of the Hungarian film industry… to centralize the heretofore widely scattered governmental administration and control of the industry and to facilitate the preparation of…regulations for the reorganization plan'.[109]

With alacrity, the IMB enhanced the spider's web of ties binding state and private enterprise. This government-appointed triumvirate not only sat on the board of Hunnia but also appointed the Film Fund's Supervisory Committee, which, in turn, became a *de facto* censorship board. The effect was a semi-nationalization of Hungarian film production:

> There existed… independent film production in Hungary. The producer could make what he chose; whether his film was a flop or a success was his affair, his luck or misfortune. On the other hand, there was a state film company which rented studios. And no other studios were available because there was no other company. In close connection with this company [Hunnia] there was a Film Industry Fund, in the last analysis a government organization, which actually censored, accepted, or rejected scripts. Films could only be made on the basis of scripts approved by the state. Yet film production was not nationalized, the state itself did not produce films….Of course, anybody who decided to go to work without applying … for a[n FIF] subsidy could produce what he liked, but there was no producer who commanded such capital.[110]

In their attempts to rationalize, those steeped in a culture of étatism ironically rationalized their own expansive reach. '…[T]here was so much and often diffuse regulation, that it gave the bureaucracy more and more reason to believe that it should be [more] involved in the film profession'.[111] Regulators held that rules and decrees could build the film industry and secure its future. Many in private industry concurred, sometimes requesting even greater state incursions. Stymied by industry stagnation through 1933, private entities preferred to defer to the state, requesting that it settle disputes, stabilize market positions, provide credits, and negotiate access to new markets. The new IMB, they hoped, would devise uniform, consistent policies which would bridge the divides the film industry had been unable to address on its own.

It swiftly became apparent that the IMB's initiatives would impose no greater coherence on the politics of film than previously existed nor bring the industry peace and stability. Rather, it created confusion. As the features previously discussed illustrate, the IMB, through its control of the Censorship Committee and the FIF, sanctioned production which yielded a wide range of representations of the nation. As output grew, so too did the rifts between sectors of the film world. Producers, theater owners, and distributors all demanded industry-wide agreements and government financial and legislative assistance by claiming they represented the nation's best interests, rational commerce, social progress, and the advancement of culture. The OMME, representing film producers and distributors, regularly called for the reduction of the fees assessed on films, particularly censorship, import, and FIF surcharges. It lobbied for legislative limits on double features and the amount of time a film could be shown in a single theater, and called for exhibition quotas, legal obligations guaranteeing that movie houses buy and show their products. Opposition to these plans most often came from the MMOE, the advocate for cinema owners. Because of the cultural significance of film, the MMOE maintained that just as producers received state subventions from the Film Fund, theaters deserved aid as well. They suggested everything from utility discounts to tax reform, specifically targeting the 'luxury tax' assessed on ticket sales.[112] Further, cinema owners felt aggrieved that the state required much of them, obliging even struggling small theaters to pay for and show Hungarian newsreels, but offered little in return.

Despite sporadic cooperation, disagreement was the hallmark of the mid-1930s movie world. Nearly every policy consideration became a politicized skirmish. The IMB proved incapable of resolving even the most basic questions, such as what roles should the state assume and who best represented the interests of the nation. Subsidiary questions concerning exhibition quotas, theater cartels, and professional training set the film establishment afire. Debates over the nature of production – should it be oriented for export or the domestic market, should quality trump quantity or vice-versa – continued to rage.[113] Between 1933 and 1938, these conflicts not only pitted distributors, producers, theater owners, and government authorities against each other, but they also shattered the precarious solidarities within those constituencies. Premier theater and small theater owners clashed over the matter of trusts, exploding all prospects of a unified association of motion picture exhibitors. Within the realm of distribution, schisms formed between distributors of domestic and foreign film. These industry-wide intra-organizational differences became increasingly significant, as they opened into fissures through which populism, antisemitism, and other forms of radicalism penetrated the profession.

The battle over exhibition quotas was symptomatic of the ways industry constituents gussied up their self-interests in the clothes of rationalization, order, and the national interest. For years, domestic filmmakers had unsuccessfully lobbied for the adoption of compulsory Hungarian language movie showings. Once their products proved they could attract audiences and domestic sources of financial support in 1933–4, Hungarian producers and foreign film translation firms began to argue in favor of quotas in order to insure the wide, and guaranteed, distribution of their works. Quota proponents were adamant that this would attract new audiences, particularly in rural Hungary where the population shunned movie houses because of limited access to films in its mother tongue.[114] Audiences with new or rejuvenated passions for film would mean better business opportunities for all, be they producers, distributors, or theater owners. These were the general outlines of a discussion that began in earnest in 1933 and peaked in the summer of 1935, when the Lower House of Parliament considered the questions of Hungarian language production and foreign film translation. In 1935, the government granted the wishes of the quota constituency, passing the desired legislation.

Cleverly, in all public forums, film producers and their allies phrased their pro-quota arguments in terms of the industry's financial integrity and of Hungarian culture's struggle for survival. Language was the heart of Hungarian culture, and Hungarian language film production, not to mention the dubbing of foreign films into Hungarian, was fundamental to preserving and fostering the national essence. Since the programming of most Hungarian theaters amounted to 'an exercise in [foreign] language instruction that cost HP 3.7 million', the obligatory showing of Hungarian film could only have a positive effect, argued proponents.[115] 'It is our prize interest, indeed it is our obligation to spread Hungarian language', wrote OMB member and editor of the *Hungarian World News* newsreel, Géza Ágotai.[116] Only through Hungarian language could film promote education, moral improvement, and the 'productivity of Hungarian thought', above all in the countryside.[117] Quotas would thus help shear film of its international nature, argued parliamentary quota advocates Rudolf Meskó and Tibor Törs, allowing motion pictures to build the nation, its culture, and economy. No patriot could oppose a minimal quota requirement, nor could anyone with a shred of business sense. Parliamentarians in favor of quotas bombastically suggested that 25 percent of all films shown be Hungarian made, an outrageously high number considering that at the time, less than five percent of the films shown in Hungary were domestic products. Others proposed that within three years, all foreign language works be dubbed into Hungarian, despite the fact that dubbing technology was rarely used compared to subtitling.[118] Few ministers objected outright to quotas, and if they did, they couched their objections as concerns about the technical difficulties of dubbing or protests that one need not force theaters to behave in a patriotic and 'nation supporting' manner, since that behavior was of course instinctual.

Outside of Parliament, however, there was opposition aplenty. The MMOE, on behalf of cinema owners, came out strongly against quota legislation. If assured showings in all Hungarian theaters, suggested MMOE officials, film producers would be incentivized to make the cheapest, not the most technically and artistically sound, films. Lack of competition could have a 'catastrophic impact', wrote MMOE vice-president Pál Morvay, on the quality of Hungarian productions.[119] Rather than encouraging an orderly, efficient, and nation-nurturing process, quotas would enfeeble the national culture.

Despite its members' misgivings, the MMOE eventually acceded to the inevitable. Focusing instead on restricting the size of the quota, they were able to pare the numbers down. When Interior Minister Miklós Kozma issued decrees to 'preserve the Hungarian language in motion picture performances' in July 1935, he required only ten percent of all films exhibited be in Hungarian.[120] Up to half of that number could be Hungarian-dubbed foreign film. In January 1936, the quota would rise to 15 percent, two-thirds of which could be foreign works dubbed in Hungary. To protect the virtue of the Hungarian language, the decrees empowered the censors to refuse exhibition permits to films which misused the Hungarian language.[121] Politicians, cultural commentators, and censors, it seems, saw poor grammar and incorrect usage as grave threats to their culture's integrity.

The quota legislation marked an important shift. First, it reified the government's right, if not its obligation, to intervene in film in order to advance 'national' goals of preserving the Hungarian language and promoting Hungarian filmmaking. Second, quota legislation altered the orientation of Hungarian production. The industry's early 1930s output was geared toward international film markets. By 1935, Hungarian producers could concentrate on the domestic market. Films produced in Hungarian primarily (although not entirely) for Hungarian consumption became conceivable, if not desirable, triggering a qualitative transformation.

Whereas quotas could fix the national interests for producers, sorting out national priorities regarding the complicated rental arrangements for movies was another matter entirely. Sound forced theater owners to pay far more than they had during the silent era, not merely for equipment, but also for the higher distribution costs the talkies commanded. The early 1930s Hungarian response to this new economic landscape was the formation of two film theater groups, one dominated by the theaters owned by the German Ufa film company, and another by the mogul István Gerő.[122] Several hundred cartels such as these existed in the mid-1930s Hungarian business world, some powerfully influencing the Hungarian economy.[123] Many were more influential than those involved in the movie business, but few became as symbolic of Hungary's larger domestic struggles with capitalism and modernity. Through 1938, no issue created more rancor, especially among Christian nationalist forces in the film world, than movie house trusts.

In essence, these trusts were unions of theaters whose owners joined forces to improve their negotiating power. As collectives they possessed more clout than individual theaters, enabling them to demand lower rates for film licenses from distributors. Both Gerő's trust and the Ufa trust tended to attract urban premier and second-week theaters, meaning they were the largest, best-managed, most financially secure theaters. When the distribution companies began to charge higher rates to non-trust theaters in order to maintain profit margins, it was small theaters, late-run theaters, and rural theaters that bore the brunt of the price increases. This, in short, shattered all sense of unity among theater owners. Larger theaters felt they could only bring in profits if they played hardball with distributors and smaller theaters felt the larger theaters were getting rich at their expense.

The trust issue was representative of Hungary's difficult struggle with its own backwardness. Although there were Jewish-owned and Jewish-run theaters on both sides of the battle, the rhetoric soon pitted rationalization against morality; foreign, Jewish, unbridled capitalism against a notion of the collective good described as 'Christian' and 'Hungarian'. Istvan Gerő, a practicing assimilated Jew and the man who introduced sound to Hungary, became the lightning rod in this debate. Gerő opened the country's first sound cinema in Budapest in 1929, and soon began buying theaters and organizing the Movie Theater Trust [*Mozgóképüzemi tröszt*], also known as the Royal Theater Trust or simply the Gerő Trust. Designed to prevent distributors from setting prices for films unilaterally, the Gerő Trust quickly gained power. In short order, Gerő moved into production, spurring contemporaries to dub him the 'uncrowned czar' of the Hungarian film industry.[124]

His cartel was potent, yet it was also more vulnerable than the smaller Ufa Trust, which had the heft of the Third Reich behind it. Gerő understood that this made his trust the only viable target of the anti-trust movement he always suspected was essentially an anti-Jewish, anti-big capital campaign. In Gerő's view, the anger of the small theaters was misplaced. The responsibility for high distribution fees was not his; the only way to combat those fees was to unify. In other words, all theaters should join trusts, which would then negotiate for the best films, foreign and domestic, at affordable prices. If theater owners could merely turn their attentions toward the true enemy, the distributors, then their precarious finances would improve, and their businesses become more rational. Distributors, themselves split between importers of foreign pictures and promoters of domestic products, were keen enough to comprehend the threat Gerő represented, and did their best to sow division among theaters. They petitioned the Interior Minister in May 1934 to limit theater unions to groups whose combined capacity could not exceed 2,500 seats. They purposely stirred discontent among struggling small theaters, especially those rural cinemas run by veterans or war widows, against larger urban theaters. The distributors claimed they had no choice but to charge smaller theaters higher fees because of the unwillingness of the larger 'trust' theaters to make reasonable concessions. The trusts, in their words, were mutilating Hungary.

By June 1934, the rift between small and large theaters had grown vast. For the first time, a group of premier and second-run theaters, in effect those belonging to the Gerő and Ufa trusts, walked out of the MMOE. They formed their own professional organization, the Representatives of the Interests of Premier Film Theaters [*Bemutató Filmszinházak Érdekképviselete*]. While this association would occasionally reach accommodation with the MMOE, most contentious issues remained unresolved through the 1930s. Relations between these adversaries worsened in August 1935, when Gábor Bornemisza, editor of the radical right journal *Daybreak* [*Virradat*], won the presidency of the MMOE running on an anti-trust, anti-Jewish platform.[125] His election effectively ruptured the MMOE, resulting in a clash that once again saw film concerns become stepping-off points for arguments about the essential nature of Hungary.

By foregrounding urban-rural splits and further intertwining them with the problems of antisemitism and anti-capitalism, Bornemisza and his supporters – the smaller city theaters, including second week and third run cinemas, and the smaller theaters of the countryside – reconfigured the familiar trope about 'Christian Hungary' to fit their priorities. They denounced urban Hungary, that of premier theaters and monopoly capitalists, that of 'sinful' Budapest, as Jewish and un-Hungarian.[126] Rural theater owners, on the other hand, were the true purveyors of Hungarian culture. If not for their theaters, the pastoral folk of Hungary stood to lose their connections to the Hungarian language and culture. In addition, rural theater owners saw themselves as integral to the long-term health of the industry, claiming that figurative procreation occurred in their seats. If the film industry wanted to breed loyal and larger audiences, it had to insure the solvency of rural theaters and design its products with countryside appeal. This was the rural vision of rationalization and the national interest: theaters of the people not of the bourgeois, providing bread for the table rather than champagne for the glass. If free markets could not support this vision, state aid should.

Helping to propel Bornemisza's ascension was a deep, industry-wide concern about the continuing closures of theaters. As of November 1934, there were only 285 regularly operating sound theaters outside of Budapest. Eighty-eight theaters had licenses but did not operate, an additional 46 were officially bankrupt, and another 38 operated silently.[127] Despite the presence of the trusts, Budapest's theaters had undergone a thinning out process as well. As of May 1934, their numbers had declined to the lowest total since 1918, although the capacity of the average theater was higher than it had ever been. This mimicked international trends. The cost of sound equipment made small theaters untenable; larger modern cinemas were the way forward. Despite this, there were six percent fewer seats for movie-goers to fill in 1934 Budapest than there had been before the advent of sound in 1929.[128]

This situation, Hungary's film establishment almost unanimously believed, was a grave threat to Hungary's film and national cultures. Bornemisza became the spokesman for a loose alliance of producers, distributors, and exhibitors who argued it was the role of the government to provide the plans and the means to license theaters and to keep them open, especially in the countryside. As of early 1933, only 99 of Hungary's 2,993 villages with fewer than 3,000 inhabitants possessed movie houses.[129] This meant that nearly two-thirds of Hungary's population had no access to sound theaters.[130] Even with rapid growth between 1935 and 1937, when the number of rural theaters climbed from 302 to between 384 and 410 (while Budapest's cinema count rose from 74 to 79), millions of Hungarians missed the sound experience.[131] While no one seriously considered building cinemas in every village, some film elites expressed concern that the isolated countryside would suffer cultural stagnation. This paternalist argument was typically circular: failure to provide 'civilization' to the outlying fringes of Hungary where the heart of Hungarian culture lay would threaten the well-being of the nation.[132]

The national interests – given form in the rationalization, quota, and trust debates – boiled down to avoidance of responsibility, scapegoating, simple solutions, and appeals for state assistance. Build more rural theaters, provide state aid for production, mandate domestic showings, and regulate wages, proposed the distributors and producers of the OMME, and Hungary's film culture will flourish. Hungarian language production will triple and imports decrease. Cart before the horse, countered members of the MMOE. Hungary must first manufacture more Hungarian films, providing theaters the grist needed to mill successful programs and attract rural audiences.[133] In other words, producers and distributors wanted theater owners or the state to make the primary cash outlay, while theater owners believed the producers and distributors should act first. In the end, the industry proved unable to resolve essential disagreements over who would provide the capital investment assuring the industry's long-term health; who was to blame for the industry's failures; and who was primarily responsible for nurturing a national film culture.

Christian Nationalism in the Studios: The Turuls' Antisemitic Attacks

As calls for state action and the rhetoric of antisemitic populism echoed among theater owners starting in the early 1930s, those same ideas, held at bay longer in the production sector, soon began to percolate through, chiefly among young Christian film professionals. In the summer of 1933, the Ministry of Religion and Education officially recognized the curriculum of the National Film Association Film Arts School run by the movie director Béla Gaál. Since all but one or two of Hungary's private film schools had failed in the late twenties and early thirties, state certification of Gaál's program marked an important point in the revival, standardization, and professionalization of film industry training. Of course, there were costs to this element of industry rationalization. Theater owners balked at the new taxes the state slapped on movie tickets to fund this and future government-sanctioned film schools. In addition, some of the products of the program, the self-described 'third generation' of Hungarian film professionals, found themselves educated but unemployed.[134] This was to have serious repercussions.

Because Hungarian filmmaking was so precarious, investors rarely risked using unknown quantities, whether they were actors, directors, scriptwriters, or cameramen. Between 1931 and December 1938, Hungary produced 132 films. Nearly half, 63 films in total, featured one of three leading men: Pál Jávor (28 films), Antal Páger (21 films), or Imre Ráday (14 films). A handful of talented actors, including Sándor Pethes, Gyula Kabos, Gyula Gózon, Gerő Mály and Gyula Csortos, played either leading or large supporting roles. Pethes appeared, on average, in every other film and Kabos in every third film. Over the same period of time, nine women starred in 68 of Hungary's 132 films. Irén Ágai led the way with 11 featured roles, followed by Ida Turai, Margit Dayka, Klari Tolnay, Zita Szeleczky, Éva Szörényi, Mária Lázár, Rózsi Bársony, and Zita Perczel. Seven screenwriters wrote 45 percent of Hungary's films (60 total), and many more were credited to others but likely penned by one of those seven. Ten directors made 101 of Hungary's 132 films, with two men, István Székely and Béla Gaál, accounting for 44 films, or fully one-third of the era's bounty.[135] These statistics make two facts clear. First, a limited circle of recognizable and productive talent held sway in the world of Hungarian filmmaking.[136] Second, the crews behind the casts in this small circle were predominantly Jewish. Jews commanded the production segment of the profession, in both numbers and influence. Through the mid-1930s, nearly all of Hungary's top producers, directors, screenwriters, and many of the top players listed above were Budapest Jews. Many were internationally well-connected, having trained in Weimar Germany, London, Paris, Vienna, or Hollywood. Nearly all believed they were engaged in a vanguard industry, serving the nation through a new mass entertainment and education medium. This mix of a cosmopolitan pedigree and hypermodern optimism set them apart from the newly trained but underutilized 'third-generation' of film professionals. Frustrated with their inability to break into the business, many among the third generation gravitated to the industry's radical populist wing, calling on the state to assure them gainful employment and endorsing 'de-Jewification' to secure their places.[137]

Within the production sector, populist and antisemitic rumblings began to surface in 1934 and coalesced by 1936. The realization that a 'Jewish conspiracy' existed in the Hungarian film world came to Viktor Bánky, a film editor, assistant director, and the brother of the renowned actress Vilma Bánky, in 1935. Bánky had worked in Berlin until early 1933, when he returned to Hungary as part of the second wave of Hungarian film professionals ousted from Germany. He attributed his struggle to find work in Hungary to lingering resentment against Germany on the part of expelled Jews. Bánky believed that in retaliation for Germany's treatment of them, the Jewish producers Joe Pasternak and Ernő Gál had made a pact to deny employment to all Christian film professionals.[138] Pasternak, in Bánky's mind, imagined even further revenge: a purging of all Christians from Hungarian film, conducted with the aid of Gál, the family of director István Székely, and contacts at the MFI.[139] In Bánky's estimation, the final conspirator was István Gerő. Bánky wildly charged that Gerő influenced nearly 80 percent of the films made in Hungary through provision of credit for production, and naturally had no interest in employing Christian professionals.[140]

Around the time of Bánky's epiphany, a Hunnia employee named Kálmán Zsabka began recruiting his fellow Christian employees into the Turul Society [*Turul Szövetség*]. The Turul Society was one of Hungary's most influential and politically active radical student organizations. Originating immediately after the Great War, it had agitated in favor of the first antisemitic legislation passed in any European country, the 1920 *numerus clausus*, which limited Jewish enrollment at Hungarian universities.[141] A growing Turul faction became more rightist, nationalist, antisemitic, and populist in orientation in the later 1920s and 30s.[142] In 1935, the Turuls turned toward the film industry, where they found a ringleader in Zsabka. While the number of Christian co-workers Zsabka succeeded in converting is unknown, anecdotal evidence suggests that he won a substantial following among young and disaffected film personnel. He marshaled them into the Turuls' cultural wing, the Fraternal Society of Fine Arts [*Szépmives Bajtársi Egyesület*]. There they stirred resentment of Jews and oligarchs and spearheaded a movement for 'all Christian youth' to purify Hungarian cinema from within. Bánky was one of the early adherents, and in the Turul Society he found like-minded film comrades.

Just as the fight over trusts, with its clear antisemitic overtones, was heating up in 1935 and early 1936, organs and spokesmen on the far and populist right took up the Turul cause. They demanded that Jews be expelled from the entire Hungarian film establishment, its production side in particular. The István Milotay-edited radical rightist newspaper, *New Hungary* [*Uj Magyarság*], first issued this ultimatum in January 1936, forcefully suggesting all practising Jews and Jews by birth be extricated however possible.[143] The mouthpiece of the Turul Society, the journal *Comrade-in-Arms* [*Bajtárs*], soon joined the fray, calling for a 'changing of the guard' in the world of Hungarian film. *Bajtárs* urged an end to what it perceived to be the destructive influence of the 'cosmopolitan' Jews and their repression of those creative Christian talents whose goal it was to manufacture true Hungarian products. It called on the government and the film industry to aid these legitimate Hungarians and curtail the corrosive influence of those inauthentic Hungarians presently in control. Without naming names, it was clear the article's author was implicating precisely the 'cabal' of Pasternak, Székely, Gál, and Gerő fingered by Viktor Bánky. *Bajtárs* then went a step further, denouncing the false, international nature of Hungarian film. 'Why must every domestic film have a role which pillories our society, binding it with tasteless forms of life?' asked the author. No longer should Hungarian film be made by 'occasional capitalists' or 'trumped up foreign directors'. It should be made by Hungarian artists and talents instead. There was no need for film production if it disregarded these precepts.[144]

In February 1936, the Turul offensive continued, reiterating the familiar thunder of anti-capitalist, pro-countryside populism, wrapped in an acknowledgment of film's critical national-cultural significance:

> It is undeniable that film is the most powerful propaganda tool for the national culture. Today in Hungary film is not utilized for that goal. A well-constructed national politics of culture, while rooted in Hungarian racial uniqueness, in Christian morals, and a Christian worldview, must draw film into its propagation. The education of the wide masses and the cultivation of the nation make it all the more justified and urgent. Silent film had an international character. The task of sound film is to express the racial character of the country that produces it.
>
> The film production that today is called Hungarian stands far from the spirit of the Hungarian people [and] from their culture. It is the greatest opponent of the conservative politics of culture in the country and from the perspective of mass education [it exerts] a destructive influence.[145]

Funded by 'cosmopolitan capital' which 'inundated the country with jargon films and dream productions' filled with a 'destructive, nation-splitting spirit', Hungarian film could not represent the nation's essential culture.

The new generation, continued the article's author István Jóny, 'is obliged to rid the nation of this cosmopolitan spirit'. To build '100 percent national' film production Jóny exhorted the youth of Hungary to join the national camp and refuse to patronize Jewish films and theaters. This would spur a wave of bankruptcies, which the government could rectify by awarding the insolvent companies to Christian Hungarians and underwriting the risk involved in filmmaking through increased subsidies. Christian investors would then enlist 'in the service of nation-educating cultural politics' and real national film production would ensue. Hungary would be best served, concluded Jóny, if it mimicked Italy and Germany. It should create a government-supervised Film Chamber empowered to control the entire cinema realm.

Action soon followed these rhetorical fusillades. In mid-1937, the Turul Fraternal Society of Fine Arts dispatched delegations to speak with the heads of the two Hungarian film studios, Hunnia and the MFI, on behalf of Christian professionals. They achieved some success. Hunnia immediately hired a member of the Fraternal Society delegation, as did Ernő Gál on behalf of the MFI.[146] According to Viktor Bánky, the impact of these actions was far-reaching, but not in the way he expected. Jewish and non-Jewish film elites recognized the threat posed by the Turul Society. They negotiated with the associations representing more moderate Christian film professionals and rewarded those who distanced themselves from the movement.[147] Christians found jobs, but preference went to those who were not Turuls.

This resistance at the studio level ran counter to opinion at higher government levels, where support for the Turuls was growing. As early as 1936, some of Hungary's film regulators were already moving in the direction the anti-Jewish right desired, actively promoting the cultivation of Christian filmmakers. Yet they had not reached the point where ideology would trump economics, a point they would reach only two years hence. The background to the film *Pagans* [*Pogányok*] is illustrative. In early 1936 the Hungarian Stylus Film company submitted plans to Hunnia and the government authorities to do an adaptation of the Ferenc Herczeg's novel by the same name. The book and film featured the early eleventh-century battles between King István and the pagan Hungarian tribes of Danubian plain, potent mythological material concerning Hungary's Christian origin. Stylus requested Hunnia provide two weeks of studio time as well as an unprecedented HP 50,000 of production credit. Noting the film's cultural and historical importance, the two delegates of the Ministry of Religion and Education serving on the Hunnia Directorate, State Secretary Baron Gyula Wlassics Jr. and László Balogh, also agreed that funding the film afforded Hungary the opportunity to launch the career of a 'Christian Hungarian director',[148] thus advancing two high priorities.[149] The notion of using films such as *Pagans* to catapult Christian directors to prominence appealed to the Hunnia board, but the difficulty inherent in transforming Herczeg's novel into screen fare convinced the board to discourage Stylus' pursuit of a novice director. Hunnia ultimately decided to back *Pagans*, but with less credit than Stylus Film requested. Hunnia also approved of the choice of Emil Martonffy, already a well-known and experienced Christian Hungarian director, to make the movie. Hunnia's choices implied that in 1936, promoting Christian film production might be laudable, but not if it jeopardized financial success.

Over the next several years, between 1936 and 1939, the Turul Society attracted new and prominent members, men and women frustrated by false starts such as these.[150] The Society also expanded its anti-Jewish/anti-cosmopolitan campaign in print, on the streets, and in parliament. These actions soon turned violent. On 11 February 1937, Gerő's Forum theater premiered the István Székely's directed, Ernő Gál produced picture *Affair of Honor* [*Lovagias Ügy*] to great critical acclaim.[151] Eleven days later, from its tent encampment along the Danube, the Turul Society organized a large protest, calling, in this case, for government provision of jobs.[152] After the gathering, the Turuls headed to the Forum to protest *Affair of Honor*, inspired by a similar Turul-organized rally in Pécs the previous day. Turul actions in Budapest, in turn, ignited antisemitic demonstrations in Debrecen, Szeged, and throughout the countryside. These rallies continued through March and drew participation from what soon became Hungary's largest national socialist party, the Arrow Cross. The protests forced theater owners to cut the film's run short, and created enough upheaval to attract the attention of the German and American consulates.[153] Despite denunciations from Prime Minister Kálmán Darányi and Minister of Religion and Education Bálint Hóman, the Turuls loudly and forcefully proclaimed the need to cleanse Hungarian film and theater, to rid Hungary of the three-headed Gerő-Gál-Székely monster.

As a play, *Affair of Honor* had been a stage smash in Budapest and had not attracted the ire of the radical right, so what was it that made the feature so objectionable to the Turuls? Both the film's content and its production symbolized the problems the right had with Hungary's budding mass culture industry and Hungarian society more broadly. The filmmakers decided to re-write much of Sandór Hunyady's play from the middle of the second act, adding a poor girl-rich boy romance that had not appeared in the stage version. However, the insertion of romance was not the only change to the play's text. Hunyady's original work was based on a dispute between two Jews named Fried and Geld, who nearly have a duel before settling their differences.[154] The screenwriter, László Vadnai, altered the storyline so that the film's central dispute occurred not between two Jews, but between a Jewish accountant, Mr. Virág, and Pál Milkó, the nephew of the Christian director of the company the Jewish accountant serves. Gyula Kabos played Virág, and the film expresses clear sympathy for this Jewish protagonist while exhibiting distaste for custom.[155]

The Turuls found these alterations and sentiments offensive. The bulk of the movie now revolved around the nephew of a Christian Hungarian seeking to make restitution for insulting a Jew. For the Turuls, this was all that was wrong with Hungarian society. Not only were assimilated Jews controlling Hungary, they were demanding public subservience. That a Christian would even deign to consider an apology was an affront to nearly all on the Hungarian right. Further, as reconciliation between the film's protagonists results in the marriage of the Christian Pál Milkó (Imre Ráday) and the Jewish Baba Virág (Zita Perczel), the film envisages a future for Hungary that the Turuls and most on the right abhorred: that of a physical union between Christian and Jew.

It was also the making of the film, as much as its content, which aggrieved the right. *Affair of Honor* had a significant Jewish creative component. All of the key figures involved in the film's production and premier – Székely, Gál, and Gerő – were of Jewish origin. Both of the movie's male protagonists (Gyula Kabos and Imre Ráday), as well as several of those who played significant supporting roles (Gyula Gózon, Zita Perczel) were also Hungarian Jews. For the Turuls, this was intolerable, and one of the far-right newspapers noted with exaggeration that the purpose of the student protests was to point out that 'without exception, every one of the actors was Jewish'.[156] The actual numbers of Jews in the film was beside the point. For the Turuls, Jewish involvement in film was symbolic of what accounted for Hungary's weakened economic, military, and spiritual state. Jewish domination of Hungary's culture and its critical industries needed to be rolled back.

Reaction to the movie spread like wildfire on the right. *Virradat*, the journal that MMOE president Gábor Bornemisza edited, and the Centrum publishing house, which he owned, gave wide distribution to an outpouring of antisemitic bile. Unsigned letters in *Virradat* attacked the director 'István Székely-Schwarcz' as a man of 'Galician origins' who 'had nothing to do with the Hungarian race' and could not possibly create films true to the Hungarian spirit.[157] Centrum published a pamphlet on the 'ghettofilm' style that saturated and emasculated Hungarian filmmaking.[158] This response to *Affair of Honor* energized the movement to eliminate Jews from the Hungarian film profession. In July 1937, the Chief Secretary of the Turul Society, József Végváry, submitted a memorandum to several cabinet-level officials, asserting that 'nearly all films traded [in Hungary] served Jewish business and racial interests exclusively' and as a result, all Hungarian film was tainted by a 'ghetto-flavored immorality'.[159] Only the employment of more Christian-Hungarians – writers, directors, cameramen, musicians, set designers, actors, and actresses – could purge the ghetto flavor and provide Hungary's predominantly Christian audiences with a suitable product. Further, Végváry argued, government ministries needed to provide Christian filmmakers easier access to production licenses and material aid.[160] The explicit corollary to Végváry's argument was that Jews had to be prohibited from engaging in all film activity.

Végváry's petition encouraged and legitimized the practice of antisemitism in film production. Coming on the heels of the *Affair of Honor* protests, the petition seemed a palpable turning point to the Jewish community. The leading neolog Jewish daily, the pro-assimilation, politically conservative, and anti-Zionist *Equality* [*Egyenlőség*], which sometimes pooh-poohed instances of antisemitism in Hungarian society, nervously wrote of a 'nascent wave of antisemitism in the Hunnia film factory'. The paper cited a recent decline in the number of Jewish actors and actresses involved in films, a trend that had accelerated since the Turul memorandum's submission. Since that point, wrote László Palásti, antisemitism had become customary in the selection of casts for films, and the number of Jewish film technicians was waning as well.[161] By October, there was concern among Jews and liberals that the state-controlled Hunnia film studio had adopted an 'Aryan paragraph' for productions.[162]

While some of Hungary's film journals, such as Imre Somló's *The Film* [*A Film*], claimed it was ludicrous to believe that Hunnia boss János Bingert would condone such a change,[163] a series of December 1937 *Egyenlőség* articles lent credence to the assertions of rising bias. In an early December interview, a Jewish actor identified only as Emil F. remarked that antisemitism and Turul domination of the arts had made the situation intolerable for Jews. He and others were now non-entities, unable to find employment on the stage or before the camera. 'The Turul rules; it dictates and reviews [all contract decisions] in the theaters…[and] the situation is the same [in film]'.[164] Others charged that Jewish musicians found the doors to film studios and top theaters shut, forcing them to eke out livings working in smaller theaters.[165] One of Hungary's most prominent actors of Jewish descent, Lajos Gárdonyi, ruefully noted that by December 1937, the Turul Society had secured for itself permanent places alongside the chiefs of the two Hungarian studios, ostensibly supervising the casting of all Hungarian films. 'They not only kicked out Jews from among the actors, but they did not permit [Jewish] support professionals or statisticians'.[166] Remarks about the Aryanization of film casts and worries about the well-being of Christian companies, unmentioned prior to 1937, regularly appeared in Hunnia's Executive Committee minutes in 1938.[167] It was apparent that Turul concerns had spread to the highest levels of Hungarian filmmaking and by early 1938, the society's vision of a Christian, organic Hungarian culture threatened to become the norm. The debate over whether Jews could have a role in creating or promoting that culture and whether they belonged to the Hungarian nation had reached a new stage.

Conclusion

The new cultural and political implications of language-specific films amplified society-wide concerns with the images of Hungary and Hungarianness just as Hungary's film industry became self-sustaining. Through the 1930s, rural theaters opened or resumed long-suspended operations and audiences viewing Hungarian film expanded to include greater numbers of the peasant and working classes. Hungary's film establishment began to produce far greater numbers of films designed primarily for the domestic market, rather than for foreign markets, yet continued to work to mold images of Hungary seen abroad. As a result, the industry engaged in an increasingly complex struggle to define itself and to forge an attractive national identity suitable to interwar realities. The results were a multitude of arguments and alternatives, often competing and contradictory. These notions ruptured the film community, but also forged unexpected unities. From censors to diplomats, urbanists to populists, producers to exhibitors – all defined Hungarian, national, and the national culture according to parochial self-interests determined by competing social, political, economic, artistic, and ideological agendas. In many cases, their definitions were malleable, allowing censors to disregard their statutory guidance, official image makers to offer incongruous visions of Hungary, and filmmakers to cooperate with government bureaucrats to challenge mainstream illiberal ideas and devise syncretic, modernist visions of the future. This chaos and contradiction confounded many, especially the populists and Christian nationalists of the right, whose frustration about filmmakers' inability to draw sharper lines demarcating Hungarian national identity burst through the surface in the mid-1930s. Their attacks on internationalism, capitalism, and the Jewishness of the filmmaking enterprise, combined with the movie community's own inability to reconcile its festering disputes, produced the industry's single consensus: the government must be more engaged. State intervention legislating exhibition quotas, extending government-directed film enterprises, and arbitrating disagreements lent the film business a degree of security it had not previously enjoyed. However, if film was ever to achieve its sacred cultural mission, more sweeping government involvement, something more powerful than the IMB, would be necessary.

Notes

1. Balázs Trencsényi, '"Imposed Authenticity": Approaching Eastern European National Characterologies in the Inter-war Period', *Central Europe* 8/1 (May 2010), 34, 36, 46.
2. Andrew Lass, 'Aesthetics of Distinction in Czechoslovakia', in Verdery & Banac (eds): *National Character and National Ideology in Interwar Eastern Europe*, 43.

3. Lajta, 'A magyar film története', 12.
4. István György, 'Vitam et sanguinem – a magyar filmért!' *FK* VI/4 (1 April 1933), 2–3.
5. Sándor Jeszenszky, 'A magyar film – A magyar kultura egyik létkérdése', *FK* VII/4 (1 April 1934), 1–2. Jeszenszky was also an advisor to the Minister of Religion and Education. As Nicholas Reeves notes, Jeszenszky's view was shared the world over during the interwar and wartime period. Reeves, *The Power of Film Propaganda: Myth or Reality?* (London, 1999), 4.
6. *MOL-Ó* – Hunnia, Z1124-r1-d20. Német-Magyar filmcsere…, 1933–42. J. Széll to J. Goebbels, Nr.157 789/1933 (undated, probably 30 November 1933).
7. Jeszenszky, 'A magyar film', 1–2.
8. Castiglione was also an advisor in the Trade Ministry and Budapest's Corso theater director. Henrik Castiglione, 'Magyarország film-világpiaci jelentősége', *FK* VII/7–8 (July-August 1934), 1–4.
9. István György, 'Magyar Film', *Nyugat* 27/21 (1934), 458.
10. Anna Manchin, 'Fables of Modernity. Entertainment Films and the Social Imaginary in Interwar Hungary', (Brown University: Unpublished Ph.D. dissertation, 2008), 5–6.
11. On the nation as discursive concept, see esp. Craig Calhoon *Nationalism*. Minneapolis, MN, 1997); R. Suny & M. Kennedy (eds), *Intellectuals and the Articulation of the Nation*. (Ann Arbor, MI, 2001); Bhabha, *Nation & Narration*.
12. Verdery, 'Introduction', in Verdery & Banac (eds), *National Character and National Ideology in Interwar Eastern Europe*, xxii.
13. 'A Magyar Külügyi Társaság megalakulása', *Külügyi Szemle* 1/1–2 (July-October, 1920), 1. See also Kuno Klebelsberg, *Neonacionalizmus* (Budapest, 1928), 247.
14. An excellent discussion of Klebelsberg and Hungarian interwar cultural diplomacy is contained in Zsolt Nagy, 'Grand Delusions: Interwar Hungarian Cultural Diplomacy, 1918–1941', (University of North Carolina: Unpublished Ph.D. dissertation, 2012). See also parliamentary hearings throughout the 1930s. For example, Gyula Petrovácz, in a parliamentary session, claimed that: 'In appreciation of the situation confronting the Hungarian nation, it can be declared…that although they humbled us at Trianon, the majority of our relations can be rectified not using diplomacy, not using politics, rather with the weapons of literature and art'. *Az 1935.évi április hó 27-ére hirdetett Országgyülés Képviselőházának naplója,* Harmadik kötet. Hiteles kiadás (Budapest, 1935), 93. Erik Ingebrigsten, 'Privileged Origins. "National Models" and the Reform of Public Health in Hungary', in György Péteri (ed), *Imagining the West in Eastern Europe and the Soviet Union* (Pittsburgh, PA, 2010), 54, 38.
15. G.C. Paikert, 'Hungarian Foreign Policy in Intercultural Relations, 1919–1944', *American Slavic and East European Review*, 11/1 (February 1952), 42–65; esp. 43–5. Paikert, a Secretary in the Ministry of Religion and Education, was

intimately involved in the cultural political efforts to convince the West of the validity of this equation. He wrote of Hungary's interwar and wartime cultural diplomacy, particularly the demonstration of Hungarian 'cultural superiority', as a matter of 'national life and death'.

16. Quoted in Langer, 'Fejezetek', 27.
17. Záhonyi-Ábel, 'Filmcenzúra Magyarországon a Horthy-korszak', http://www.mediakutato.hu/cikk/2012_02_nyar/12_filmcenzura_magyarorszagon/. The Film Council retained a minor censorship role examining weekly newsreels until the VKM disbanded the Council in July 1933.
18. *MOL* – BM, K158-cs2, OMB, 1932. Folio – 272/1928.eln. – 'A m.kir. OMB elvi jelentőségü határozatainak gyüjteménye'.
19. *MOL* – BM, K158-cs14, 1944. Minutes of the 18–19 January 1928 plenary session of the OMB, 'Nemzetépítő irányelvek a filmcenzúrában'.
20. Ibid. The Interior Ministry oversaw the OMB, but the OMB included representatives from the Justice, Army, Religion and Education, Foreign Affairs, Treasury, Trade, and the Prime Ministries. A minority (33 percent or less of every reviewing committee) of its members were private film company representatives, actors, and actresses. From 1932 on, the Hungarian Evangelical, Lutheran, and Catholic churches all had the right to name representatives to the Censorship Committee.
21. Jill Forbes & Sarah Street, 'European Cinema, an Overview', in Forbes & Street (eds), *European Cinema*, 17, 11. Scitovszky admitted that determining the 'preferred expression' of what was Hungarian was a difficult task. 'We all know that the national interest and ethics, the national- and social moral order demands the enforcement of considerations so wide-ranging and far reaching that cannot possibly be enlisted in austere paragraphs. But I am convinced that the members of the committee, by their moral and intellectual qualities, will enforce in their work with the right subjective considerations even principles that cannot be contained in written laws'. Quoted in Manchin, 'Fables of Modernity', 54.
22. Lists of films prohibited found in *MOL* – BM, K158-cs1-8, but these files include few of the censors' explanations.
23. 'A magyar filmszakma nyolc éve számokban', *FK* V/3 (1 March 1932), 5–6. See also *MOL* – BM, K158-cs7, OMB, 1936. 'Nyilvántartás – az OMB által megvizsgált filmek évi vizsgálatának részletes statisztikai adatairól'.
24. *Filmkultúra* statistics show censorship, as measured by the banning of film, dropped off in 1930 to around three percent but then rebounded to a high of over five percent of all film examined (12.2 percent of all sound film by meter) in 1931. While industry statistics are littered with errors and inconsistencies, that censors became much more stringent in late 1930 and 1931 is undeniable. In 1931, Hungarian authorities forbid approximately 30 of 214 sound feature films they reviewed, over twice the number and three times the length of censored silent film. *MOL* – BM, K158-cs2, 1932 and K158-cs7, OMB, 1936.

25. For more detailed descriptions of Hungarian censorship in the early 1930s, see Frey, 'National Cinema, World Stage', 73–90 and Frey, 'Aristocrats, Gypsies, and Cowboys All: Film Stereotypes and Hungarian National Identity in the 1930s', *Nationalities Papers* 30/2 (2002), 383–8.
26. *MOL* – BM, K158-cs2, OMB, 1932. 'A Filmcenzura tiz éve. Megjelent az elvi döntések tára', [Undated, probably December 1930].
27. On controlling stereotype content for explicitly political purposes, see S. Reicher, N. Hopkins, and S. Condor, 'Stereotype construction as a strategy of influence', in R. Spears and P. J. Oakes, et al. (eds), *The Social Psychology of Stereotyping and Group Life* (Malden, MA, 1997), 94–118. For analysis of a similar example of censorship and national culture from the interwar period, see Kevin Robins and Asu Askoy, 'Deep Nation. The national question and Turkish cinema culture', in M. Hjort & S. MacKenzie, (eds), *Cinema & Nation*, 203–21.
28. Béla Scitovszky wrote that 'Censorship is a necessary evil, and should be applied only to the degree necessary, because as is the case with medicine, excessive use of it can have a poisonous and destructive effect'. See Scitovszky, 'Nation-building principles in film censorship', *Rendőr* (21 January 1928), 6. An example of this light touch was recounted by the writer Zsigmond Móricz, who wrote in his diary that Film Industry Fund (and likely OMB members) told him that during the filming his *I Can't Live without Music* [*Nem élhetek muzsikaszó nélkül*], the gentry could not carouse so much, as it presented them too negatively. The filmmakers responded accordingly. Balogh & Király, *'Czak egy nap a világ…'* 396.
29. Péter Hanák, 'The Image of Neighbors in the Hungarian Mirror', in A. Gerrits & N. Adler (eds), *Vampires Unstaked: National Images, Stereotypes and Myths in East Central Europe* (Amsterdam, 1995), 55–67.
30. On the problem of limited identifiers that small countries possess, see O'Regan, *Australian National Cinema*, 92.
31. László Marácz, 'Western Images and Stereotypes of the Hungarians' in Gerrits & Adler (eds), *Vampires*, 35.
32. While this use of stereotypes may appear problematic, it indicates that many contemporaries were aware of how context and function impacted representation. On the varied functions of stereotypes, see Richard Dyer, *The Matter of Images. Essays on the Role of Representation* (London, 1993), 11–17.
33. This approach traces back to the beginning of the Horthy era. See Nagy, 'Grand Delusions', ch.5.
34. The term *magyarság* appears repeatedly in contemporary discussion of Hungarian culture. This amorphous and ultimately vacuous concept of a cultural essence has analogies in other European cultures. See Núria Triana-Toribio's discussion of 'Spanishness' in *Spanish National Cinema* (New York, 2003), 6–7.

35. *MOL* – KM, K66-cs164, 1930, III-6/c: Film. Gerevich to KM, 5 May 1930.
36. As Attila Pók has explained books and articles, this was a form of scapegoating or allocating blame. See Pók, 'Atonement and Sacrifice: Scapegoats in Modern Eastern and Central Europe', *East European Quarterly* XXXII/4 (January 1999), 531–48; Pók, 'Scapegoating and Antisemitism after World War I', *CEU Jewish Studies Yearbook* III (2002–03), 125–34; and Pók, *The Politics of Hatred in the Middle of Europe. Scapegoating in Twentieth Century Hungary: History and Historiography*. (Szombathely, 2009).
37. Jenő Gábor, 'Erdélyi, Felvidéki és Bácskai filmhiradókat a revizió érdekében', *FK* VII/1 (1 January 1934), 14–15.
38. 'Hirek', *Mozivilág* II (XIX)/12 (23 March 1930), 5. The plan was not enacted.
39. 'Scitovszky Béla belügyminiszter és Bud János kereskedelmi miniszter hangosfilm-üzenete a magyar közönséghez a magyar film támogatása érdekében', *FK* IV/5 (1 May 1931), 1.
40. *MOL* – VKM, K507-cs91-t12, 1937–41. 'Jgykv.', 545–53.
41. *Pesti Hirlap* estimated that in Germany and Switzerland alone, more than 70 million viewers would see the film. 'Magyar Filmpropaganda Németországban', *Pesti Hirlap* (14 March 1930), 19. For additional reports on the success of *Hungária*, see *MOL* - KM, K66-cs164, III-6-c Film, szinház 1930–32514.
42. *MOL* - KM, K66-cs164, III-6-c Film, szinház 1930–32514. MKK Stockholm, 15 February 1930, 200/1930. Underline in original.
43. *MOL* - KM, K66-cs378, 1938, III-6/c: Film, szinházügyek. MFI to Lajos Villáni, 6 April 1934. For evidence that some of the promised funds were received late if at all, see *MOL* - MFI, K675-cs1-t3, 1928–44. MFI VBi ülésének napirendje és jgykv-e, 14 March 1934.
44. 'Hangos Hungária', *FK* VII/4 (1 April 1934), 5.
45. *MOL* - KM, K66-cs240, 1934, III-6/c: Film, szinházügyek. Bevezetés – 'Az uj hangos 'Hungária' – film vázlata'. My summary of the film comes directly from this eight page document and the extant copy of *Hungária*. I am grateful to Zsolt Nagy for sharing his copy of *Hungária* with me.
46. 'Két új magyar film', *Budapesti Hirlap* (24 March 1934).
47. 'Hungária', *Pesti Hirlap* (24 March 1934).
48. 'Hangos Hungária', *FK* VII/4 (1 April 1934), 5.
49. *MOL* - MFI, K675-cs1-t3, 1928–44. MFI VBi ülésének napirendje és jgykv-e, 14 March 1934.
50. *MOL* - KM, K66-cs240, 1934, III-6/c: Film, szinházügyek. All above references to reviews come from 'Melléklet az 1600/1934.sz jelentéshez. Holland sajtóhangok a Hungária hangosfilmről', 7 June 1934. Reports from December 1934 indicate the film continued to help lessen the 'socialist anti-Hungarian atmosphere' in the Hague. *MOL* - KM, K66-cs264, 1935, III-6: Kulturális ügyek. KÜM 34.659/11–1934, 14 December 1934.

51. Linde-Laursen, 'Taking the National Family to the Movies', 26. Linde-Laursen claims these characteristics make a film 'national'.
52. István Radó, 'Miért juthat csődbe?' *Mozivilág* XXX/25–26 (25 December 1936).
53. Tivadar Csáky, 'A magyarfilmgyártás gazdasági jelentősége', *Külkereskedelmi Figyelő* II (15 February 1935), 1.
54. János Bingert, 'A magyar film útja a világpiac felé', *FK* VIII/2 (1 February 1935), 4–5; Géza Ágotai, 'A Velencei II. Nemzetközi Filmmüvészeti kiállitás', *FK* VII/9 (1 September 1934), 6–9.
55. Ágotai, 'A Velencei II. Nemzetközi Filmmüvészeti kiállitás', 6–9.
56. Hungary negotiated bilateral cultural/intellectual agreements with Austria (1934), Poland (1934, 1935), Italy (1935), Germany (1936), Estonia (1938), Finland (1938), Japan (1938, 1940) & Bulgaria (1941).
57. Protochronistic films stressed that the given nation was first in its technological or cultural achievements, that it had a special claim on 'civilization' or 'progress'.
58. Elemér Boross, quoted in Nemeskürty, 'A magyar hangosfilm első évei 1931–1938', 43–4. Underline in original.
59. Romsics, *Hungary in the Twentieth Century*, 178.
60. Béla Kempelen, 'A filmgyártás jövő feladatai', *Film Élet* 1933/1 (19–26 December 1933), 4–5.
61. 'És a felbujtó?' *Fejérmegyei Napló* XXXVIII/241 (24 October 1931), 1–2.
62. István György, 'Vitam et sanguinem', op. cit., 2–3. The concept of the 'sacred task' is Kempelen's in 'A filmgyártás jövő feladatai', 4–5.
63. Andor Lajta, 'Magyar történelmi filmeket!' *FK* VII/4 (1 April 1934), 12.
64. Kempelen, 'A filmgyártás jövő feladatai', 4–5
65. Jeszenszky, 'A Magyar film', 1–2. Italics in original.
66. Borbándi, *Der ungarische Populismus*, 101. Hungarian populism was known as the *népies* or *népi* movement. The terms derive from the root *nép*, whose meaning is very similar to the German *Volk*.
67. Tamás Hofer, 'The "Hungarian Soul" and the "Historic Layers of National Heritage": Conceptualizations of Hungarian Folk Culture, 1880–1944', in Banac & Verdery (eds), *National Character*, esp. 67, 70, 77, 80.
68. Populism was a European and perhaps worldwide phenomenon. It is important to note that the variant was simultaneously conservative, revolutionary, and divided, with strong left and right-wing components. After a brief period in which leftist, anti-Nazi populism was ascendant in 1937, Hungarian populism became increasingly racist into the 1940s. Sz. Péter Nagy, ed., *A népi-urbánus vita dokumentumai* (Budapest, 1990); Charles Gati, 'The Populist Current'; Pál Romány, 'Tanyavilág és falukutatás', *Magyar Tudomány* 163/9 (September 2002), 1187–8. Miklós Lackó, 'Populism in Hungary: Yesterday and Today', in Joseph Held (ed), *Populism in Eastern Europe* (Boulder, CO, 1996), 107–27.

69. Hanebrink, *In Defense,* 126; András Sipos, 'Who is a "True Hungarian"? The Movement of "Spiritual Defence of the Fatherland" and the Image of the Enemy', *CEU History Department, Working Paper Series 2* (Budapest, 1995), 119–31. Sipos rightly points out that anti-German and anti-Jewish populism often fit hand-in-glove.
70. Hanebrink, *In Defense,* 124–5.
71. The latter is the core argument of Vera Ranki's *The Politics of Inclusion and Exclusion: Jews and Nationalism in Hungary* (Teaneck, NJ, 1999).
72. Trencsényi, '"Imposed Authenticity"', 34–44.
73. 'Urbanists' is the term used by Pál Ignotus.
74. I discuss the nature of the term Jewish in chapter four. Many of the urbanists were born into Jewish or recently converted families, but many had, as Richard Esbenshade points out, 'given up all connection with the organized Jewish community and any over Jewish identity'. Esbenshade, 'The Radical Assimilated: Hungarian "Urbanists" and Jewish Identity in the 1930s', in L. Greenspoon, R. Simkins & B. Horowitz (eds), *The Jews of Eastern Europe, Studies in Jewish Civilization,* v. 16 (Omaha, NE, 2005), 117.
75. Unnamed interwar Hungarian architect, quoted in Hofer, 'The 'Hungarian Soul', 1880–1944', 73.
76. Manchin, 'Fables of Modernity', 61–110. Manchin argues (84): 'All of the more prominent Hungarian directors of the 1930s, with the exception of Marton Keleti, learned their trade in Berlin, Vienna, or Paris from an international group of filmmakers…Their personal experiences as Hungarians becoming highly successful abroad were mirrored in the cosmopolitan view of modern Hungary in their films'.
77. This is the conclusion of Manchin's analysis of 1930s Hungarian sound film. See Manchin, 'Fables of Modernity'. Manchin draws from Balogh & Király, '*Csak egy nap a világ…*', 79–80.
78. István Székely, 'A magyar filmgyártás eddigi müvészi és erkölcsi mérlege', *FK* VI/11 (1 November 1933), 1–2.
79. Some may have adopted a more expansive view of the national interest. Others who sat on the boards of both the regulatory agencies and Hunnia may have decided lining their own pockets was the most important national interest.
80. 'A magyarországi filmszinházak helyzete', *FK* VI/12 (1 December 1933), 1–2.
81. Statistics vary. According to Tibor Törs, MP, Hungarian moviegoers purchased over 15 million tickets annually. Other estimates for 1935 claim that theaters sold 18.5 million tickets. Törs quoted in Az országgyűlés képviselőházának 35. ülése. 1935 évi június hó 18-án kedden. *Az 1935. évi április hó 27-ére hirdetett Országgyülés Képviselőházának naplója,* 102. For the higher estimate, see *Cultural Policy in Hungary* (Paris, 1974), 78. Government statistics are rare, but 1934 numbers indicate that Hungarians spent over HP 10 million on movie

tickets in 1934. MKK Statisztikai Hivatal, *Magyarország községeinek háztartási viszonyai az 1934.évi községi költségelő iranyozatok szerint* (Budapest, 1935), 21.

82. Hake, *Popular Cinema of the Third Reich*, 9–11.
83. Manchin, 'Fables of Modernity', 17–18; Cunningham, *Hungarian Cinema*, 42.
84. On cars and nation, see Tim Edensor, *National Identity, Popular Culture and Everyday Life* (Oxford, UK, 2002), 118–37.
85. G. Balogh, V. Gyürey & P. Honffy, *A magyar játékfilm története a kezdetektől 1990-ig* (Budapest, 2004), 48; Cunningham, *Hungarian Cinema*, 41.
86. Manchin, 'Fables of Modernity', esp. 19; 88–9.
87. János Füki, 'Hatósági jóindulat és kedvezmények szükségesek az uj vidéki mozik üzembehelyezésénél', *FK* VIII/3 (1 March 1935), 4–5.
88. MP Ernő Brody, quoted 18–19 June 1935, Lower House hearings excerpt in *FK* VIII/7–8 (1 July 1935), 10.
89. András Komor, 'Egy mozijáró naplójából', *Magyar Filmkurír* X/49–52 (24 December 1936).
90. György, 'Magyar Film', *Nyugat* 27/21 (1934), 458. György blamed not only filmmakers, but Hungarian audiences in particular for lacking any sort of 'Hungarian consciousness[ses]'.
91. Balogh and Király assert that prior to the mid-1930s, Hungarian films generally appealed to the lower and middle ranks of society or youth, relying on love stories or early life crises to do so. They define 'senior' films as mid-30s works which addressed themes aimed at more mature audiences, featuring older men and often mixing nostalgia for a mythical past with some form of family loss or generational conflict. These films aimed to build audiences and bridge the era's social divides. Balogh & Király, *'Czak egy nap a világ…'*, 191, 289–96.
92. Ibid., 74.
93. On films and ideological messaging through 'diffuse emotional effects' and 'symbolic transformations of socially relevant issues', see Stephen Lowry, 'Ideology and Excess in Nazi Melodrama: *The Golden City*', *New German Critique* 74 (Spring-Summer 1998), Stephen Lowry, 'Ideology and Excess in Nazi Melodrama: The Golden City', *New German Critique* 74 (Spring-Summer 1998), 125–49 and *Pathos und Politik: Ideologie in Spielfilmen des Nationalsozialismus* (Tübingen, 1991).
94. *MOL* – KM, K66-cs296, 1936, III-6/c: Film, szinházügyek. Lits to KÜM, Munich, 14 March 1936.
95. *MOL* – KKM-i Levéltár Külkereskedelmi Hivatal, K734-cs3-t12. Kiállitás, vásár, film. 'Feljegyzés a vadászfilm németnyelvü kópiájának Göring német miniszterelnöknek ajándékozása tárgyában', 18 February 1938.
96. *MOL* – KMI, K429-cs59-t2. Vegyes MTI levelezés, 1938–9. Ambassador Pelényi to Kozma, Washington, DC, 19 September 1938.

97. Manchin, 'Interwar Hungarian Entertainment Films and the Reinvention of Rural Modernity', *Rural History* 21/2 (October 2010), 199.
98. Manchin, 'Interwar Hungarian Entertainment Films', 210.
99. Sonja Simonyi, '*Modernity and Tradition. Film in Interwar Central Europe*', National Gallery of Art exhibit guide (Washington, DC, 2007). Depending upon who used the term, the search for a 'third way' or 'third road' could mean a medium between capitalism and communism, Nazism/Fascism and Bolshevism, and even democracy and Bolshevism.
100. Balogh & Király, '*Czak egy nap a világ…*', 673.
101. Anna Manchin, 'Jewish Humor and the Cabaret Tradition in Interwar Hungarian Entertainment Films', in Lina Khatib (ed), *Story Telling in World Cinema: Forms* v.1 (New York, 2012), 37–8.
102. Cunningham, *Hungarian Cinema,* 45.
103. Miklós Lackó, *Korszellem és tudomány 1910–1945* (Budapest, 1988), 181–210.
104. Verdery, 'Introduction', in Verdery & Banac (eds), *National Character and National Ideology,* xxiii.
105. Mary Nolan, *Visions of Modernity. American Business and the Modernization of Germany* (Oxford, UK, 1994), 6.
106. Quoted in Kőháti, 'Magyar film hangot keres (1931–1938)', *Filmspirál* II/2-1 (1996), 100–1.
107. 'Kozma Miklós lesz az átszervezett magyar filmgyártás vezetője', *Filmújság. Zsolnai László Hétilapja* II/1 (7 January 1933), 1, 6.
108. 5200/1933 M.E. sz, 12 May 1933.
109. *NARA* – RG59 [State], M-973, Roll 211, 864.4061/Motion Pictures 42. Report from Consul General John B. Osborne to Sec State, 'Modification of the Decree No. 6292/1925 M.E. Regarding the Promotion of Hungarian Film Production', 17 May 1933.
110. Nemeskürty, *Word and Image,* 73. Nemeskürty, 'A magyar hangosfilm első évei 1931–1938', 31. Unfortunately, few of the archives of the FIF exist today and thus it is impossible to fully evaluate the accuracy of Nemeskürty's conclusion.
111. Lajta, 'A magyar film története', 42.
112. Depending on the locality, this tax could be as low as three or as high as 17 percent of gross ticket sales.
113. See, e.g., the plans proposed by Hunnia's Director Dr. János Bingert, the producers Ernő Gál and the Berlin-based, Arzén Cserépy, the cinema journalist Andor Lajta, groups of parliamentarians, and numerous others. Bingert, 'A racionális magyar filmgyártás felépitése és fejlesztése', *FK* VII/1 (1 January 1934), 3; Ernő Gál, 'A magyar film eddigi mérlege and jövő rentabilitása', *FK* VII/1 (1 January 1934), 4–5; 'A filmgyártás racionalizálásáról', *FK* VII/2 (1 February 1934), 1; Lajta, 'A magyar film története', 50.

114. Rudolf Meskó, quoted in 'Az országgyűlés képviselőházának 35. ülése. 1935 évi június hó 18-án kedden', *Az 1935. évi április hó 27-ére hirdetett Országgyülés Képviselőházának naplója*. Harmadik kötet, (Budapest, 1935), 101.
115. Miklós Kozma, quoted in 'Az országgyűlés képviselőházának 35. ülése', op. cit., 103. At the time, HP 3.7 million were equivalent to approximately US$ 740,000.
116. Géza Ágotai, 'Az utószinkron kérdése Magyarországon', *FK* VIII/6 (1 June 1935), 5–7.
117. MP József Szabó, quoted in the 18–19 June 1935 Lower House hearings excerpted in *FK* VIII/7–8 (1 July 1935), 2–20, esp. 16. Hunnia head Bingert predicted that dubbing would bring significant new audiences. Bingert, 'Okos és józan filmgyártás, valamint céltudatos szinkronizálás uj tömegeket hoz a magyar filmszinházaknak', *FK* VIII/9 (1 September 1935), 2–3.
118. With the passage of quota legislation, dubbing increased substantially. Szilvia Dallos, *Magyar hangja…a szikronizálás története* (Budapest, 1999), 17–26.
119. Pál Morvay, 'A magyarországi filmszinházak helyzete 1933-ban', *FK* VI/12 (1 December 1933), 1–2.
120. 180.000/1935 BM and 180.100/1935 BM, 26 July 1935.
121. *MOL* –BM, K158-cs6, OMB, 1934–5, Alapsz. 319/1935; K158-cs7 OMB, 1936, 402/1936; K158-cs8, OMB, 1937, 338/1937; K158-cs9, 1938, 362/1938.
122. According to László Zsolnai, nine of Budapest's 12 premier theaters were either members of Gerő's Mozgóképüzemi trust or the Ufa trust. By 1934, according to Henrik Castiglione, that number had decreased to nine of 14 premier theaters in Budapest, but still represented 69 percent of all of Budapest's premier theater seating capacity. Castiglione himself was a theater owner and member of Gerő's trust. György Guthy, the editor of *Mozi- és Filmvilág* and a consistent proponent of the anti-trust position, claimed that Gerő determined the programming for all but two of the 14 premier theaters in 1938 Budapest. See 'A budapesti filmkölcsönzők a mozitrösztök feloszlatását kérik a kereskedelemügyi minisztertől', *Filmújság. Zsolnai László Hétilapja* I/1 (10 December 1932), 5; Henrik Castiglione, 'A Budapesti mozièlet erőviszonyai', *FK* VII/5 (1 May 1934), 1–4; and György Guthy, 'Gerő István uralkodik pénze és összeköttetése révén a belföldi filmpiacon', *Mozi- és Filmvilág* II/3 (25 March 1938), 4.
123. The Gömbös government promised to confront the trusts and place Hungary on a 'Christian and national course' but never did so. Jenő Gergely, *Gömbös Gyula* (Budapest, 2001), 257.
124. Langer, 'Fejezetek', 104. István Borhegyi, the *Mozivilág* editor, regularly demeaned Gerő using this term.
125. Gábor Bornemisza was the brother of Géza Bornemisza, Gömbös' Minster for Manufacturing and one of the leaders of the 'reform generation' of radical right student activists. Gergely, *Gömbös*, 288.

126. Horthy used the term 'bűnös' meaning both guilty and sinful. See Lukács, *Budapest 1900*, 179.
127. 'Magyarország mozgóképszinházai 1934. végén', *FK* VII/12 (1 December 1934), 4–5.
128. Henrik Castiglione, 'A budapesti moziélet erőviszonyai', 1–4.
129. Henrik Castiglione, 'A vidéki filmszinházak az új népszámlálás tükrében', *FK* VI/5 (1 March 1933), 1–4.
130. Pál Morvay, 'Magyarország filmszinházákultúrája a statisztika bonckése allat', *FK* VI/7–8 (1 July 1933), 3.
131. Henrik Castiglione, 'A budapesti mozipark adattükre', *FK* XI/1 (1 January 1938), 4–8. In the 1936/37 year alone, nearly 40 new theaters opened their doors. Castiglione's figure of 302 operating theaters does not seem to match exactly with the figure of 285 he cited in 1934. If one uses 285, the increase in theater numbers was 35 percent. Official government statistics from 1934/35 indicate that there were 310 operational rural theaters and 75 operational Budapest theaters in early 1935. MKK Statisztikai Hivatal, *Magyar Statisztikai Évkönyv, Uj Folyam XLIII, 1935* (Budapest, 1936), 347. Endre Hevesi, estimated that by 1936, there were at least 409 working rural sound theaters. Hevesi, 'The Year in Hungary', *1937–38 International Motion Picture Almanac* (New York, 1938), 1171.
132. The term culture is used by István Erdélyi precisely in this sense. See Dr. István Erdélyi, 'A szünetelő vidéki mozgó[kép]szinházak megnyitásának kérdése', *FK* VII/12 (1 December 1935), 5–6. On intellectuals, culture, nation, and state expansion, see R. Suny & M. Kennedy, 'Introduction', in Suny & Kennedy (eds), *Intellectuals and the Articulation of Nation* (Ann Arbor, MI, 2001), esp. 25–33.
133. A series of sometimes acrimonious articles appeared in *Filmkultúra* in 1935, pitting Henrik Castiglione, representing producers, distributors, and large theater owners against Pál Morvay, representing the non-trust theaters. Castiglione's position was that better education for the countryside, a better economy, and more theaters was the best path to forging a stable and successful Hungarian film culture. In other words, make Hungary more modern and less agricultural, like Austria, and its film culture would flourish. Morvay disagreed, arguing that more films were necessary to assure the existence of rural theaters. It would be wonderful if the Hungarian economy were to improve, Morvay believed, but the reason farmers do not patronize the cinema is not a lack of cash, rather the lack of attractive and comprehensible products. Increased Hungarian language production, then, was the obvious path to a healthy, growing Hungarian film culture.
134. The term 'third generation', referenced Gyula Szekfű's famous history of Hungary, *Három nemzedék és ami utána következik*. Hungary's younger, domestically-trained film professionals used this designation to self-identify.

It differentiated them from the founding fathers of Hungarian film, active during the Habsburg era, and the second generation of the 1920s and 30s. Announcing its presence in June 1937, the third generation declared itself imbued with 'richer spirits and ideas' than the 'old and tired' second generation. It insisted on the rights to make films and earn livings. 'A harmadik filmnemzedék élniakarása', *FK* X/6 (1 June 1937), 4–5.

135. Statistics from Nemeskürty, 'A magyar hangosfilm első évei 1931–1938', 58–91. The number of films each actor participated in either as a star or as a lead supporting actor is as follows: Sándor Pethes (approximately 65), Gyula Kabos (45), Gyula Gózon (42), Gerő Mály (34) and Gyula Csortos (24). Leading actresses were, in addition to Ágai, Ida Turai (9), Margit Dayka (8), Klari Tolnay (8), Zita Szeleczky (7), Éva Szörényi (7), Mária Lázár (6), Rózsi Bársony (6) and Zita Perczel (6). Kabos, Gózon, Mály, Ágai, and Perczel were Jews.
136. In comparison with the Hollywood star system, Hungarian stars appeared in a far greater percentage of the country's overall output The United States made hundreds of films each year, and American stars, even at the height of their popularity, appeared in no more than a handful of films annually.
137. While the terms de-Jewification and de-Jewify do not exist in English, they are literal translations of several Hungarian [*zsidótlan, zsidótlanítás*] and the German word *Entjudung*. I have used this phraseology connoting complete removal of Jews from the community because it accurately imparts the malignant intentions of contemporaries who used the terms, far more than the benign English phrase 'removal of the Jews' and the inaccurate 'expulsion of the Jews'.
138. The rightist paper *Uj Magyarság* made these same charges in mid-1937. According to the Jewish daily *Egyenlőség, Új Magyarság* charged that István Székely and Joe Pasternak secretly agreed to ensure Jewish domination of Hungarian film. See 'Árja filmek kellenek?' *Egyenlőség* (9 September 1937), 5.
139. *BFL*– Bánky [Bánki] Gyula Viktor, Nb. 2540/1945; 70–100, quotes from 71–2. [*...itt valóban a keresztény filmszakemberek terv-szerü* [*sic*] *kiszoritásáról van szó*]. Bánky, like many film establishment figures active during the war, was later accused of 'crimes against the people' and/or war crimes. Bánky himself was sentenced to six months in jail and five years loss of political rights for his roles in the production of antisemitic films and in the expulsion of Jews from the movie business.
140. *BFL* – Bánky [Bánki] Gyula Viktor, Nb. 2540/1945; 93.
141. For Turul involvement in the *numerus clausus,* see Iván Berend, 'The Road toward the Holocaust: The Ideological and Political Background', in R. Braham & R. Vagó (eds), *The Holocaust in Hungary: Forty Years Later* (New York, 1985), 34–5.

142. Borbándi, *Der ungarische Populismus,* 172; Lackó, 'Populism in Hungary', 114.
143. Milotay was one of the most important voices among Hungarian racial and antisemitic nationalists. János Gyurgyák, *A Zsidókérdés Magyarországon* (Budapest, 2001), 440–6.
144. 'Turulhivatás 1936-ben: az őrségváltás kiharcolása', *Bajtárs* (17 January 1936), 1–3.
145. István Jóny, 'Hozzászólás a nemzeti filmgyártáshoz', *Bajtárs* (20 February 1936). Italics in original.
146. *MOL* – KM, K66-cs335, 1937, III-6/c: Film, szinházügyek. MFI records indicate that in a June 1937 meeting, the Turul group requested the MFI 'provide the means and opportunities for young Christians, after being trained accordingly, to receive roles in film production'. Tőrey cleverly told the Turuls that opportunity-wise, the MFI will support them (Christians) in all cases exactly the same way as Hunnia does. 1 July 1937, 'A MFI VBi ülésének napirendje és 28 számú jgykv-e'.
147. One of the organizations they turned to was the National Alliance of Hungarian University and Polytechnic School Students [MEFHOSz].
148. *MOL-Ó* – Hunnia, Z1123-r1-d1, 1936. 26 June 1936 Hunnia Igazg. jgyek., 5–6.
149. Langer, 'Fejezetek', 145.
150. *BFL* – Testimony of Dr. Géza Staud, in Ferenc Kiss, Nb. 4077/45; 19–20. Kiss, one of Hungary's best known actors, was perhaps the most renowned Turul.
151. 'Aus der Filmwelt', *Pester Lloyd* 84/34 (12 February 1937), morning edition, 8; Imre Somló, 'A lovagias ügy', *A Film. A Magyar Mozi és Filmszakma Lapja* IV (March 1937), 10.
152. 'Diaktüntetés a Kossuth Lajos utcában', *Friss Újság* (23 February 1937), 5.
153. 'Ujább diaktüntetések Pécsett', *Friss Újság* (23 February 1937), 5; 'Studentendemonstrationen auch in Debrecen', *Pester Lloyd* 84/45 (25 February 1937), morning edition, 4; GsB (Werkmeister) report to the AA, 1847/420 892–901, 10 March 1937, reprinted in György Ránki (ed), *A Wilhelmstrasse és Magyarország* (Budapest, 1968), 206; and *NARA* – RG59 [State], M-1206, 864.4016/103, John Montgomery to the Secretary of State, No.878, 19 November 1937. See also István Székely, *A Hyppolittól a Lila Akácig* (Budapest, 1978), 153.
154. Tibor Sándor, 13 December 1998 presentation, 'Zsidó sorsok magyar filmen', Budapest, Puskin Theater.
155. László Kelecsényi, *A magyar hangosfilm hét évtizede, 1931–2000. Hyppolittól Werckmeisterig* (Budapest, 2003), 20.
156. 'Az ifjúság tüntetése a 'Lovagias ügy' cimű film ellen', *Új Magyarság* (23 February 1937), 4.
157. Quoted in Sándor, *Őrségváltás,* 93–4.

158. Sándor, *Őrségváltás*, 97–8.
159. The memorandum was addressed to Hungary's Minister for Manufacturing, Géza Bornemisza and Minister of Religion and Education, Bálint Hóman. The first quote in this sentence comes from László Zsolnai, who quotes Végváry's memorandum; the remaining quotations are from *Új Magyarság*. I was unable to find the actual memorandum in government records. László Zsolnai, 'Sürgősen állítsák be a filmgyártásba a keresztény erőket. Hozzászólás a Turul Szövetség gettófilm memorandumához', *Filmújság. Zsolnai László Hétilapja* (17 July 1937), 1–4; 'Sürgősen állítsák be a filmgyártásba a keresztény erőket!' *Új Magyarság* (14 July 1937).
160. Zsolnai, 'Sürgősen állítsák...' op. cit.
161. László Palásti, 'Miért kapnak a zsidó filmszinészek egyre kevesebb szerepet?, *Egyenlőség* (22 July 1937), 7.
162. 'Árjaparagrafus a filmgyártásban', *Demokrata Ujság* (2 October 1937).
163. 'Három árja sir magában', *A Film. A Magyar Mozi és Filmszakma Lapja* IV (October 1937), 1–2.
164. 'Magyar zsidó szinészek kálváriája', *Egyenlőség* (9 December 1937), 7. Emil F. was likely Emil Fenyvessy.
165. Gábor K., quoted in 'Magyar zsidó szinészek kálváriája', *Egyenlőség* (9 December 1937), 7. Both Emil F. and Gábor K. urged Jews to develop greater unity and greater 'senses of self'. They also asked audiences to consider boycotting theaters and films which refused Jews roles.
166. 'Miért bocsátják el a zsidó alkalmazottakat a filmgyárakban', *Egyenlőség* (16 December 1937), 8.
167. *MOL-Ó* – Hunnia, Z1123-r1-d1, 1936–45. See 1937–8 Hunnia Igazg. jgyek.: 23 April 1937, 20 November 1937, 24 January 1938, 9 February 1938, 14 March 1938, 20 April 1938, 5 July 1938, 6 September 1938.

3

National Cinema, International Stage: Film Trade and Foreign Relations

Introduction

Every national film industry develops within an international framework, a product of transnational cultural, financial, political, and aesthetic interactions. The Hungarian example is emblematic. Between 1933 and 1938, Hungarian production grew by leaps and bounds, and export and Hungary's place in global cinema became existential matters. Greater numbers of Hungarian sound features forced the cultivation of larger audiences to insure continued viability. There were two possible options to guarantee the needed revenue: expand the domestic market and/or win access to foreign screens. While theaters were (re)opening at a reasonable rate, the mid-1930s proved that autarky was not feasible for a small, linguistically-trapped country such as Hungary. To have a completely self-sustaining domestic market similar to Germany, France, England, the United States and even Czechoslovakia, one that could assure profits for the average Hungarian-made picture, Hungary's premier film statistician guessed that the country required over 900 regularly operating theaters.[1] As of late 1937 Hungary remained a 'mere' 300–400 short.[2] Thus, it was no surprise that at precisely the same time as Hungary's film establishment energetically engaged in aesthetic and political debates over the purpose, creation, and content of its films, even conservatives, populists, and the most radical nationalists could not resist the allure of foreign sales.

The export question, like all film questions, was enmeshed in broader considerations of Hungary's future. Accepting that export was not a luxury but a necessity, when Hungarians film figures discussed capturing foreign audiences, the specific audiences they targeted often spoke volumes about their politics and their corresponding concepts of Hungarian identity. The choices generally boiled down to whether Hungary's focus should be closer relations with its proximate neighbors, Germany, or the United States. Determining where the Hungarian state should seek markets, economic assistance, diplomatic leverage, or enduring partnerships for its film industry were politicized decisions nested in wider discussions about mass culture and Hungary's place in Europe and beyond. Hungarian authorities and some producers favored pursuit of Germany, salivating at the thought of their films being seen by a population which exceeded Hungary's by a factor of nearly ten. Ideas then prevalent in Germany about the creation and protection of a 'national' film industry spawned by a unique race-based culture appealed in Hungary, particularly in bureaucratic and right-leaning nationalist circles.[3] However, some of Hungary's oligarchs feared Nazi radicalism more than they appreciated its nationalist orientation. Thus, segments of officialdom, and nearly every one of Hungary's privately-owned production and distribution companies, took a different tack. Almost every Hungarian distributor was of Jewish origin, and since 1933 these men and women had distanced themselves from Germany. Having a hand in both imports and exports, they looked beyond Germany for possible markets while simultaneously importing fewer German films.

By the mid-1930s, these conflicting orientations created new breaches that would split government officials and divide them from many of their movie business compatriots. While Hungarian bureaucrats ambivalently courted Berlin, Hungarian importers and filmmakers instead turned toward Hollywood, Vienna, and even to Hungary's Little Entente neighbors for markets and goods. The most promising developments were collaborative Austrian-Hungarian German-language productions. These transnational innovations represented a distinct Central European form until German aggression rendered them impossible in the later 1930s. As the decade progressed, developments such as these clarified the extent to which the Hungarian film establishment was not fully in control of its own decisions. External pressures heavily influenced the transformation of domestic film culture. By 1938, for example, Hungary's sometimes reluctant tilt toward Germany appeared as if it were becoming irreversible, as other options vanished. To the chagrin of many of Hungary's film professionals, their industry increasingly found its fortunes shackled to the Nazi giant. Yet this incline toward Germany was belied by the behavior of a large portion of Hungary's film establishment and Hungarian moviegoers. Through the 1930s, Hollywood products, not German ones, dominated the country's cinema programming. Hungary's most profitable movie export market was not Germany, but the United States. So once again the story lines re-emerged: priorities and philosophies clashed, the nature of the medium caused conflicts, Jewish 'questions' surfaced in a variety of forms, and paradox abounded as Hungary sought its niche, buying and selling film in the global marketplace.

The Regional Context: The Film Cultures of Europe's Center and the Politics of Culture

As Hungarians peered outward at the small to medium-sized states nearby, they observed a range of cinematic development. From the Balkans to the Soviet border, the process of building sound cinema cultures took similar but regionally distinct forms when compared to interwar Hungary. In a literal and figurative sense, Hungary found itself in the middle. It was not the most developed sound culture, nor the most primitive. In Central and Southeastern Europe, the correlation between national wealth and the vitality of domestic sound production was clear, although this alone did not determine whether states succeeded in fostering a sound film culture characterized by indigenous language production. As one might expect, the northern and western regions of Central Europe, meaning Austria, Czechoslovakia, and Poland, were home to the healthiest film cultures, with the southern and eastern lands of Yugoslavia, Romania, and Bulgaria trailing well behind.

By far the most rapid and seamless transition to sound occurred in Czechoslovakia, the Central European state with the highest per capita income and most developed network of theaters.[4] Czechoslovak producers had struggled in the early days of the Republic, but during the last years of the silent age production was surprisingly vibrant, with its filmmakers averaging 30 to 35 features per year.[5] Thus, unlike in Hungary, a viable industry existed at the time of the conversion to sound. With a population eager for Czech language production and a government with the means and mechanisms to support it, Czech sound filmmaking developed rapidly.[6] Bolstered by an exhibition network which by 1938 expanded to over 1,800 theaters with a capacity to hold more than 600,000 viewers, Czechoslovakia supported the most mature cinema culture in Central Europe outside of Germany. Professionals led by those working at the famed Barrandov studios (owned by the father and uncle of the future President of the post-communist Czech Republic, Vaclav Havel) created avant-garde animation and 30 to 50 feature films annually from 1933 through 1938.[7] All of this, of course, collapsed in 1938 and 1939 with Germany's crippling and eventual destruction of the Czechoslovak state. With much of its capacity appropriated by its Nazi overlords, Czech film production never exceeded more than 11 films annually during World War II.[8] The new Slovak state, although it did produce domestic newsreels, made no feature film during its short wartime existence.[9]

Austrian sound film production, which will be considered in greater detail later in this chapter, did not match Czechoslovakia's early vigor. Similar to Hungary, Austria's silent filmmaking had gone quiet in the late 1920s, and the studios of Vienna were slow to adapt to sound. Like Hungary, Austria was hampered by the disappearance of investment capital and the failure of the majority of its silent era production companies. The country's first sound film, Georg Jacoby's *Money on the Street* [*Geld auf der Strasse*] was realized only with German support.[10] However, despite increasing state intervention during the rule of Engelbert Dollfuss and a stifling 1934 treaty with Germany that made exports into the German market subject to racial restrictions, Austria ultimately proved to be a multinational hub, a place where, at least through the mid-1930s, filmmakers from across Central Europe flocked, often for international co-productions.[11] Because of their vibrant film cultures and sheer number of theaters which could potentially screen Hungarian film, both Czechoslovakia and Austria appealed to Hungarian filmmakers.

In Poland, sound film took multiple paths, as both Polish and Yiddish talkies gradually replaced silent film.[12] Until the arrival of sound, the history of the motion picture in Poland was highly politicized, often along ethno-religious lines. Even in the silent era, fights and competition resulted from the language used for intertitles. Polish exhibitors publicly berated cinema-goers who frequented German or Jewish theaters. 'In large part', writes Shiela Skaff, the 'history of Polish cinema is a narrative of people alternately participating in and negotiating ways to avoid the linguistic and class tensions with which they lived on a daily basis'.[13] As in Hungary, domestic silent film production had virtually ground to a halt in the latter 1920s. The first sound film to premier in Poland was an Al Jolson film, as it was in Hungary. In 1930, native filmmakers completed the country's first Polish-made sound film, *A Dangerous Love Affair*, although it took several more years for Warsaw's main studio, Falanga, to be fully equipped for sound.[14] By 1935, seven out of eight Polish theaters possessed sound projection equipment.[15] Hungary's neighbor to the north found that using film as a mechanism for forging a national culture created the same possibilities and problems as Hungary experienced. With sound, 'issues of historical remembrance arose, and debates over the (in)accuracy of cinematic representations of the nation's past and present' became contentious.[16] The economic struggles of filmmakers in Warsaw mirrored those in Budapest. Both industries relied primarily on established theatrical actors rather than risking capital on unknowns, and ticket receipts in the two capital cities made up an enormous percentage of each country's overall box office sales.[17]

At the other end of the spectrum stood the Romanian, Bulgarian, and Yugoslav cinema industries. In Romania, the technical and economic transition to sound proved 'catastrophic' through the early 1940s.[18] Unable to mobilize private or state funds, Romania lacked a domestic studio designed for sound feature manufacture. As a result, Romanian sound film production was entirely dependent on studios outside its borders, particularly those in Paris, Prague, Berlin, and, ironically, Budapest. Two of the top grossing Romanian language films of the 1930s, *The Phantom Train* and *Marie, Légende Hongroise,* were filmed with Romanian casts in Hungary's Hunnia studios.[19] As late as 1938, Romania possessed a scant 372 sound theaters and the managers of these theaters showed a marked preference for American film, a phenomenon consistent with many other European states.[20] Despite the establishment of a National Fund for Cinema in 1934 and a National Cinema Office in 1938, national production was limited to newsreels, documentaries, and culture films. Romania did not finish its first home-made sound feature, *A Stormy Night,* until director Jean Georgescu completed it in a dank documentary studio in late 1941. Even with significant Italian aid, just two more full-length Romanian language features came to fruition prior to 1945.[21]

Yugoslavia found itself in comparable straits. In the late 1930s, the number of Yugoslav cinemas was less than half the total possessed by Hungary, even though the population of Yugoslavia was nearly twice that of its northern neighbor.[22] Yugoslavia lacked the capability for sound film production until the early 1940s, when Hungarian experts helped build sound facilities. Prior to 1942, Serbo-Croatian language films did appear, but all were co-productions assembled in studios outside the geographical limits of Yugoslavia and its wartime successor states. A Yugoslav cast, for example, produced the Serbo-Croatian film *Love and Passion* in New Jersey, but never screened it in Yugoslavia.[23] The first Serbian sound fiction film made at home, *Innocence Unprotected,* was not completed until 1942, and regular domestic production did not exist until after the war. Prior to 1940, Hollywood imports dominated the Yugoslav market, with German products a distant second.[24]

Bulgaria's post-Great War producers of culture, like Hungary's, actively strove for the creation of 'native, specific' national forms of literature, music, and the visual arts. Although Bulgaria lacked the resources to incorporate film in this endeavor, its leaders identified motion pictures as a mechanism for nationalization and for 'contributing to the national Bulgarian cause abroad', mirroring contemporary beliefs prevalent throughout Europe.[25] Bulgaria's first feature with sound, the musical *Song of the Mountains,* premiered in 1934 and, like its second sound feature, was made in Hungary. Of the 16 Bulgarian entertainment features made between 1934 and 1944, several were silent, while most were little more than vehicles for patriotism, folklore, or literary adaptations.[26] One of the more intriguing patriotic products of this era was the 1943 Hungarian-Bulgarian sound co-production, *Seaside Encounter* [*Tengerparti randevú/Bŭlgarski-ungarska rapsodia/* Българско-унгарската рапсодия]. Filmed and situated in Budapest, the film provided a metaphor for Hungarian-Bulgarian diplomatic relations through a music academy melodrama featuring an international love affair. Despite this and other efforts of Bulgaria's creative class, 'film stagnated on the periphery of cultural and economic activity' prior to the end of World War II.[27]

For Hungary's political and cultural elites, the appeal of exporting locally and facilitating the development of the film cultures of its less advanced neighbors was obvious. In the states surrounding Hungary there were over three million ethnic Hungarians and many more Hungarian-speakers. Films sold to these countries would not have to be dubbed or translated, and they would serve the crucial political purpose of keeping the Hungarian culture alive in the lands severed by Trianon. Károly Guttmann, MGM's Budapest representative, expressed a variant of this philosophy as early as 1933. He argued that Hungarian creative talents should make apolitical movies. These need not be expensive, multiple language version films; they could be in Hungarian and dubbed in other languages post-production. These pictures should then be exported to Hungary's closest neighbors, including Yugoslavia, Romania, and Czechoslovakia. These markets, implied Guttmann, were the key to the future of Hungarian production. The addition of several million Hungarian speakers and approximately 250 theaters could almost guarantee Hungary a language market large enough to support continuous, profitable Hungarian-language movie making. In addition, following the American dictum that trade follows film, Hungary's neighbors would be more apt to import other Hungarian products, thus promoting Hungary's ulterior national interests.[28] Better cultural relations and better trade would undoubtedly improve the political situation over time.

Regrettably, the decade of the 1930s was not the time for such a plan. Territorial revision was not only the ultimate goal in the minds of Hungary's political elite, it was the one goal shared by virtually all of Hungary's populace. Hungary's neighbors had little incentive to allow uninhibited distribution of Hungarian cultural products.[29] Rightly, they suspected that every initiative coming out of Hungary was motivated by irredentism. Likewise, Hungarian officials showed little willingness to open their country to the cultural products of its neighbors. Over the course of the 1930s, therefore, no real film trade developed between Hungary and Romania or Yugoslavia.

In fact, Yugoslavia prohibited the import of all Hungarian cultural goods until 1937. Even relatively simple matters of film trade, involving the reciprocal exchange of newsreels, remained problematic until the last years of the decade. Due to disputes over possession of Transylvania, relations with Romania were so antagonistic that Baron Lajos Villány, head of the Cultural Department of the Foreign Ministry, threatened in 1939 that he would 'utilize international organizations and friendly contacts' to prevent other European nations from showing Romanian film once Romania developed sound production capability.[30]

Film relations with Romania and Yugoslavia's Little Entente ally, Czechoslovakia, were better, but not notably. Rarely did Hungary import more than five Czechoslovak features a year, and not until the second half of the decade did it export that number. Czechoslovak censors regularly forbid the showing of Hungarian films, films approved elsewhere in Europe, for reasons the Hungarians found spurious. This prompted the Hungarians to respond in kind or to refuse Czech films import licenses. This tit-for-tat mentality was so pervasive it affected virtually every film exchange. When Hunnia decided to import and dub the Czechoslovak hit *Hej Rup* in 1937, it did so with the understanding that it would be able to export films to Czechoslovakia. When Czechoslovak authorities then refused Hunnia's exports, a minor 'film war' ensued and Hungarian exports sunk from twelve in 1937 to five in 1938.[31]

Until the dismemberment of Czechoslovakia and the outbreak of World War II, Hungary could not gain access to the audiences it so desperately desired: Hungarian-speakers in neighboring states. When searching for other alternatives, Hungary's film elites eliminated many other European states without fully exploring their feasibility. They dismissed English-language exports to Britain for fear of British import quotas. They generally believed exports to France and Italy to be doomed endeavors, as neither country expressed much interest in Hungary's wares. Even if they had, both countries subsidized their native producers, a fact which convinced Hungarians that their products would never receive fair screenings. As wounds from the Osso controversies of 1933 were still fresh, most of Hungary's top film officials remained skeptical of any French plans for production in Hungary.[32] Plus, in terms of national politics, many Hungarian elites perceived France to be anti-Hungarian, Italy pro-Romanian. With all of these options closed off – France, Italy, Romania, Yugoslavia, and Czechoslovakia – the Hungarian film establishment looked elsewhere. Its gaze fell primarily on Germany, America, and to a lesser degree Austria. Hungary's film elites saw these countries as more than markets for Hungarian features. They were also sources of capital for film fabrication and organizational models for the evolving Hungarian national film industry. Each option represented a distinct direction, afforded Hungary a greater or lesser degree of sovereignty over its own film affairs, and came with unique risks and rewards.

German-Hungarian Film Relations: The National Idea and the Centrality of Antisemitism

When Gyula Gömbös became Prime Minister in 1932, he immediately signaled his desire to expand Hungary's economic relations with Germany. On 1 February 1933, only two days after Hitler's takeover in Germany, Gömbös dispatched his emissary Sándor Khuen-Hédervåry and days later his Foreign Minister, Kálmán Kánya, to reiterate this wish. Although his initial preference was to strengthen ties between Hungary, Italy, and Austria, Gömbös quickly realized Hitler and the military and economic might of Germany were the best means of leveraging Hungary's revisionist aims and rebuilding its economy.[33] In early 1933, Hungarian and German government officials discussed the expansion of both trade and political relations.[34] In mid-June, Gömbös became the first foreign head of state to visit Hitler, ending Germany's diplomatic isolation. Among other things, the two leaders spoke about the intensification of political cooperation between their two states as well as the creation of a revisionist bloc with Italy.[35]

The government's wish to broaden its relations with Germany incorporated greater cultural exchange, film products included. Even before Goebbels' Ministry for Enlightenment and Propaganda was fully established, Hunnia chief János Bingert wrote to the German Ministry of Interior expressing his aspiration for a broad film trade agreement. In this March 1933 correspondence, Bingert proposed two possibilities. First, extend the agreement between Germany and Hungary, worked out unofficially in 1929, allowing Hungary to export four films to Germany yearly, exempt from quota considerations. Second, immediately commence discussions with the aim of creating a new agreement regularizing the export of Hungarian film to Germany. Without the opportunity to recoup its film investments through sales in the German market, Bingert asserted, Hungarian national film production, then in a vulnerable nascent stage, would be stillborn.[36]

Other high-ranking Hungarian officials voiced similar opinions. In December 1933, *Reichsminister* Joseph Goebbels and Hungarian Interior Minister József Széll corresponded regarding Hungarian film. Even as the non-official Hungarian film world clamored against German racial nationalism, Széll expressed his approval. Hungary, he claimed, realized the artificiality of internationalism, and would fabricate only those pictures that could be branded as cultural products of Hungary. Széll reiterated the widely shared belief that profit was highly unlikely on production of Hungarian version only films in Hungary. Hungary needed the German market to generate money-making goods, specifically the ability to export six to eight films annually. Shrewdly, Hungary's Interior Minister contended that these films would not represent direct competition for German films. They would be distinct Hungarian products and would attract new audiences, not divert them from German films.[37] Germany's Propaganda Minister responded positively to Széll's appeal. Pleased that Hungary, like Germany, had decided to give its cinema 'a definite national face', Goebbels promised collaboration between Germany and Hungary. He mentioned German interest in Hungarian films and potential for profit in the German market. He proposed that negotiations for quota free imports start immediately.[38]

Of course, negotiations did not begin forthwith. Rather, both sides made rhetorical statements or took symbolic steps to create the impression of improved relations. János Bingert publicly declared his country's affinity for Germany and noted that the time was ripe for more Hungarian-German co-productions because of the 'ideological and moral' proximity between the two peoples. The reason the Hungarian-French Osso films were not runaway hits in 1932–3 was that the French and Hungarians had totally different intellectual mindsets, Bingert suggested. The Osso films had been written primarily for French audiences. The Hungarians and Germans understood each other's national nuances and thus their joint productions would undoubtedly be more successful. But co-productions and German films made in Hungary would not be enough, he concluded. Germany must open its market to Hungarian features, as only this would permit Hungary to achieve its creative, nation-nurturing potential and establish its production on a sound footing.[39]

In short, Hungary needed Germany. No other European country possessed Germany's market size. Hungary, believed its bureaucrats, had a historic affinity for the German language, not to mention a large pool of German-speaking film professionals, making German a natural language for Hungarian film made specifically for export.[40] Although Hungary's need for Germany would never be reciprocated, in early 1934 Goebbels made a symbolic gesture of goodwill, implying that Hungary might have some role to play in Germany's cultural markets. Goebbels extended the unofficial 1929 agreement permitting Hungary to export four films annually to Germany, quota free, for one year. In addition, he proposed broad bilateral cultural negotiations.

Despite assurances from Goebbels' office that the German side would attach 'the greatest value on the import of national Hungarian [film] products' the agreement had no impact whatsoever, greatly frustrating Hungarian authorities.[41] Correspondence from the Ufa branch in Budapest in 1934 indicates that Bingert and other Hungarian officials were well aware of Nazi resistance to the showing of Hungarian features in Germany. These letters also reported rumors that Hungarian journalists were considering unleashing a retaliatory propaganda campaign against German film.[42] A *Reichsfilmkammer* official, however, denied that Germany was hindering the screening of Hungarian film. We 'have nothing against' Hungarian film, he remarked. In fact, continued the official, relations with Hungary are 'exceptionally happy'.[43]

Few took this assertion seriously. Hungarian distributors, due to German refusals of import licenses, exported no films to Germany between 1934 and May 1936, hardly indicative of exceptionally happy relations.[44] For some members of the Hungarian government, the reason Hungarian-German film exchange had become inert was obvious: Germany's perception of Jewish domination of Hungarian film. In 1933, István Székely directed the dual German-Hungarian language film *Scandal in Budapest* [*Skandal in Budapest/Pardon, Tévedtem*]. Produced by the German branch of Universal Films, *Scandal* was, to the vexation of the Nazis, a sensation in Germany.[45] From the movie's inception, however, Hungarian government officials expressed concern over the overwhelming presence of Jews in the cast and crew, fearing that German authorities would forbid the picture. The Hungarian Ambassador to Germany, Szilárd Masirevich, viewed the picture as a total success, except for the elephant in the room. Of the 'Hungarians who occupied leading positions in this film, [all] were Jewish (production [Joe Pasternak], script [Károly Nóti], music [Miklós Brodszky], direction [Székely] and leading actors/actresses [Franciska Gaál, Gyula Gózon, and Szőke Szakáll])'. All, in fact, had been chased out of Berlin by the German government only months earlier. Considering the 'cultural and economic value' represented by the Hungarian-German film trade, '...if the above quoted proportion of Jewish/Non-Jewish collaboration were to stabilize – a danger which by chance exists due to the domestic invasion of Hungarian film professionals expelled from Germany – this would indeed come with unfavorable consequences for us as Hungarians'. Masirevich reported hearing Germans derisively complain that Jews were 'taking over Hungarian cultural products and film'. Reality aside, wrote Masirevich, films such as *Scandal* represented Hungarian culture in a 'totally slanted way' and reinforced German misconceptions. The impact, he warned, could be disastrous. If Germany came to believe that Hungarian culture had become totally 'Jewified' [*elzsidósodik*] and divested of its *völkisch* character, Hungarian cultural prestige would undoubtedly be lost. Germany would have no interest in purchasing 'precisely that Hungarian-packaged film culture which they so radically endeavored to root out at home'. From an economic standpoint, Masirevich cautioned, this 'might seriously harm Hungary's neophyte film production'.[46]

The Hungarian government could solve this dilemma easily, wrote another Foreign Ministry official, by exercising its powers of oversight through the Inter-Ministerial Film Committee established by Prime Minister Gömbös. This would ensure that Jews did not play such prominent roles in films designed for export. In Székely's next film, *Rákóczi March* [*Rákóczi induló*], the IMB did take a more active role, and the number of credited Jews dropped, eliminating their 'undesirable preponderance'.[47]

As these letters indicate, by the end of 1933 Hungarian government officials concerned with the film trade already had made several crucial suppositions. First, they convinced themselves that the dissemination of

Hungarian film in Germany was absolutely essential to future success. Without that exposure, Hungarian film production would never establish a viable commercial foundation. Second, some Hungarian officials were beginning to view the significant Jewish element of Hungarian film establishment as a threat to the long-term well-being of the film industry. At the same time, they were cognizant of the fact that without this Jewish segment, there would be no Hungarian film production at all. These assumptions, particularly the Jewish conundrum, weighed on the minds of both the German and Hungarian sides as they began to discuss a bilateral cultural agreement and a separate film compact in late 1934.[48]

The Nazi takeover and the Gömbös government's handling of issues related to the rights and autonomy of Hungary's German minority made the matter of cultural exchange more pressing for Germany's leaders. Germany had long utilized trade and economic agreements to exert influence in Hungary, and in the mid-1930s it began to augment its direct investments in Hungarian industry and agriculture.[49] In October 1934, Germany's Minister of Science, Health, and Education Bernhard Rust traveled to Budapest specifically to add culture to the equation.[50] Film accord negotiations began in earnest after Rust's visit. An additional, perhaps more important motivation for the movie talks was that Film Fund officials had floated the idea that, in order to underwrite the construction of a third Hunnia studio, they would raise the charge assessed on all feature film imports from HP 1,000 to 1,500. On the heels of this proposal came word that Hungary would institute a law requiring that theater programs include a fixed percentage of Hungarian-made or Hungarian-dubbed films, as discussed in the previous chapter. When news of these potential changes began to circulate, representatives of the German giant Ufa and the American majors asked their embassies to intercede. Ufa, which had seen its bottom line suffer in Hungary due to an abrupt fall in German market share, was particularly worried, especially because it appeared that the American companies were able to win exemptions while German firms were not.[51]

Unofficial reports indicate that Hungarian authorities intended their quota and surcharge legislation, in part, to be a 'counterstroke against the [perceived] German boycott of Hungarian films'.[52] If it meant to coerce Germany to negotiate, it succeeded. According to German embassy documents, if Hungary were to forgo raising the FIF surcharge, Germany would concede to importing films that German authorities deemed 'of national value' from Hungary. German officials also promised international distribution for all films made by German companies in Hungary.[53] These offers formed the basis of years of episodic negotiations between German and Hungarian film authorities. After talks failed to produce a deal in late 1934, cultural negotiations did not restart until after September 1935, when Prime Minister Gömbös and his favorite Nazi leader, Hermann Göring, authored a secret pact. Gömbös promised to make Hungary a Nazi-style dictatorship within two years, signaling that a cultural rapprochement might follow his planned political re-alignment. Yet even then, no such change occurred. Gömbös died before he could affect his revolution and the secret agreement apparently had no bearing on Hungarian-German cultural relations, which remained icy.

This distressed Hungarian officials in the Foreign Ministry and the Ministry of Religion and Education. They believed Nazi lack of interest in signing cultural and film agreements to be detrimental to both countries. A Councilor in the Hungarian Embassy in Berlin outlined this concern in a letter to Foreign Minister Kálmán Kánya. The absence of agreements for the exchange of literature, arts, and film resulted in the precipitous decline of German-language culture in Hungary. Councilor Bóbrik argued that this left a vacuum whereby the 'excessive spread' of other languages displaced German, generating commensurate gains in influence for non-German countries. A cultural agreement would help to stanch these losses and restore cultural and, by implication, political order in Hungary. Specifically related to film, Bóbrik warned that Hungarian insistence on the distribution in Germany of movies made by artists of Jewish-origin would needlessly poison existing and future Hungarian-German cultural relations.[54] Bóbrik was likely informed by a July 1935 speech by Goebbels' lieutenant Hans Hinkel. Hinkel clearly stated that Germany had no interest in importing 'destructive' films made by Jewish artists in foreign lands, particularly those that were 'damaging to the Volk' and made in Hungary, Austria, Poland, France, and the United States. Germany wished to strengthen its cinematic ties with Hungary, but only if Hungary understood this viewpoint.[55]

Not until late February and early March 1936 were moribund Hungarian-German cultural discussions reinvigorated. First, Germany signed a cultural agreement with Austria and turned its energies toward concluding a similar pact with Hungary, which reciprocated, anxious not to be left behind as Germany reformulated its Central European relations. Second, Karl Melzer and Oswald Lehnich, two important Nazi film bureaucrats, visited Hungary. Melzer, the business head of the *Reichsfilmkammer*, and Lehnich, the chief of the foreign trade section of the Chamber, traveled to Hungary to revitalize German-Hungarian film talks. After touring the newly modernized Hunnia studios, both men met separately with state officials and representatives of Hungary's two most powerful film associations, the MMOE and the OMME, raising the hopes of the Hungarian film establishment that its products might finally be granted entry to hallowed German theaters.[56]

Germany's premier film magazine, *Licht, Bild, Bühne,* officially announced that relations with Hungary had indeed improved, a conclusion reached by Lehnich in the course of his visit. Ties with Hungary were strong, declared the magazine, despite Jewish-sponsored boycotts of German film and direct Jewish attempts to ruin Hungarian-German cultural relations. In the article, Lehnich, like Goebbels several years earlier, promised closer contacts with Hungary and future collaboration. He expressed German pleasure with Hungarian efforts to create a national film industry and agreed that Hungary must export to Germany in order to better amortize production costs. Since there was great interest on the part of the German public in Hungarian products, including Hungarian history, comedies, and musicals, Germany would gladly help the Hungarian film industry. The caveat, suggested Lehnich, was that Germany would only import Hungarian films that suited the 'German mentality'. In Lehnich's view, the exchange of film must take into account the larger context of the political, economic, and cultural relations between the two lands.[57]

What Lehnich and other Nazi authorities labeled the larger context contained the obvious ideological component – antisemitism. When Germany and Austria signed a cultural agreement in March 1936, the new Hungarian Ambassador to Germany, Döme Sztójay, informed his superiors that the most important aspect of the accord was that Austria consented to the 'Aryan paragraph'. It was crucial that Hungary do the same, demanded

Sztójay, since Hungary's failure to do so was the sole reason Hungarian film had not shown in Germany over the last two years.[58] The American Minister to Hungary, John F. Montgomery, confirmed Sztójay's recognition of the obvious: the Jewish question was responsible for the chill that had characterized Hungarian-German film relations. Montgomery informed his superiors of the visits of Lehnich and Melzer to Budapest and directly addressed what prompted the trips. He noted that:

> ...Dr. Lehnich would not have come to Budapest merely to urge the export of a few Hungarian films for German consumption, and his exhortations to better relations between the German and Hungarian motion picture trades are of themselves an admission that the German films have during the recent season found steadily increasing difficulties in the Hungarian market. The [American] Legation has privately been informed that the difficulties have arisen because the motion picture industry in Hungary is almost entirely in the hands of persons of the Jewish race, with the result that many of the actors and actresses are of that race. This has led to the exclusion of such films from Germany owing to the anti-Jewish laws and regulations, and consequently to reprisals on the Hungarian side adversely affecting the importation of German films into Hungary.[59]

A mid-April 1936 article in the *Völkischer Beobachter* presented a slight twist to the argument. A German-Hungarian Film Agreement would make it possible for 'the best Hungarian films to come on to the German market...giving expanded Hungarian film production a more secure financial footing'. Naturally, a film treaty would also mean more German films in Hungarian cinemas. At the same time, *Völkischer Beobachter* could not hide the Nazi leadership's ultimate justification for a film concord. The Jews of Budapest had conducted an effective boycott of German film, particularly its premier theater owners. *Völkischer Beobachter*, the mouthpiece of the Nazi Party, voiced the hope that a film alliance would essentially oblige Hungarian authorities to break the boycott.[60]

Hungarian and German authorities concluded the first bilateral agreement signed by either country specifically to regulate film exchange in October 1936, although it would be fifteen additional months before Hungary's parliament approved it. For Germany, the agreement guaranteed a minimum of 50 feature exports to Hungary exempt from import fees. In exchange, Hungary would be permitted to send five films to Germany. Both sides consented to outlaw films deemed inflammatory by the other and to coordinate action on movie issues of mutual concern. Despite the fact that Hungarian authorities felt that it would help attract German filmmakers to Hungary's studios and allow Hungarians to act in German film, the Film Agreement heavily favored the Nazi behemoth. It stipulated that all films Hungary exported to Germany must abide by German law. Abiding by German law meant doing exactly as Ambassador Sztójay and his successors had repeated *ad nauseum:* accept the Nazi 'Aryan paragraph'. A secret Interior Ministry memo and later German diplomatic correspondence leave no doubt that this was the understanding reached by Hungarian and German film officials. Hungary had consented to a 'confidential administrative agreement' or 'secret protocol' accepting the Nazi requirement that all German-language movies made by Hungary for export to Germany be 'Jew-free'.[61] One of the small victories Hungarian diplomats won from the Nazi negotiators was that in theory, Hungarian officials belonging to the IMB were to certify whether films met the German requirements.[62] This arrangement was certainly preferable to direct examination by a German authority, because it preserved Hungarian film sovereignty and because, based on the judgment of the Prime Minister's Office, full German investigation of the Aryan question could be far-reaching and problematic.[63] While designed to retain an element of independence, this decision created the legislative imperative to consider Jewish involvement in filmmaking, a problem with long range ramifications.

Knowledge of the secret protocol was not well concealed, and members of the film establishment, Jewish ones in particular, expressed unease even as negotiations with Germany were under way. *Magyar Filmkurír* quoted an anonymous source as saying, 'In the future we must…strive to create Aryan film production, because only this way can we succeed in placing Hungarian film on the German market'. Nervously proclaiming that it could not believe these rumors and injurious ideas had substance, *Magyar Filmkurír* demanded they be denounced. The journal requested clarification from Interior Minister Miklós Kozma.[64] That neither Kozma nor any other high government official made any public denial of the rumors could be read as assent, but more likely it illustrated the ambivalence Hungary's conservative political elite had concerning the 'Aryan paragraph'. On both sides were bureaucrats who thought of the film industry's Jewish question in economic terms. One side believed the presence of Jews to be detrimental to the financial health of the industry, the other that Jews were the economic engines, like it or not, of Hungarian culture. Nearly a year prior to the conclusion of the German-Hungarian film compact, in January 1936, the radical right newspaper *Uj Magyarság*, outlined the former position. *Uj Magyarság* editor István Milotay began calling for the de-Jewification of Hungarian film, connecting the long-term survival of the film industry with exports to Germany. *Uj Magyarság* quoted the renowned Berlin-based Hungarian director Géza Cziffra, who warned that if current Hungarian films were to be shown in Germany, they could touch off an explosion of antisemitism there. Far more preferable, Cziffra suggested, were Hungarian produced Jew-free exports, which would reap revenues of up to RM 70–80,000 apiece.[65] The process would feed on itself. Aryan filmmakers crafting Aryan subject films cast with Aryan players would then have the money to make more 'true' Hungarian films.

A Hunnia memorandum written in early 1937 reached conclusions similar to Cziffra's. The memo's author proclaimed that future Hungarian exports to Germany needed to select casts and crew in accordance with the 'Aryan paragraph'. He noted that all other recent cultural agreements negotiated by Germany with Austria, Poland, Czechoslovakia, and Italy included this same stipulation. Naturally, remarked the writer, abiding by the 'Aryan paragraph' would cause the costs of making Hungarian films to rise, but it was the only real possibility open to Hungary. The waiver of export fees on four to five films annually, agreed to by Germany, represented savings of approximately HP 75,000. Delaying ratification of the agreement would jeopardize currency exchange contracts the FIF and Hunnia were negotiating with Germany, representing a potential loss of nearly RM 130,000. Failure to adopt the agreement would totally destroy any chance of making foreign version films in Hungary, for if Hungary rejected Germany, all other suitors would lose heart. The consequences of this rejection, noted the memorandum writer, could be a net yearly loss of two million *pengő*.[66]

Thus, radicals, bureaucrats, and Hunnia employees, later joined by an influential portion of the production segment of the film realm, spoke in favor of tighter relations with Germany, premised upon regulation of Jews, in order to grow Hungary's national film industry. These individuals believed that export of film to Germany, collection of duties on imported German films, and encouragement of German film-making and German capital investment in Hungary could guarantee Hungary a profitable film industry and win it first-tier international status.[67] Proponents of this view found encouragement in 1936, when German companies invested in a number of filmmaking endeavors in Hungary, including three quite successful works. Director Willy Reiber and his Berlin-based Cinephon Film made a musical/tourism feature film titled *Danubian Melodies* [*Donaumelodien*] that yielded reasonable returns in Germany, although it never appeared in Hungary. Georg Höllering's *Hortobágy* was financed by the German Bioscop company and became one of Hungary's top draws in 1936. Goebbels later awarded it the predicate *künstlerisch wertvoll*, one of the six special designations for film in Germany.[68] The dual version *Sister Maria* [*Sein Letztes Modell/Mária nővér*], the third Hungarian-German movie made in 1936, very nearly achieved the same box office success as *Hortobágy*. Although it did not receive a German predicate, it won recognition at the Venice Biennale. A joint effort of the Hungarian Pallas and Phőbus Film companies, and the German conglomerate Bavaria Films A.G., the project was actually two entirely separate versions of the same script. Both Hunnia-filmed versions featured the Hungarian baritone Sándor Svéd, a Vienna *Staatsoper* star. Rudolf van der Noss directed the German version, which starred Camilla Horn, Pál Jávor, Svéd, Julia Serta, Otto Tressler, and Hilde von Stolz. The Hungarian version featured some of the same actors and actresses, including Jávor, Svéd, and von Stolz, but other leading roles were played by the Hungarians Éva Szörényi, Lili Berky, Olga Eszenyi, Gerő Mály, and Gyula Gózon, a Hungarian Jew. Viktor Gertler, also a Hungarian Jew, directed, not van der Noss. So, while they insisted that the films that entered their market be Jew-free, Nazi authorities, probably unintentionally and due to Hungarian IMB prevarication, allowed German money to be spent on creative efforts that included Jews.

These joint undertakings and a more welcoming attitude on the part of the Nazi censors seemed to signify better Hungarian-German film relations, but they were mirages. Although the three co-productions were a step forward, they were less than Hungarian authorities desired, which was the placement of a significant number of Hungarian films on the German market and real investment in German-language film made in Hunnia studios.[69] Further, these 1936 gains were lost in 1937, when only one German-Hungarian feature reached Reich screens. *Young Noszty*, distributed in Germany as *Ihr Leibhusar*, and previously discussed in chapter two, was a love story featuring a gambling aristocratic officer and a bright, *nouveau riche* American woman of Hungarian descent. Made by Märkische Film it was a hit in both countries. Other German-language features made in Hungary with hopes for sale in Germany, particularly the Austrian-Hungarian co-productions finished after 1936, were not so lucky. *Catherine, the Last* [*Katharina, Die Letzte/3:1 A Szerelem Javára*] did not win the German censor's approval, likely because it was funded by the Jewish-Hungarian producer Imre Hirsch. With the exception of *Young Noszty*, German companies seemed to have lost their desire to work in Budapest, and Germany's censors again wielded a quick red pen when it came to German-language features made by Jewish Hungarians. And for Hungarian-language films subtitled or dubbed in German, the German border became virtually impenetrable. By 1937, Germany's behavior agitated even those Hungarians who advocated for a closer relationship. The film settlement, not yet ratified by the Hungarian government, was already crumbling.

The responsibility for the agreement's unraveling cannot be attributed to Germany alone. Jews and other anti-German constituencies within Hungary's film establishment reacted to German efforts to directly influence the Hungarian movie business with distaste and distrust. Instead of increasing the number of German film imports, Hungarian distributors continued to show a marked preference for Hollywood. Rather than look to the German market to export their products, these Hungarians turned to Austria and the United States. Of even greater significance was the stance taken by Interior Minister Miklós Kozma. Kozma, who remained a commanding film world figure, signed the 1936 film agreement with Germany on behalf of the Hungarian government. Thus, he possessed the authority to implement the agreement, but it appears he had little intention of doing so. Despite his wartime and immediate post-World War I antisemitic credentials, Kozma was a pragmatist who came under increasing attack by more radical members of his own party and factions within the Gömbös government.[70] In 1935, when he had arranged for his company, the Hungarian Film Office, to begin making features, he contracted Ernő Gál, a Jewish producer, to launch this effort. As Interior Minister he refused to bust the theater trust run by István Gerő and, in addition, made arrangements through Gerő to exhibit films and dub foreign films in Hungarian.[71] His position on Jewish involvement in film relative to German relations was no different. When he discussed the Jewish question with Joseph Goebbels in a private meeting in late 1936, Goebbels dismissed his ideas as 'shortsighted'.[72] While Kozma certainly believed trade with Germany would bring financial resilience to the film industry, he had no wish to dismiss many of Hungary's best artists and business minds. Realizing this dilemma and distracted by the larger problem of the Reich's interventions regarding Hungary's treatment of its German minority, Kozma essentially tabled the 1936 German-Hungarian film concord during his tenure as Interior Minister.

Between 1936 and 1938, German-Hungarian cultural relations, film relations in particular, foundered.[73] Despite regularly touting the 'natural and historical' ties constituting their shared 'union of fate' [*Schicksalsgemeinschaft*], Hungary and Germany both became dissatisfied with the 1936 agreement.[74] If the increasingly tight German border and shallow German pockets aggravated Hungarian officials, Hungary's unwillingness to consent to German film supremacy irritated Nazi authorities even more. By 1936/37, Germany's share of Hungary's premier film market had slipped to 17 percent, approximately equal to Hungary's segment of its own market, and nearly a 70 percent drop from 1932/33.[75] Making matters worse, in the eyes of Nazi bureaucrats, through 1938 the Hungarian state did little to reduce the roles Jews played in the film industry. German *Filmpolitik* toward Hungary thus became focused not only on diminishing Hollywood's portion of Hungary's market and expanding the possibilities for exhibition of German film, but on compelling Hungary to enact domestic reform to reduce or eliminate the influence of 'film Jews'. Nazi authorities used various levers, from diplomatic channels to film journal articles to the direct involvement of German citizens and government offices in 'assisting' Hungary's Aryanization of its film industry. The latter efforts, which began in earnest in 1938 and continued through the early 1940s, illustrate the degree to which Nazi officials were willing to export their ideology by intervening directly in the political economy of Hungarian culture.

To accelerate the slow pace of Aryanization, Ufa covertly underwrote two domestic Hungarian productions during the 1937/38 film season, funding two Christian-cast only works by director János Vaszary and the Objektív Film Company.[76] More significantly and successfully, Nazi bureaucrats authorized the German filmmaker Fritz Kreisle to bankroll the newly formed Mester Film Company, in July 1938. Founded and run by Miklós Mester, a former Ministry of Religion and Education official who would later regain his state appointment and assume high-ranking positions in the Hungarian Film Chamber and Hunnia through the first months of the German occupation of Hungary in 1944, the company would have folded had Kreisle not provided HP 300,000 for the purposes of making four films.[77] This cash convinced Hunnia of Mester Film's viability, and it extended credit to cover remaining expenses. In short order, Mester became one of the few stable Christian-owned, Christian-run Hungarian film production and distribution companies. Two of the company's first four films, *Bence Uz* [*Uz Bence*] and *Pista Dankó* [*Dankó Pista*], were extremely popular. The early beneficiary of increasingly invasive Nazi intrusions into the Hungarian film world, Mester Film remained one of Hungary's most successful Christian film companies through the early 1940s.

In response to other frustrations, the Nazis made a second attempt to advance their interests in Hungary in 1938. Hungary's Parliament failed to ratify the 1936 Film Agreement and German companies had great difficulty obtaining Hungarian import licenses. Thus, Joseph Goebbels' deputy, Oswald Lehnich, dispatched Günther Schwartz, the chief of the Foreign Office of the *Reichsfilmkammer*, to Hungary to negotiate solutions. Schwartz and L. Levente Kádár of the Hungarian Interior Ministry produced a revision of the 1936 agreement, to which the presidents of the Hungarian IMB and the German *Reichsfilmkammer* later assented. According to the German embassy, the 1938 revision restated the articles of the earlier accord and included several new confidential addenda. In the addenda, each side pledged to take concrete steps to address the other's grievances, particularly by easing import of each other's films.[78]

The revision had a formal and legalistic feel, tweaking technical concerns of minimal significance. The very fact that it had to be worked out at all demonstrates the gulf between the German and Hungarian film communities. Unlike the 1936 agreement, however, the 1938 concord held the promise of success. The Hungarian signatory, József Széll, in his second go-round as Interior Minister, pledged to push the document through Parliament, something his predecessor Miklós Kozma had purposely neglected to do. More pro-German and anti-Jewish than Kozma, Széll had begun developing plans for limiting Jewish involvement in the Hungarian film industry and was in the process of dismantling the Gerő Trust. He convinced the Lower House of Parliament to adopt the 1938 Film Agreement alteration, signaling a pro-German shift.[79]

While the Nazis welcomed these steps, they remained dissatisfied with the evolution of Hungarian film culture, which had proven far less docile, more fractious, and more independent than they had hoped. Although clash and competition, rather than symbiotic cooperation, remained the relational norm through 1938, German pressure and domestic Hungarian trends now converged. By 1938, the racial nationalism endorsed by the Nazis, factions within the Hungarian bureaucracy, the Turul Society, and Hungary's growing Arrow Cross fascist movement, among others, coalesced to produce an unprecedented round of state-supported initiatives meant to strip Jews from the leading positions they occupied in the Hungarian film establishment.

The Dual Monarchy Sequel: Co-productions and Austrian-Hungarian Film Relations

Through the mid-1930s, as Hungary vainly sought access to the screens of the Reich, it had much greater success in another German-language market, Austria. Between 1934 and 1938, Austrian-Hungarian film relations progressed through two distinct phases. A 'co-production' phase ran from 1934 through 1936. Joint filmmaking projects, mostly made by Austrian companies or casts in Hungary's Hunnia studios, characterized this period, earning Budapest a reputation as the 'second center of independent German-language production' outside of Germany.[80] The 'trade' phase commenced in 1936 and lasted through March 1938. Film trade relations normalized during this period as both Austria and Hungary began to produce greater numbers of exportable films, even though Germany more overtly dictated Austrian film production. However, just as Austrian-Hungarian film exchange was growing strong, Austria ceased to exist as a result of the *Anschluß*. With it went Hungary's first real Central European export market.

In the early 1930s, Austria's cinema culture had become highly dependent on Germany. Many Austrian film companies had German stockholders and Austria was reliant on the German market for both exhibition of its films and imports to fill its own theaters. The Nazi rise completely altered the Austrian movie landscape. The Austrian film industry, like its Hungarian equivalent, became a haven for German expellees. Between March 1933 and March 1934, Nazi censors blocked the distribution or exhibition of nine of the 14 films imported from Austria, putting the health of the industry in jeopardy.[81] In March 1934 the German Film Industry Union, in practice, forced Austria to abide by the Nazi 'Aryan paragraph'. It announced that Germany would not permit the import of Austrian works made by film professionals previously excised from the German industry, and the Dollfuss government assented. These lethal pressures provided space for Austria and Hungary to forge a more cooperative relationship.[82] Oskar Pilzer, the head of Austria's Tobis-Sascha and one of the most important figures in the Austrian movie business, explained the logic of this relationship. Although German censors closely analyzed Austrian films and rejected them if they contained non-Aryans, censors were assumed to be more lax with films from Hungary.[83] Therefore, Vienna's leading players should continue to make film, but their products should be exported from Vienna as Hungarian in order to gain access to Germany.

In large part, the Austrian-Hungarian relationship developed through the person of Joe Pasternak and his employer, Universal Films. Though on paper Pasternak still worked for Universal's German branch, he had more or less relocated to Budapest and Vienna by 1934, one of the many Hungarians banished from Germany as a result of Nazification. Pasternak organized and produced a number of transnational works in 1934–5, combining Hungarian, Austrian, and German casts and production personnel. In essence, Pasternak reconstructed some of what Universal had organized in Berlin several years earlier, most notably multi-national crews that included Jews. His efforts began as vehicles for the Hungarian-Jewish actress Franciska Gaál [born Franziska Zilverstrich], whom

Pasternak claimed even Hitler adored.[84] The first of the Universal co-productions was the 1934 hit *Peter,* which Pasternak produced and Hermann Kosterlitz [Henry Koster] directed. It starred Gaál, Hans Jaray, Ludwig Stössel, Hilde von Stolz, Anton Pointner, Imre Ráday, and Hans Richter, and it garnered the Best Comedy of the Year Award at the 1935 Moscow International Film Festival. A second Universal release in 1934, *Spring Parade* [*Frühjahresparade/Tavaszi Parádé*], gathered an ensemble headed by the Hungarian-born director Géza von Bolváry and Hungary's most expert cinematographer and cameraman István Eiben. Filmed both in Budapest and Vienna, this prototypical 'Viennese film' harkened back to Biedermeier era settings, values, and music. The cast was the best of the Hungarian and Austrian film communities, including Gaál, Paul Hörbiger, Wolf Albach-Retty, Theo Lingen, Hans Moser, Anton Pointner, Annie Rosar, Adele Sandrock, Piri Vaszary, and Tibor Halmay. A smash wherever it played, *Spring Parade* competed at the Venice Biennale and helped reinforce Franciska Gaál's standing as one of Europe's most beloved actresses.[85] Universal's third co-production, *Little Mama* [*Die kleine Mutti/Kismama*], made in late 1934, utilized much of the same workforce from previous efforts. This German-language comedy shot in the Hunnia studios premiered in Budapest's Gerő-owned Forum theater with Hungarian subtitles. Universal's final Gaál effort, made in 1936 by the Austrian Universal spinoff RORA-Film AG and the Hungarian Hirsch & Tsuk company, was *Catherine, the Last* [*Katharina, Die Letzte/3:1 A Szerelem Javára*]. This soccer-based romance directed by János Vaszary presented a Hungarian national side which lost the first leg of a home-and-home to its Austrian opponents. Naturally, the Hungarians equalized in the second, allowing their star player to win the affections of the female lead. The picture appeared in Austria and Hungary with entirely different casts. The Hungarian version, due to a contract dispute, did not include Gaál, substituting instead another outcast from the Reich, Rózsi Bársony.[86]

Pasternak explained years after the war that his decision to move Universal's bilingual production from Berlin to Budapest was simple. Ignoring the Nazis' antisemitism, the German government's increasing pressure on Vienna, and the fact that Budapest was Europe's 'least expensive city for film [production]',[87] Pasternak claimed with a movieman's panache:

Figure 8: Franciska Gaál. Public domain.

> I think I'd have gone to Budapest if I'd had to make our pictures in a damp wine cellar with Gujerati-speaking actors. I loved Budapest....To me it always had a magic, a poetry like no city in the world. It was gay, but not neurotic and disordered, as was Berlin. It was small enough so that you felt the place was yours; yet it had the quality that distinguishes a city from a town: an endless variety of people and moods. The girls were beautiful and chic, which helps any city, and the men were good companions, which is almost as pleasant.[88]

These Universal projects were some of Hunnia's most important, providing the studio a degree of professionalism and crucial income in the mid-1930s. Pasternak introduced the 9am to 6pm workday, and declared that neither 'hangovers, passionate romances, [nor] a surfeit of paprikash' excused his employees.[89] His successful endeavors inspired several other independent Austrian-Hungarian co-productions, most of which were meant for export to the German-speaking world. Among these films were *I Accuse You of Love* [*Es flüstert die Liebe/Szerelemmel vádollak*], a 1935 Géza von Bolváry love chase made by Styria Film of Vienna. Styria, founded by the Austrian Heinrich Haas, also funded the László Kardos directed comedy *Three and One-half Musketeers* [*Drei und einhalb musketiere/Három és fél muskétás*].[90] This film, which starred two Hungarian actors of international repute, Pufi Huszár and Szőke Szakáll, was made only in German, though it was later dubbed in Hungarian and released in Hungary. *I Accuse You of Love,* on the other hand, played in Hungary in German.

Styria Film proved an excellent client as it made a third film at the Hunnia works, this one in 1936. Unlike Styria's previous works, the Ralph Benatzky directed *Girls' Institute* [*Mädchenpensionat/Leányintézet*] utilized a mostly Austrian and German production crew. This staffing decision won the film access to the German market, where it did quite well. This was Styria's last Budapest venture, but it was far from the last German-language co-production made in the Hunnia studios by an Austrian company. Filming of the Béla Gaál comedy *Mircha* [*Mircha/Der kleine Kavalier/Taxigavallér/Bubi*] and the Karel Lamac drama *Skylark* [*Pacsirta*], starring Márta Eggerth and Lucie Englisch, kept Hunnia's studio active for weeks in 1936. Both were projects of Vienna-based companies, RORA and Atlantis Film respectively. A second Lamac work, *Zoro and Huru in Paradise* [*Pat und Patachon im Paradies/Eine Insel wird entdeckt*], a comedy featuring the pair known as Pat and Patty in the German-speaking world and Zoro and Huru in Hungary, finished production in early 1937. It too was an Atlantis project, and it also involved Swiss and Hungarian firms. Even City Film, the vestige of the Osso/Mannheim investments in Hungarian cinema, funded Austrian-Hungarian co-productions. Included among them was the 1934 *Bad Ending, but It's All Good* [*Ende Schlecht, Alles Gut/Helyet az Öregeknek*], a dual version film directed in Hungarian by Béla Gaál and in German by Hans Schulz. City also financed the István Székely directed, Hermann Kosterlitz written, Pál Abraham scored *Ball at the Savoy* [*Ball im Savoy/Bál a Savoyban*], a German-language operetta. The film, released to great acclaim in Budapest in early 1935, starred one of the most beloved pairs in interwar Austrian cinema, Hans Jaray and the Hungarian-born Gitta Alpár.

The fourteen co-productions sired by this goulash of Austrian-Hungarian creative forces between 1934 and early 1937 represent a unique chapter in Hungarian and European filmmaking.[91] Whether described as 'independent', 'émigré', or 'non-Germany' [*Nichtdeutschland*] films, this oeuvre was clearly different from the vanilla standardization which characterized the MLVs of the early 1930s. It involved casts and crews using locations, idioms, and customs with which they were thoroughly familiar.[92] Born of the region's 'productive insecurity', this distinctive pre-World War II Central European cinema offered a transnational vision of comedy, romance, and musical drama.[93] The films presented spectacular middle-class visions of a utopian present, where liberated sophisticates caroused, sang, and danced in posh art deco Budapest hotels, on sailboats in the Balaton, and in cafes in Vienna. In other cases, they resuscitated a Habsburg idyll, a mythical age where polyglot cosmopolitans of bourgeois and aristocratic origin mixed and mingled. Their makers ignored the territorial limits of and many of the pathologies of interwar Europe while contributing to the mutual sustenance not only of Hungarian and Austrian film cultures, but the broader German-language film culture outside the Nazi Reich. Eventually, the producers of these images brought their formulas to California, where they continued to cooperate for decades and made significant contributions beyond Hollywood's golden age.[94] More immediately, their efforts paid great dividends for Hungary. International stars, prestige, and hard currency came to Pasareti Street. The films enhanced the reputations of Hungarian producers, directors, cameramen, screenwriters, and actors, stabilizing the Hungarian industry and its nascent star system. Success throughout German-speaking Europe gave Hungary a commodity tailor-made for export which offered back-door passage into markets otherwise closed to Hungarian products.

Yet these were big, urbanist films, and for precisely that reason, the Austrian-Hungarian co-productions came with risks, contributing to the industry's expanding fractures. Nationalist and populist opposition became further entrenched in response to the overt cosmopolitanism of these works. The co-productions also altered Hungarian film finance, affording greater risks and rewards for the Hungarian film establishment, especially Hunnia. Impressed by the cachet, large budgets, and potential payoff of the co-productions, Hunnia provided more services and significantly larger lines of credit to foreign producers than it typically did for Hungary's native filmmakers. In 1936, for example, Veit Harlan, infamous for his later direction of *Jud Süß*, filmed *Everything for Veronika* [*Madonna im Warenhaus – Alles für Veronika/Minden Veronikáért*] in the Hunnia atelier. This film, a joint effort of Hunnia, Atlantis Film of Vienna, and Thelka Film of Bern, Switzerland, was made only in German and later subtitled in Hungarian. Nevertheless, Hunnia contributed HP 209,000 to the project, approximately four times more than the company risked on the average 1936 Hungarian-language film made in its facilities. The combined amount Hunnia spent on five Austrian-Hungarian (in some cases Austrian-Swiss-Hungarian) co-productions – *Everything for Veronika, Pacsirta, Girls' Institute, Mircha,* and *Zoro and Huru in Paradise* – was over HP 1.2 million, or HP 242,000 per picture. This was at least five times Hunnia's capital expenditure for contemporaneous Hungarian-language films.[95]

There is little data indicating what sort of returns Hunnia received, though anecdotal accounts indicate that Hunnia became a creditor to several foreign production companies during this time. Hunnia records do show that *Péter* and *Spring Parade* were two of the studio's four most successful releases in 1934. Another of the aforementioned co-productions, *Skylark,* was one of the Hungarian market's biggest draws in 1936, and it also did extremely well in German-language markets.[96] *Three and One-half Musketeers*, however, did poorly due to less than stellar dubbing.[97] In the main, these films created revenue and won Hunnia a reputation for production quality. However, by late 1935, the Nazis had caught on to the Austrian-Hungarian conspiracy to sneak non-Aryan products into the Reich. Over time, fewer co-productions gained entry to Germany, and thus fewer reaped the receipts for which their creators had hoped. Still, Hungarian officials saw a silver lining. These films, many of which featured Hungarians, Hungarian milieus, and Hungarian subjects, could project Hungarian culture into the protected shadows of neighboring countries, such as Czechoslovakia.[98] Even if banned in Germany, the films retained an important cultural-political role.

Despite the commercial, aesthetic, and political successes of the Hungarian-Austrian filmmaking efforts, a full-fledged reciprocal film exchange between the two countries, counter-intuitively, did not develop until Austria was pulled closer to the German orbit. Prior to 1936, except for the co-productions, Austrian distributors generally did not contract to show Hungarian films domestically. Nor did Hungarian distributors show much interest in importing the products of Austria's film community. By early 1937, the cultural-political calculus had changed, prompting both sides to alter their positions. The main catalysts for this change were threefold. First, Germany began exerting authority over Austrian film production.[99] Second, news of the German-Hungarian film agreement of October 1936 became public. Third, Austria and Hungary concluded a film exchange agreement that same fall. The first development ended plans for future co-productions, while the second crushed all possibility that films with high numbers of Jewish professionals would be allowed into Germany. The third was the only positive development, permitting any Hungarian film dubbed in German to be exported to Austria, provided it cleared the hurdle of the censor and its distributors paid the perfunctory fees.[100] This agreement and the implicit realization by Austrian and Hungarian officials that the German option was no more paradoxically caused the Austrian-Hungarian film trade to thrive. In the 1936/37 film year, Austria provided 17 films to Hungary, giving it the fourth largest presence on the Hungarian feature market, one picture behind France. Unlike France, Austria reciprocated, finding a place for Hungarian film in its theaters.[101] In 1936, ten Hungarian pictures premiered in Austria, a number nearly equivalent to the three previous years combined.[102] Hungarian features matched that number in 1937, indicating that Hungarian film was on the cusp of securing a stable niche when the *Anschluß* occurred in March 1938.[103] The Austrian market became virtually impenetrable for Hungarian film once it became part of the Reich. In a matter of days a once promising export market evaporated, and Hungary lost a critical partner.

The disappearance of Austria, its market, and its co-production support magnified the instability and sense of gloom that suddenly permeated the Hungarian film industry. As the Viennese welcomed German troops, Hungary's parliamentarians began discussing comprehensive anti-Jewish legislation. This cast a pall over Hungary's film industry, whose future had appeared so promising just months earlier. With the forces of 'Christian Hungary' arrayed against them internally and the Nazi menace now on Hungary's borders, the film industry, especially Hungary's Jewish filmmakers, found themselves besieged. Their choices increasingly limited, they stubbornly turned away from the Nazi ogre, clinging to their preferences for products from across the Atlantic.

Hollywood Reasserts its Dominance: Hungary Takes a Sliver of the American Pie

The five years prior to 1938 were the most favorable in interwar Hungarian-American film relations, despite the occasional bitter dispute. During this period, American film companies re-established their control over the Hungarian feature film market. Some, in fact, relocated their Central European operations from Berlin to Budapest or expanded their existing Budapest offices, and many hired Hungarian firms to copy Hollywood products for European distribution.[104] As Thomas Elsaesser reminds us, there can be no consideration of a national film culture without consideration of audience tastes and the exhibition sector, and in Hungary this sector was ruled by Hollywood.[105] Throughout this era, American film represented at minimum one of every two new features projected on Hungarian screens. As the German share of Hungary's premier film market shrank between 1933 and 1937, nearly all of Germany's loss became Hollywood's gain. America's portion of Hungary's premier film market rose from 36 percent in 1932/33 to over 52 percent in 1933/34 to nearly 60 percent in 1936/37.[106] The American share remained between two and three times its German competition's through 1939.[107] Already in 1934, Hungary represented a more lucrative market for American products than Sweden, Austria, Poland, and even India.[108] Clearly, Hungarian distributors, if not Hungarian audiences and exhibitors, preferred the products of Jack Warner's dictatorship to those of Joseph Goebbels'. Among the European states, only Italian and Scandinavian agents and cinema-goers expressed similar devotion to Hollywood fare.[109]

How to interpret the success of Hollywood in Hungary remains puzzling. Decades of film scholarship have refuted the notion that films project clear ideological visions that audiences passively absorb.[110] Instead, it is evident that many Hollywood films were overtly universal and thus sources of fantasy, fun, and a wide range of audience interpretations, translations, and reconfigurations.[111] That individual films were hits may have been a function of the appeal of an upwardly mobile, consumerist ideal transfigured in a Hungarian context.[112] Success may also have been due to the simple fact that the Hollywood works were technically, artistically, or thematically superior to the competition, or that Hungary's audiences were less nationalistic or more indifferent than elites presumed.[113] Perhaps the reason for Hollywood's triumph was more crass and serendipitous: Hungary too was subject to Hollywood's 'overseas campaign' and the result was that the law of averages favored American films.[114] In the absence of audience surveys, coherent box office statistics, and film company records, it is difficult to know if distributors or audiences were more responsible for the proliferation of American movies. Rankings from 1935 typify the lack of clarity. Of the ten most profitable films, two of the top three were Hungarian-made. Overall, however, four of ten were American, three of ten were German, and one was a British adaptation of *The Scarlet Pimpernel*.[115] Most likely, a combination of audience tastes, particularly the tastes of Budapest audiences, and the preferences of distributors who most certainly favored American stock, were the primary drivers behind Hollywood's command of the Hungarian market.

Contemporary Hungarians seemed just as confused by the American presence, reading into it far more significance than mere ticket sales. Although *Magyar Filmkurír* had proclaimed the defeat of American film by Europe in 1932, only five years later the country's official film trade journal, *Filmkultúra*, pronounced the opposite to be true.[116] Hollywood had conquered Hungary, and Hungary's film establishment had come to see cooperation and collaboration with American companies as the best way to further the interests of theater owners, film distributors, and producers alike. Clearly, Hungary's elites, like their Italian fascist counterparts, saw Americanism not as antithetical to the development of their 'self-consciously national culture', but essential to it.[117] With their usual bravado, Hungary's elites once again urged foreign film companies, specifically American ones, to invest in Hungarian film production. 'The "genius train" of Hungarians does not travel only outside of Hungary', wrote Pál Kende. On board were many natives laboring in Hungary, anxiously awaiting their opportunity to contribute to international filmmaking endeavors.[118]

American companies brushed aside these pleas, choosing to abandon the path of co-productions with Hungary in the 1930s.[119] Instead, the best Hungarian film professionals were gradually absorbed by Hollywood, some by invitation, others because they were eventually forced out of Hungary.[120] This increased integration may have helped make Hollywood pictures more 'authentic', and thus more palatable, in Hungary.[121] Certainly, it augmented the perception of a cultural kinship shared by the two countries' film establishments. Film and popular culture magazines such as *Film, Szinház, Irodalom* and *Sziházi Élet* religiously reported on Hollywood developments, trends, and gossip. In the imaginations of some of Hungary's cultural commentators, Hungary had become one of the favored feeders for the American industry. Newspapers and trade journals assumed every visit made by Hollywood executives to be a search for Hungarian talent or for Hungarian subjects. 'Americans come to Hungary for themes for films, just as they go to Paris for painters and music', claimed the critic Imre Fazekas in December 1935.[122]

Fantastic reveries such as this were indicative of a mainstream affinity for America felt by Hungary's private film establishment, overwhelming the voices of those who viewed the American presence in Hungary with distaste. The latter included film professionals who saw competition from the United States was the most dangerous threat to Hungary's national production.[123] They were, of course, joined by partisans of Germany. Together they voiced generic reactions to Americanization: there was too much Hollywood in Hungary, and worse than that, people actually seemed to like the amoral, non-national values Hollywood espoused. Their viewpoints, however, remained in the minority into the early 1940s. Hungarians harbored far less fear of American cultural imperialism than their continental counterparts and defined themselves less against Hollywood than in convergence with it.

Most Hungarian film professionals, in fact, embraced America and its market potential, envisioning cinema globalization as nuanced exchange and interaction rather than a one-way process of homogenization. Michael Curtiz and Menyhért Lengyel urged the Hungarian movie establishment to send its creations to America as early as 1931.[124] Curtiz, already a Hollywood stalwart, and Lengyel, a Hungarian writer/screenwriter who emigrated to America in 1937, felt there were natural cultural and political links between the two states worth nurturing, and that Hungary could influence America with its creative talents and their works. Specifically, Curtiz and Lengyel pointed out, Hungarian speakers in America were fairly concentrated in a restricted geographic area and in a handful of American cities. These Hungarian-Americans numbered nearly one million and were sure to consume quality Hungarian products. While Hungary's revisionist politics blocked its access to larger and more concentrated Hungarian-speaking audiences in Europe, it posed no problem vis-à-vis the United States.

Initially, Hungarian film professionals ignored the advice of two of the country's most famous exports to Hollywood, seeking, during the early 1930s, to break into German-language markets instead. But by the mid-30s, export to the US became a realistic and desirable option, as Hungary's domestic production flowered. Shipments of Hungarian film to the States boomed. Hungarian companies targeted the Hungarian-speaking populations of New York, Ohio, Pennsylvania, Illinois, and West Virginia. Their products did not compete for top billing nor yield substantial profits, yet they developed America into Hungary's largest export market by the mid-1930s. Over the course of the decade, nearly every significant Hungarian sound film found its way to American screens. Two New York City-based companies run by Hungarian immigrants, Danubia Pictures and Hungária Films, quite capably distributed Hungarian movies to theaters in at least 13 states, from New York to Virginia to Louisiana to California. Seven Hungarian feature films premiered in the US during the 1932/33 season. This total rose quickly, numbering 16 features and two feature length culture films (including *Hungária*) in the 1935 calendar year. Audiences queued outside New York's Tobis theater for hours to get tickets to the premier of István Székely's *Rakoczi March* [*Rákóczi Induló*].[125] In 1937, 25 Hungarian features appeared across America and in the first four months of 1938 *alone*, 22 new pictures opened.[126] Based on official trade estimates, Hungary's film exports across the Atlantic made it the fourth ranking exporter to the United States, behind only Germany, England, and France.[127] However, because currency exchange was difficult in the 1930s and many trades were completed in kind or not reported at all, the official statistics are inconsistent and unreliable. If the correct numbers of films were used, as determined by Hungary's own diplomats in America, Hungary would displace France as the number three provider of film to the United States by 1937. This was an eye-popping achievement for a country that had not been a player in the international film trade three years earlier.

According to the director Béla Gaál, this orientation toward the American market transformed its movies. Gaál hypothesized that as sources of foreign investment in Hungarian-made film dried up in the mid-1930s, Hungary's producers increasingly concentrated on American markets. Combined with the fact that nearly all Hungarian films were pieced together on shoestring budgets, the result was that in the majority, Hungarian features were of low quality and characterized by 'superficiality' and 'dilettantism', excepting, of course, Gaál's own works. Gaál implied, however, that this superficiality also was due to the necessity of creating a product that could be exported to America and understood by American audiences. Of Jewish origin and often the target of rightist and populist venom for producing anational films, Gaál would never use the word 'cosmopolitan' because of the political baggage attached to the term. Ironically, however, he agreed with many of those who criticized Hungarian film as such. He too felt many of Hungary's products were inaccurate representations of Hungarian culture.[128] They were slick, sugary, kitschy, homogenized products that, with appropriate subtitles, any audience could digest. Gaál's critique was a sardonic reversal of the typical nativist condemnation of Hollywood film heard throughout Europe in the 1930s. It may have been a rehashing of the well-worn European trope of Hollywood as other and European as craft. In reality, it was an unexpected confirmation that most Hungarian filmmakers envisioned Hollywood more as model than alterity.

While America seemed willing to import Hungary's 'shallow' pictures for limited regional showings, American officials remained hypersensitive to any attempts to restrict the sale of their own 'superficial' features in Hungary. They became quite irritated when Hungary's bureaucrats made efforts to boost domestic production and the quality of that production by taxing foreign films. American diplomatic correspondence records regular Hollywood protests any time Hungary sought to amend its film regulations.[129] Any increase in censorship costs, import fees, or fees levied for the promotion of domestic Hungarian production earned an immediate rebuke from the Motion Picture Producers and Distributors of America [MPPDA], relayed directly to the State Department or the American Consul in Hungary. Most worrisome of all were Hungary's efforts to install and later increase quotas regulating the number of Hungarian-made films exhibited in theaters. Hungary's 1935 legislation requiring that a minimum of ten percent of film theater programs be Hungarian provoked American movie executives to voice stout opposition. Even quotas requiring only that a percentage of films be dubbed by Hungarian firms led Frederick Herron of the MPPDA to fret over prospective market share declines. Herron derisively remarked that he could not 'see how it will be possible for us to carry on there if they enforce such import regulations on our people'.[130] The majors considered threatening 'to completely withdraw from Hungary and strip the market of American films', according to the American *Chargé d'Affaires*, but ultimately rejected this approach, believing it would play into the hands of German interests and Hungarian fascists who wished to see American film pushed from the continent.[131] Constrained by their fear of Germany, the majors eventually acquiesced to every change in Hungarian law, albeit with dramatic bluster. Ultimately, their fears proved prescient. Germany did want to force Hollywood from the continent and in 1942 appeared to succeed.

Through the 1930s, however, American imports dominated the Hungarian market. American majors became firmly ensconced in Budapest and their studios did steal some of Hungary's best professionals. On the other hand, Hungarian exports found audiences across America, Hungarian films incorporated American motifs and relations between the two movie establishments proved synergistic. Hungarian-American connections in film should give scholars pause, as they offer an alternative to the narratives of European discomfort with Hollywood's 'dollar imperialism' and to the common notion that the 'othering' of 1930s national film industries occurred primarily in contrast to Hollywood.[132] Additionally, this cooperation provides evidence of a pre-Cold War attraction to 'America' and all that this term represented in the heart of Europe. It reminds us of the significant role that distributors played in shaping Hungarian audience preferences and complicates our understanding of the evolution of Hungary's national film culture.

Conclusion

Hungary's quest for a place in the global film market, prior to 1938, illustrates that options were available and different paths possible. To the chagrin of the larger powers, Germany and the United States, Hungary exercised film sovereignty, expressed through its decisions to enact exhibition quotas, its resistance to demands that it reduce the roles of 'film Jews', and its active pursuit of independent transnational production, primarily with Austrian companies. More than state bureaucrats, Hungarian distributors and audiences determined what appeared in Hungarian movie houses, and both preferred American cinema despite Germany's efforts to break Hollywood's grip. The Hungarian film establishment convinced itself that it was a wellspring of subject matter themes and talent. The great film powers would inevitably recognize this, and then decide to exploit and grow Hungary's national filmmaking capacity. This self-assurance that Hungary would help provision the world grew as its movies succeeded domestically and, to a lesser degree, abroad.

This ballooning belief in a global desire for Hungarian production should be mistaken neither for a unity of vision within the film establishment nor freedom of action abroad. In fact, the more Hungarian cinema achieved, the more the industry's options became limited and Hungary's film sovereignty constricted. As the 1930s unfolded, external influences, especially those originating in Germany, exerted increasing pressure, exacerbating existing conflicts within the Hungarian film world. Variations on the 'Jewish question' emerged everywhere. The centrality of Jews to German-Hungarian film relations was clear. Hungarian-Austrian movie matters were primarily determined by Jewish film professionals, most of whom had been expelled from the Reich. Jewish-Hungarian movie men, especially distributors, and their Jewish-American counterparts, charted the course of Hungarian-American cinema relations. Germany, partially in retaliation for Hungarian distributors' preferences, refused to open its market or fund German-language production in Hungary. This encouraged the Hungarian search for alternatives, such as collaboration with Austria. When this partnership yielded a non-Nazi, popular, distinctly Central European movie product that purposely took up transnational themes and utilized a border crossing *lingua franca* rather than Hungary's national vernacular, Nazi authorities quashed the experiment. Thereafter, given Hungary's poor relations with most of its neighbors, far fewer opportunities for Hungarian-led international production remained. In addition, from 1936 on, Nazi manipulation of Hungarian cinema became more overt, and business as usual became more difficult. As the threats of Nazi expansion, racial nationalism, fascism, and antisemitism became acute, these actions and ideologies literally pushed Hungarians to Hollywood, bolstering ties and reinforcing Hungary's over-inflated sense of importance to America. At the same time, Nazi actions meant that access to markets in Central Europe other than Germany, and even continued access to the United States, became less likely.

For Hungarian film to maintain its viability in the late 1930s, it appeared to have little choice but to hew closer to the Nazi model of a race-based culture propagated by an activist state. Within Hungary, there were growing numbers who believed this change to be long overdue. There were legions, however, who opposed this development, including those in the upper reaches of the Hungarian government. They resisted German pressure not out of philosemitism, but because of distaste for mass radicalism and a desire to protect Hungary's economic and political sovereignty. By 1938, these film industry fissures, cracked by stressors emanating from Germany and America, were about to become gaping chasms.

Notes

1. Henrik Castiglione, 'A magyar mozipark fejlesztése', *FK* VIII/2 (1 February 1935), 4–10.
2. Contemporary estimates assumed sales of around 250,000 tickets to break even. Access to foreign markets meant a minimum of tens of thousands of potential ticket sales. István Radó, *A Magyar filmgyártás a kapitalizmus alatt* (Budapest, 1975), 139.
3. This statement is a generalization, characterizing a trend in Hungarian officialdom. There was considerable opposition on both the Hungarian right and left to closer ties with Germany, but in the main, Hungary's ruling elites viewed improved relations with Germany as necessary, if not positive.
4. G. Heiss & I. Klimeš, 'Kulturindustrie und Politik', 399–400.
5. Larry Langman, *Destination Hollywood: the influence of Europeans on American filmmaking* (Jefferson, NC, 2000), 21.
6. Wingfield, 'When Film Became National', 113–38, esp. 115. Czechoslovakia funded domestic film production through a quota system.
7. Martin Votruba, in 'Historical and National Background of Slovak Filmmaking', claims nearly all production was Czech, with less than one movie per year identified as Slovak. See Votruba, 'Historical and National Background of Slovak Filmmaking', *KinoKultura*, Spec. issue #3 (December 2005), 6. http://www.kinokultura.com/specials/3/votruba.pdf
8. Peter Hames, *Czech and Slovak cinema: theme and tradition* (Edinburgh, 2009), 10.
9. Votruba, 'Historical and National Background of Slovak Filmmaking', 7.

10. Robert von Dassanowsky, *Austrian Cinema: A History* (Jefferson, NC, 2005), 42.
11. von Dassanowsky, *Austrian Cinema,* 53–7, 62–4. For Austrian-Hungarian co-productions, see the documentary by Petrus van der Let and Armin Loacker, *Unerwünschtes Kino*, 2005. Petrus van der Let Filmproduktion in cooperation with Robert Koster, ORF/Film/Fernsehabkommen, Filmfund Wien, Filmarchiv Austria, Jewish Broadcaster, Új Budapest Filmstúdió, and Duna-TV. See also Walter Fritz, *Kino in Österreich 1929–1945. Der Tonfilm* (Vienna, 1991), 59.
12. W. Stradomski, 'The Jewish Cinema in Inter-war Poland', trans. A. Rodzinska, *Polish Art Studies* 10 (1989), 173–7.
13. Skaff, *The Law…*, 74, 130 & 47. Skaff explains the problem of sound thus: 'The transition from silent to sound cinema was…a Pandora's box that unleashed innumerable domestic and international problems. It was welcomed, to a certain extent. Sound cinema allowed filmmakers in Poland, like filmmakers around the world, to make films in their national language. It opened the door to musicals and other genre films. Its promise to better replicate reality excited some audiences. It encouraged some filmmakers and exhibitors to develop a national industry. Still, its proponents did not face an easy road to acceptance of the new technology. The transition from silent to sound film - which was, in effect, mandated by the film industry of the United States - occurred during a time of great change in the political landscape of Europe. …The problem with sound film was as intangible as it was obvious, and it was rooted as much in what language hides as in what it reveals'. Skaff, *The Law…*, 135.
14. *The Jazz Singer* was the first sound film in Poland, and its sequel the *The Singing Fool* the first in Hungary.
15. Haltof, *Polish National Cinema*, 24; Skaff, *The Law…*, 109–10, 122; Charles Ford & Robert Hammond, *Polish Film: A Twentieth Century History* (Jefferson, NC, 2005), 65.
16. Skaff, *The Law…*, 138.
17. Haltof, *Polish National Cinema*, 26. According to Haltof, Warsaw accounted for 33 percent of Poland's box office. Hungary's capital city, with over one-fifth of the country's population, accounted for even more.
18. Manuela Gheorghiu-Cernat, *A concise history of the Romanian film* (Bucharest, 1982), 30.
19. Between 1930 and 1939, only 16 Romanian language films were completed, the majority of which were foreign films dubbed in Romanian.
20. Barbara A. Nelson, 'Hollywood's Struggle for Romania, 1938–1945', *Historical Journal of Film, Radio and Television* 29/3 (September 2009), 295.
21. Gheorghiu-Cernat, *A concise history…*, 36–7; 'Rumäniens Filmschaffen', *Interfilm (Blätter der IFK)* 5 (1943), 52–4.
22. Mira & A.J. Liehm, *The Most Important Art*, 19–20.
23. Ibid., 20.

24. Daniel J. Goulding, *Liberated Cinema: The Yugoslav Experience, 1945–2001*, revised edition (Bloomington, IN, 2002), 1. Hungarian Censorship Committee statistics confirm this, indicating that as of the late 1930s, American films made up 56–63 percent of all film imported into Yugoslavia. See *MOL* – BM, K158-cs8-10, OMB 1937–9.
25. Evelina Kelbecheva, 'Creation of a New National Style in Bulgarian Art after the First World War', *Bulgarian Historical Review* 21/2–3 (June 1993), 108, 123–4.
26. Ronald Holloway, *The Bulgarian Cinema* (Rutherford, NJ, 1986), 77–8.
27. Liehm & Leihm, *The Most Important Art*, 21.
28. Károly Guttmann, 'Kereskedelmi szerződésekben biztosítsunk piacot a magyar hangosfilmeknek', *FK* VI/1 (1 January 1933), 4.
29. Endre Hevesi, 'The Year in Hungary', in *1936–37 International Motion Picture Almanac* (New York, 1937), 1119.
30. *MOL* – KM, K66, 420 cs, 1939, III-6/c: Film, szinházügyek. Baron Lajos Villany to MKK Bukarest (Vörnle), 6 September 1939, 33045/39.
31. Langer, 'Fejezetek', 156–7; Heiss & Klimeš, 'Kulturindustrie und Politika…', 402. One should be skeptical of these numbers. Export numbers were inflated. In 1937, for example, Hungary sold 12 films to Czechoslovakia, but several were censored and never exhibited. Further, the decline was clearly influenced by the Nazi destruction of Czechoslovakia.
32. *MOL-Ó* – Hunnia, Z1123-r1-d1. 1937. 27 August 1937, Hunnia Igazg. jgyek, 3.
33. M. Ádám, Gy. Juhász, & L. Kerekes (eds), *Allianz Hitler-Horthy-Mussolini: Dokumente zur ungarischen Außenpolitik (1933–1944)*, trans. J. Till, (Budapest, 1966), 15–16. See also *MOL* – KM K63, 180 cs, 1933, 21 – Általános külpolitika. Gömbös to Sándor Khuen-Héderváry, Budpaset, 1 February 1933, 525/T.1933; Kánya to Gömbös, 8 February 1933, 24/pol.-1933.
34. For greater detail regarding deepening German-Hungarian ties, see Michael Riemenschneider, *Die deutsche Wirtschaftpolitik gegenüber Ungarn 1933–1944. Ein Beitrag zur Interdependenz von Wirtschaft und Politik unter dem Nationalsozialismus* (Frankfurt-am-main, Germany, 1987), 60–2 and György Ránki, *Economy and Foreign Policy: The Struggle of the Great Powers for Hegemony in the Danube Valley, 1919–39* (Boulder, CO, 1983), 135ff.
35. Gergely, *Gömbös Gyula*, 273–7. Although Gömbös came to believe that Hungary had more in common with Mussolini's fascism than Hitler's Nazism, he also concluded that Germany was Hungary's closest ally in searching for solutions to the Jewish question. Ormos, *Egy magyar médiavezér*, 432–3.
36. *AAA* – GsB, Fach 218, Akt. VII, 16; Filmwesen, Bd. 1 – 1935–8. Bingert to Dr. v.Völger, 20 March 1933. Abschrift zu VI C 2313, Königl. ungarischer Minister des Innern. No. 157789/1933.
37. *MOL-Ó* – Hunnia, Z1124-r1-d20. Német-Magyar filmcsere…, 1933–42. Széll to Goebbels, Nr.157 789/1933 (undated, probably 30 November 1933).

38. *MOL-Ó* – Hunnia, Z1124-r1-d20. Német-Magyar filmcsere…, 1933–42. Goebbels to Széll, 16 December 1933, v 5513/30.11.33.III.
39. János Bingert, 'A racionális magyar filmgyártás felépitése és fejlesztése', *FK* VII/1 (1 January 1934), 3.
40. *MOL-Ó* – Hunnia, Z1124-r1-d20. Német-Magyar filmcsere…, 1933–42. Pro Memoria: a magyar-német filmegyezményhez, undated, (likely early 1937).
41. *AAA* – GsB, Fach 218, Akt. VII, 16; Filmwesen, Bd. 2 – 1935–8. Ufa to RFK, 20 September 1934. Addendum, RMVP, V 5513/1.10.34, 9 Oktober 1934.
42. *AAA* – GsB, Fach 218, Akt. VII, 16; Filmwesen, Bd. 2 – 1935–8. Ufa to RFK, 20 September 1934.
43. *AAA* – GsB, Fach 218, Akt. VII, 16; Filmwesen, Bd. 2 – 1935–8. RMVP (Scheuermann), V 5513/1.10.34, 9 Oktober 1934. Scheuermann was responding to Ufa's contention that there was no reason to import any Hungarian film. He used the phrase '*außerordentlich glücklich*'.
44. In May 1936, the Hungarian-German co-production *Donaumelodien* reached German screens, the first Hungarian made film to do so since Hungary exported *Rákóczi March* to Germany in December 1933. Several other Hungarian-Austrian co-productions were exported to Germany during this period, but by Austrian companies, not Hungarian ones.
45. Because the film was made by Deutsche Universal, the Nazis permitted it. Soon after, the Nazis Aryanized, then dissolved DU.
46. *MOL* – KM, K66, 218 cs, 1933, III-6/c – Film, szinházügyek. MKK Berlin (Masirevich) to Kánya, 8759/1933, 19 November 1933. Masirevich wrote that the film 'had taken on a yellowish spirit'.
47. *MOL* – KM, K66, 218 cs, 1933, III-6/c: Film, szinházügyek. Csatary to Kánya, 9394/1933.res.-VIII, 6 December 1933. Csatary's conclusion is debatable. While none of the lead actors or actresses in *Rákóczi March* were Jewish, nearly all the key production figures were of Jewish origin, including Székely, his assistant Béla Gaál, the producer Ernő Gál, the composer Pál Ábrahám, the screenwriter Andor Zsoldos.
48. Hunnia leaders attempted to demonstrate goodwill towards the Nazis by agreeing, in July 1934, that the film *Ich und Du* would be made without any Jews who had fled Germany. They then reversed course and irritated Nazi leaders as high-ranking as Goebbels, by pushing the export of the Franci Gaál film *Frühjahresparade* to Germany. See Geoffroy, *Ungarn als Zufluchtsort*, 280–2.
49. Braham, *The Politics of Genocide*, vol. 1, 54–6; Gy. Ránki, *Economy and Foreign Policy*, 135–44.
50. Jan-Pieter Barbian, '"Kulturwerte im Zeitkampf". Die Kulturabkommen des "Dritten Reiches" als Instrument nationalsozialistischer Auβenpolitik', *Achiv für Kulturgeschichte* 74/2 (1992), 418.

51. *AAA* - GsB, Fach 218, Akt. VII, 16; Filmwesen, Bd. 2 - 1935–8. Ufa Filmipari és Filmkereskedelmi Résvénytársaság an die GsB, 26 October 1934.
52. *NARA* - RG59 [State], M-1206, Roll 6, 864.4061/Motion Pictures 57. Enc. to Desp. No. 364 of 3 April 1936, AmLeg Budapest. Summary of Legate's conversation with Dr. Ödön Ruttkay, BM Ministerial Secretary. Ruttkay requested that his comments be kept confidential, since they 'would create difficulty if he were quoted'.
53. *AAA* - GsB, Fach 218, Akt. VII, 16; Filmwesen, Bd. 1 - 1935–8. GsB (Schnurre) to RMVP, 30 October 1934, B.Nr.784 XIV.a.
54. *MOL* -VKM, K636, 605 cs, 1932–6. MKK Berlin (Bóbrik) to Kánya, 322/biz.-1935, 30 October 1935.
55. 'Schutz dem deutschen Kulturschaffen!' *Nationalsozialistishe Partei-Korrespondenz* [*NSK*] 183 (8 August 1935). The article is the text of Hinkel's speech. A ranking *Reichskulturkammer* official and Goebbels' newly appointed special commissioner tasked with investigating non-Aryans in German culture, Hans Hinkel later took over the *Reichsfilmkammer* and in 1944 became *Reichsfilmintendant*.
56. 'A magyar-német', *FK* IX/3 (1 March 1936), 2.
57. 'Die deutsche-ungarischen Filmbeziehungen', *Licht, Bild, Bühne* 29/56 (6 March 1936), 1–2.
58. *MOL* - KM, K66, 296 cs, 1936, III-6/c: Film, szinházügyek, Alapsz. 911/1936. MKK Berlin (Sztójay) to Kánya, 1014/1936, 7 March 1936.
59. *NARA* - RG59 [State], M-1206, Roll 6, 864.4061/Motion Pictures 57. J.F. Montgomery to Sec State, No.364, 3 April 1936.
60. 'Vormarsch des deutschen Films in Ungarn', *Völkischer Beobachter* 107 (16 April 1936). For evidence of financial damage caused by the boycott, see *BFL* - UFA Filmipari Rt. - Cg23564 - 3951. 'Ufa Filmipari és Filmkereskedelmi Rt Igazgatóságának Jelentése a Rendes Közgyüléshez as 1937/38 Üzleti Évről', 14 January 1939.
61. *MOL* - BM, K150, V.Kútfő, 15.t, 3588 cs - Mozgoképüzemi ügyek. Filmgyártás. OMB ügyei. M. kir. BM, 109412 sz., 1937, V.Kútfő, 22.t, Alapsz. 108166; *AAA* - AA, Kult - Gen., Sig. R 61230, B.Nr. 5.adh.II Ungarn; Deutsch-ungarisches Kulturabkommen, Bd. 1 - 1938. GsB (Erdmannsdorff) to AA, B.Nr. 83 VII 16, 1 February 1938.
62. VKM Bálint Hóman, who took part in the negotiations, later claimed he had prevented the Germans from exercising 'quasi-oversight' of Hungarian film production. Hóman's People's Court testimony, cited in Gábor Ujváry, *A harmincharmadik Nemzedék* (Budapest, 2010), 152–3, ft.86.
63. *MOL-Ó* - Hunnia, Z1124-r1-d20. Pro Memoria: a magyar-német filmegyezményhez, undated (likely early 1937), unsigned (likely János Bingert).
64. 'Nyugtalanitó hiresztelések a magyar-német filmegyezményről', *Magyar Filmkurír* X/43–4 (11 November 1936), 2–3.

65. Quotes from *Uj Magyarság* come from an article in the premier Jewish newspaper *Egyenlőség*. 'Az Uj Magyarság zsidómentes magyar film akar [*sic*]', *Egyenlőség* (18 January 1936), 6.
66. *MOL-Ó* – Hunnia, Z1124-r1-d20. Pro Memoria: a magyar-német filmegyezményhez, op. cit.
67. *MOL* – BM, K150, V.Kútfő, 15.t, 3588 cs – Mozgoképüzemi ügyek. Filmgyártás. OMB ügyei. Dr. Ödön Ruttkay-Nedeczky memo to BM, BM 178965 sz., 1936, V.Kútfő, 22.t.
68. In this case, the predicate meant the film should get the widest possible distribution and was eligible to be shown on the holidays of Holy Friday, the Day of Repentance, and hero remembrance days.
69. *MOL* –BM, K150, V.Kútfő, 15.t, 3588 cs – Mozgoképüzemi ügyek. Filmgyártás. OMB ügyei. Ruttkay-Nedeczky to Kozma, 16 September 1936, 232/1936.
70. Ormos, *Egy magyar médiavezér,* 440–2.
71. *MOL* – MFI, K675, 1 cs, 3 t, 1928–44. MFI Végrehajtóbizottsági ülésének napirendje és jgykv-e, 12 December 1935. See also Langer, 'Fejezetek', 135; Ormos, *Egy magyar médiavezér,* 415–17. Gerő also owned the Magyar Utószinkron Kft, a post-production dubbing company.
72. Quoted in Ormos, *Egy magyar médiavezér,* 459.
73. Jan-Pieter Barbian underscores the importance of cultural relations, positing that 'it would not be totally groundless to connect stagnation of cultural exchange with the cooling of German-Hungarian relations between 1937 and 1939'. Barbian, 'Kulturwerte im Zeitkampf', 424.
74. 'Ungarische Kulturpolitik', *Germania* 318 (17 November 1934); *BA* – R11, 1259 – 'Deutsch-Ungarische Handelskammer in Budapest – Jahresbericht 1938', (Budapest, February 1938), 3.
75. Statistics from H. Castiglione, '1933/34 Grafikonja – Adatok a szezon mérlegéhez', *FK* VII/6 (1 June 1934), 5–7; 'A magyar filmpiacon 1934.augusztus 1.-től 1935.julius 31.-ig 237 magyar és külföldi film jelent meg', *FK* VIII/11 (1 November 1935), 10–11. Statistics for 1934/35 include films over 1,200 meters, feature and educational, and show the US share to be 58 percent of the total. Castiglione's revised statistics on German and American portions of the Hungarian feature film market, published in 1937, are as follows. Parentheses indicate country share/total number of films on Hungarian market:

	1932–3	1933–4	1935–6	1936–7
German features	49% (105/216)	29% (64/225)	22% (48/220)	17% (36/225)
American features	36% (77/216)	53% (120/225)	58% (128/220)	59% (132/225)

From H. Castiglione, 'Szinváltozások a magyar filmfogyasztás mozaikjában', *FK* X/4 (1 April 1937), 2–4.

76. *BFL* – UFA Filmipari Rt. – Cg23564 – 3951. 'Ufa Filmipari és Filmkereskedelmi Rt Igazgatóságának Jelentése a Rendes Közgyüléshez as 1937/38 Üzleti Évről', 14 January 1939. Ufa's directorate noted that these films, likely *Revenge is Sweet* [*Édes a bosszú*] and *Tokaj Rapsody* [*Tokaji Rapszódia*], 'did not entirely satisfy our business expectations'.
77. *MOL-Ó* – Hunnia, Z1123-r1-d1. 1938. 5 July 1938 Hunnia Igazg. jgyzk.
78. *AAA* – Kult – Gen., Sig. R 61230, Nr. 5.adh.II Ungarn; Deutsch-ungarisches Kulturabkommen, Bd. 1 – 1938. GsB (Erdmannsdorff) to AA, B.Nr. 83 VIII 16, 1 February 1938.
79. According to German embassy correspondence, Hungary ratified the Film Agreement and the secret Aryan protocol on 28 January 1938. *AAA*-Kult. Pol – Gen., Sig. R 61230, Nr. 5.adh.II – Ungarn: Deutsch-ungarisches Kulturabkommen, Bd.1 – 1938. GsB (Erdmannsdorff) to AA, B.Nr. 83 VIII 16, 1 February 1938.
80. Loacker & Prucha (eds), *Unerwünschtes Kino,* 47.
81. von Dassanowsky, *Austrian Cinema,* 56.
82. Loacker & Prucha (eds), *Unerwünschtes Kino,* 21–2. In 1935 and 1936, German manipulations basically made Austrian film subject to its rules, particularly proof of Aryan origin of main cast and production personnel.
83. Quoted in von Dassanowsky, *Austrian Cinema,* 57.
84. Joe Pasternak & David Chandler, *Easy the Hard Way* (New York, 1953), 128.
85. Loacker & Prucha (eds), *Unerwünschtes Kino,* 39. The authors call the cast, excepting Gaál, 'first class 'Aryan''.
86. Gaál went on to have disputes with directors and developed such a reputation as a malcontent that it drastically curtailed her film career in Vienna, Hungary, and even the US. Loacker & Prucha (eds), *Unerwünschtes Kino*, 94–7.
87. Endre Hevesi, 'The Year in Hungary', *1937–38 International Motion Picture Almanac* (New York, 1938), 1171.
88. Pasternak & Chandler, *Easy,* 135.
89. Ibid., 139.
90. This film is often identified as *Four and One-half Musketeers* [*4 ½ Musketiere*]
91. Galin Tihanov, 'Why did modern literary theory originate in Central and Eastern Europe? (And why is it now dead?)', *Common Knowledge* 10/1 (Winter 2004), 68. Included in this mix were Czechs and Italians, among others.
92. The 14 films were: *Catherine, the Last; Little Mama, Spring Parade; Peter; Three and One-half Musketeers; Bad Ending, It's All Good; Ball in the Savoy; I Accuse You of Love; Honeymoon at 50% off; Everything for Veronika; Skylark; Girls' Institute; Mircha; Zoro and Huru in Paradise*. There were also several Hirsch & Tsuk/RORA and Ludwig Vidor (Tobis-Sascha)/Hunnia projects which never came to fruition due to the *Anschluß*.
93. J.-C. Horak and H. G. Asper point out that it was this group of Austrian-Hungarians who brought the musical comedy from Central Europe to

Hollywood, not vice-versa. See Horak & Apser, 'Three Smart Guys: How a Few Penniless German Émigrés Saved Universal Studios', *Film History* 11/2 (1999), 134–53; H.G. Asper, *Filmexilanten im Universal Studio* (Berlin, 2005).

94. One should also include Austro-Czech co-productions in this generalization. Together, these German-language films represent an understudied body of work. On Austro-Czech MLVs, see Gernot Weiss, 'Aus gemeinsamen Wurzeln. Filmproduktion zwischen Prag und Wien bis 1938', in Johannes Roschlau (ed), *Zwischen Barrandov und Babelsberg. Deutsch-tschechische Filmbeziehungen im 20. Jahrhundert* (Hamburg, 2008), 36–40 and Kevin Johnson, 'Kulturelle (nicht-) Vermischung. Nation, Ort und Identität in tschechisch-deutschen Mehrsprachenversionen', in Kevin Johnson, 'Kulturelle (nicht-) Vermischung. Nation, Ort und Identität in tschechisch-deutschen Mehrsprachenversionen', 71–83.
95. Langer, 'Fejezetek', 127.
96. *MOL-Ó* – Hunnia. Z1123-r2-d5. Igazg-i iratok, 1940–3. 'Hunnia Filmgyár Rt.', 1944.
97. Hevesi, 'The Year in Hungary' (1936–7), 1111.
98. Ivan Klimeš 'Multiple-Language Versions of Czech Films and the Film Industry in Czechoslovakia in the 1930s', *Cinema & Cie* 4 (Spring 2004), 92. Czechoslovakia imported Austrian German-language works.
99. von Dassanowsky, *Austrian Cinema*, 73–4.
100. *MOL-Ó* – Hunnia, Z1124-r1-d19. Osztrák-Magyar kölcsön, filmcsere, 1936. Untitled documents.
101. Henrik Castiglione, 'A magyar film világviszonylatban', *FK* X/9 (1 September 1937), 5.
102. Henrik Castiglione, 'Szinváltozások a magyar filmfogyasztás mozaikjában', *FK* X/4 (1 April 1937), 2.
103. 'Ausztria', *FK* XI/4 (1 April 1938), 14.
104. Cunningham, *Hungarian Cinema*, 37.
105. Thomas Elsaesser, 'European Culture, National Cinema, the Auteur and Hollywood', in *European Cinema: Face to Face with Hollywood* (Amsterdam, 2005), 37–9.
106. See note 75, Castiglione statistics. American source statistics indicate that Germany captured 90 percent of the feature film market in 1932, and only two years later, Hollywood controlled 61 percent of the market. Jack Alicoate (ed), *The 1936 Film Daily Year Book of Motion Pictures* (New York, 1936), 1186–7.
107. The 3:1 ratio was replicated at Hungary's first film festival in 1939. Three Hollywood features, and only one Ufa product, received special screenings. Gyula Kolba, 'A Lillafüredi nemzeti filmhét', *Képes Kronika* XXI/23 (4 June 1939), 21–4.

108. J. Sedgwick & M. Pokorny, 'Hollywood's foreign earnings during the 1930s', *Transnational Cinemas* 1/1 (2010), 88.
109. 'A World Film Survey', *New York Times* (24 June 1934), x2.
110. Ian Jarvie, 'Dollars and Ideology: Will Hays' Economic Foreign Policy 1922–45', *Film History* 2/3 (September 1988), 207–21, esp. 216–19.
111. Miriam Hansen, 'The Mass Production of the Senses: Classical Cinema as Vernacular Modernism', *Modernism/Modernity* 6/2 (1999), 60, 68; Vasey, *The World According to Hollywood*, 226.
112. de Grazia, 'Mass Culture and Sovereignty', 59–60.
113. Robert Sklar, *Movie-made America. A Social History of American Movies* (New York, 1975), 224.
114. Ian Jarvie, *Hollywood's Overseas Campaign: The North Atlantic Movie Trade, 1920–1950* (Cambridge, 1992).
115. Hevesi, 'The Year in Hungary' (1936–7), 1120.
116. 'Miért győzte le az európai hangosfilm az amerikait?' *Magyar Filmkurír* VI/9 (9 March 1932), 4; Dezső Holitscher, 'A magyar film világpiac helyezete', *FK* X/3 (1 March 1937), 3.
117. de Grazia, 'Mass Culture and Sovereignty', 66.
118. Paul [Pál] Kende, 'An unsere Filmfreunde im Ausland' *Mozivilág - Kino Welt - Monde du Cinema - Cine World* XXX/17–18 (1936 September 12), 6. Kende, endorsing an inclusive nationalist vision, went on to say that: 'The Hungarian film industry has yet great treasures to reveal. The mountains of Hungarian history, the Hungarian village, the origin of the Hungarian people and Hungarian literature are so humanly large, so heaping, so brilliant, so stirring in form, that they evoke the interest of all nations, whether friends or enemies'.
119. Hungary's best hope for film-making partnerships, Paramount, explicitly ruled out co-productions in 1937. When asked if Paramount had plans to produce film in Hungary, Adolph Zukor responded 'Unfortunately, I must answer no to everyone'. See Zoltán Kasznár 'Beszélgetésem Zukor Adolffal amerikáról és a fejlődő magyar filmgyártásról', *FK* X/10 (1 October 1937), 7–8.
120. Among those who relocated to Hollywood were: the producers Jim Kay and Joe Pasternak; the directors Laslo Benedek and Steve Sekely; the director/assistant director Andrew Marton; the screenwriters Menyhért Lengyel, Sidney Garrick, Laslo Vadnay, Laslo Fodor, John Pen/John S. Toldy; the composer Nicholas Brodsky, cameraman Ernest Laszlo, yet another eventual Oscar winner; and a bevy of actors and actresses (see Chapter 1 note 95).
121. On the appropriation of Hungarian themes and plays by Hollywood, see Katalin Pór, *de Budapest à Hollywood: le theater hongrois à Hollywood, 1930–43* (Rennes, 2010).
122. 'Válságba került-e a magyar filmgyártás?' *Magyar Filmkurír* IX/51 (24 December 1935), 235. This quote did represent mainstream thought, but not

a consensus. Hungary's cultural establishment contained a significant anti-American minority, and many, including the allies of America, often chafed, like most European elites, at Hollywood's 'vacuousness' and dominance.

123. Holitscher, 'A magyar film világpiac helyzete', op. cit.
124. Mihály Kertész, 'A magyar film jövőjét a hangosfilm alapozza meg', *FK* IV/9 (1 September 1931), 1–2. See also Menyhért Lengyel, 'Beszéljünk az üzletről', *Szinházi Élet* XXI/41 (4–10 October 1931), 6–7.
125. H.T.S., 'A Magyar Film Romance', *New York Times* (21 November 1934), 23; Stoil, *Cinema beyond the Danube,* 51.
126. *MOL* - KM, K66, 654 cs, 1944, III-6/c: Film, szinházügyek. MKK New York to Kánya, 3275/1938, New York, 9 May 1938; and Jack Alicoate (ed), *The 1939 Film Daily Year Book of Motion Pictures* (New York, 1939), 636.
127. Nemeskürty, 'A magyar hangosfilm első évei 1931–1938', in Nemeskürty, *A magyar hangosfilm története,* 41. US film statistics, according to Nemeskürty, indicated that 67 German, 50 English, 23 French, 18 Hungarian, 17 Italian, 15 Soviet and around 10 films each for Poland and Sweden premiered in US theaters in 1937.
128. Béla Gaál, 'Különös tünetek', *FK* X/1 (1 January 1937), 6–9.
129. See also Margaret Herrick Library, Academy of Motion Picture Arts and Sciences, Beverly Hills [hereafter Herrick Library] - MPPAA Archive, MPAA General Correspondence, Reel 4, 1935–7.
130. *NARA* - RG59 [State], M-1206, Roll 6, 864.4061/Motion Pictures 44. F.L. Herron [MPPDA] to P. Culbertson, Chief, Div. of W. European Affairs, Dept of State, 3 July 1935. With flair and Hollywood hyperbole, Herron claimed that 'It seemed the country [Hungary] has gone so strongly Facist [sic] that our representatives are afraid to open their mouths about anything for fear of being clapped into jail'.
131. *NARA* - RG59 [State], M-1206, Roll 6, 864.4061/Motion Pictures 51. Chargé d'Affaires [Riggs] to Sec State, No.292, 8 July 1935, concerning 'Hungarian Law and Ministerial Decrees Regulating the Dubbing and Presentation of Foreign Motion Picture Films'.
132. Jindřiška Bláhová, 'A Tough Job for Donald Duck: Czechoslovakia and Hollywood 1945–1969', *Iluminace* 19/1 (2007), 215–16; Bogdan Barbu, 'Hollywood Movies, American Music and Cultural Policies behind the Iron Curtain. Case Study: Cold War Romania, 1945–1971', in S. Jakelic and J. Varsoke (eds), *Crossing Boundaries: From Syria to Slovakia*. Vienna: Junior Visiting Fellows' Conference XIV/2 (2003), 3. http://www.iwm.at/publ-jvc/jc-14-02.pdf

4

Confusion and Crisis: The Jewish Question, the Film Chamber, and the Construction of a Christian National Film Industry

Introduction

The two years preceding the outbreak of World War II were crisis years for all of Europe, and neither Hungary nor its film industry was immune. The *Anschluß*, the destruction of Czechoslovakia and resulting border changes, and the eventual eruption of World War II all had profound consequences for the Hungarian film industry. Domestic conditions only exacerbated the upheaval. Hungarian politics became dangerously radicalized in 1938–9, as the strength of the far-right increased and extremist groups became better organized. Simultaneously, Hungary's rulers drifted closer to Germany as it became clear that Britain and France would not resist the Reich's initial efforts to establish a new order on the continent. All this emboldened the Hungarian right to push talk of clarifying the national essence, above all the Christian nature of that essence, to the forefront of the discussion of where Hungary would be situated in a reconstructed Europe.

In the battleground of culture, Hungary's national nature, and its corollary, the Hungarian Jewish question, monopolized public discourse. From the press to the fine arts 'Jews' played disproportionately large roles. Decades of culture wars and callous debates over the causes of this so-called cultural 'dominance' had helped naturalize essentialist beliefs about who and what was Jewish.[1] Between 1937 and 1940, various forms of antisemitism, whether religious, racial, economic, or cultural, fused.[2] Hungary's leaders ultimately concluded that Jewish-produced culture could not be Hungarian. To reconstitute the 'national soul' in order to purge it of its Jewishness, they fashioned an entirely new apparatus to support a 'Christian national film industry'. As a result, the movie business hurtled toward a succession of unprecedented debacles. These developments revealed the strange ways radicalism and resistance to radicalism occurred within the industry as well as the multivalent and instrumental nature of Hungarian antisemitism. They also foreshadowed how unrealistic the fantasy of a Christian national industry was, and ironically, how its very success had to be premised on Jewish assistance in its construction.

'Jews' and 'Jewishness' in Hungarian Cinema

Data regarding Jewish involvement in the Hungarian cinema, while scarce, suggest that claims of Jewish prevalence in and Christian absence from the 1930s movie enterprise were credible. János Smolka, an industry stalwart of Jewish origin, estimated in 1938 that 'of the first 100 sound films made in Hungary, 93 were produced by Jewish companies and 65 directed by Jewish directors…' while film journalist György Guthy claimed that Jews produced all 100.[3] The Hunnia Directorate believed only one self-sustaining Christian-owned production firm existed in early 1938, the Kárpát Film Trade Company owned by István Erdélyi. László Zsolnai, the editor of *Filmújság,* wrote that in May 1938 there were 'hardly three or four recognized and competitive Christian directors' working in the Hungarian film industry and there were no Christians among Hungary's screenwriters. He also guessed that a full 80 percent of production chiefs and film set designers were Jews. Without exception, claimed Zsolnai, all private production capital came from Jewish sources. The only segments of the film profession where Christians outnumbered Jews, according to Zsolnai, were among actors and theater projectionists and, of course, among government bureaucrats.[4]

The story was largely the same in terms of theater management, with one major caveat. Zsolnai claimed that over 80 percent of cinema licenses were held by Christians. Most, however, entrusted the everyday operation of their businesses to Jewish managers.[5] The statistics regarding the supervision of Budapest's theaters, even after Hungary made it illegal for Jews to hold theater management jobs, were overwhelming. Vexed German diplomats noted that through January 1939, all but one of Budapest's twelve or thirteen premier theaters were either controlled or owned by Jews.[6] The journal *Hungarian Film* [*Magyar Film*] found that of the 112 cinema houses in Budapest and its immediate environs only 39 were entirely in Christian hands. In other words, Jewish Hungarians remained managers, license holders, or owners/silent partners in two-thirds of the theaters in Budapest and its suburbs.[7] The Hungarian Film Office's manager, Zoltán [Taubinger] Tőrey, confirmed the calculations made by *Magyar Film* for Budapest proper. In a presentation concerning licensing of the capital city's theaters, Tőrey told his boss, Miklós Kozma, that 50 of Budapest's 78 licensed theaters remained *de facto* in the hands of Jews as of mid-1939. Tőrey interpreted this to mean that Jewish professionals essentially handled the business of operating the metropolitan theaters.[8] There seems little doubt that Jews 'ruled' the Hungarian film industry in terms of numbers and influence, even into 1939.

These claims rest upon two assumptions. First, they assume that defining who was Jewish was easy. For some contemporaries, it may have been. Modern Jewish identity, at least since turn of the century Vienna mayor Karl Lueger's famous pronouncement that 'I determine who is a Jew', has had a powerful ascriptive component.[9] However, critical scholarly analysis, particularly that of the last two decades, has demonstrated that notions of Jewish identity in Hungary encompassed an enormous range of possibilities. Rhetoric about Jewish identity incorporated not only tropes of racially or religiously defined 'character' and association of Jews with ideologies of capitalism, liberalism, or communism, but entire visions of the Hungarian future.[10] The far right's categorization of motion pictures as 'Jewish', for example, was not always linked to the Jewishness of those who made the films. It could also refer to a cinematic concept, in form or content, which represented a perspective out of step with nationalist philosophies tied to rural or traditional Hungary. The right generally imagined a cosmopolitan, middle-class, and urban Hungary and its mass culture, as Jewish, but in certain cases, used the term Jew to connote a very specific meaning.[11] Scholars cannot afford to naively accept contemporary usages, since unpacking what contemporaries intended when they identified others as Jews is just as important, if not more so, than the label itself.

The second assumption required to rationalize the assertion of Jewish domination of Hungarian film is that there was some fundamental dissimilarity aesthetically, nationally, ideologically, biologically, or otherwise, that differentiated Hungarian Jews from Hungarian non-Jews. In the Hungarian culture wars that traced back through the 1920s, Hungary's leading Jewish figures reminded all who would listen that such designations were artificial. Statistics concerning Jewish 'rule', they asserted, highlighted the yeoman's work Hungarian Jews had done in service to the nation as cultivators and purveyors of Hungarian culture. Jews had provided these services since the Habsburg period, disseminating Hungarian culture to the Kingdom's farthest reaches. In leading roles in the Revisionist League in the 1920s and 1930s, through their 'heroic' commitment to Hungarian culture in the lands separated from Hungary by the Treaty of Trianon, and by means of significant contributions to Hungarian cultural and intellectual life, self-identified Jews pointed out that they had shouldered these nation-sustaining efforts throughout the Horthy era.[12] They reassured their non-Jewish compatriots that Hungarian culture could not be made Jewish because it was determined by public tastes, and public tastes did not divide by denomination. The public judged the quality of a work of culture, not whether a Jew or Christian created it.[13] Jews and their allies argued that accident of birth could not be presumed to foretell behavior. Jews did not act *as Jews,* but as individuals more attached to a Hungarian identity than a Jewish one. In private, even Hungary's Regent Horthy expressed sympathy for this position. In a November 1937 closed meeting between prominent Jewish leaders, Jewish industrialists, and Horthy, the Regent expressed unambiguous support for the Hungarian Jewish role in domestic and worldwide film production. Jews such as Adolf Zucker and others, he explained, controlled the film world and received accolades not because they were Jews, but because they were Hungarian Jews. He even praised the Jewish-Hungarian comedic star Gyula Kabos.[14]

Horthy never made these opinions public. Protestations lauding Jewish contributions to Hungarian culture persuaded a shrinking sector of the polity as the core stance taken by the antisemitic and integral nationalist right – that Jews and Hungarians possessed irreconcilable and incompatible cultures – became more mainstream.[15] In April 1937, in a speech given in Szeged, one of Hungary's largest cities, Prime Minister Kálmán Darányi threw gasoline on the fire, proposing that Hungary legislate a limit to the number of Jews involved in certain professions in order to reduce unemployment among Hungary's non-Jewish 'intellectuals' and restore Hungary's culture to its 'natural' Christian state. This public mention of a new *numerus clausus* helped galvanize antisemites, who began to champion the idea even more volubly than they had previously. One of the most well-known radical publicists, István Milotay, traced Jewish and Hungarian difference to the supposed origins of each respective culture, adapting *völkisch* ideas to the Hungarian milieu. The character of the Hungarian people, claimed Milotay, was 'agriculturalist…pastoral…mounted on horseback, grown together with nature, of strong military spirit'. The Jew, in contrast, 'had lived mainly in ghettoes and towns for thousands of years, alienated from nature. [He] pursues almost solely speculative and intellectual professions'.[16] In August-September 1937, the newspapers of the far right, in particular *New Hungary* [*Uj Magyarság*] and *Independence* [*Függetetlenség*], utilized similar language in targeting the film industry explicitly. They called for reoccupation of vital Hungarian space through 'Aryan film production', justifying this step by proclaiming an unbridgeable cultural chasm separated Jews and Hungarians. Noting growing mass support for the premise that a Jew simply could not be Hungarian, they cited anecdotes that rural Hungarians, especially German-Hungarian audiences, were avoiding 'Jewish' film.[17] These new realities dictated that all Jews, due not only to their inability to assimilate, but also to the undesirability of Jewish assimilation, be removed, outright, from all parts of the film industry.

Because Jews conquered by 'dissemination of their racial self-consciousness and popular beliefs', this change had to be effected quickly, suggested Turul Society leaders in April 1938. Jews spread their poison through culture, in particular journalism, theater, and film. Hungary must eliminate these anti-national influences by following the trail Hitler blazed in Germany. The Turuls' 'radical solution', like that proposed by Milotay, would ban all Jews from cultural production and distribution. It also included a boycott of all Jewish cultural products and vociferous lobbying for a set of government-organized chambers for the fine arts designed to propagate a Hungarian culture to replace the 'Jewish' one.[18] Cultural chambers, believed the Turuls and their associates, were the only feasible means of rooting out Jews. Other powerful segments of Hungarian society agreed upon the need to roll back Jewish influence, to make the public sphere Christian and national. The army, which increasingly saw itself as responsible for a 'Christian moral order', called upon Regent Horthy to reduce Jewish influence in the press, theater, cinema, cultural life, and the economy in general, and initiate a 'new, determined, uncompromising program, on national, Christian, and popular lines'.[19] Future Prime Minister Pál Teleki questioned whether Jewish culture makers could ever be loyal to the nation.[20] The respected populist writer László Németh damned Jews as 'shallow Hungarians' while Gyula Illyés attacked them as 'parasites of the arts'.[21] This widespread support buttressed the resolve of antisemites, whose efforts became more and more strident.

In the face of these demands, pressure increased on those tagged by rightist nationalists as Jewish. The dangers for scholars in accepting these essentialist claims of Jewishness are illustrated by briefly considering the background of one of the main targets, Hungary's most productive director, István Székely. Székely's identity was 'complex and ambiguous'.[22] He was raised in a highly educated, assimilated Budapest home by a bourgeois Jewish mother and a father whose Sabbatist Protestant religious identity was deemed 'Jewish' by Habsburg authorities when he was drafted in World War I. Székely lived in a Jewish neighborhood and among Jewish circles, but attended Catholic Sunday school of his own volition.[23] Székely even converted to Catholicism after the Great War, but in the late-1930s, rightist nationalists targeted him as the symbol of Jewishness in film. His is not the only example. *Filmújság* editor László Zsolnai converted to Christianity well before 1938, and many other 'Jews' in the film industry were non-observant or did not self-identify as Jewish. The obvious bears restating: being labeled a Jew was neither a simple matter of religious affiliation nor individual choice.

As antisemitic vitriol spread, some in the film world responded boldly. The Motion Picture Producers and Distributors Association instructed its Budapest subsidiaries to resist state meddling in the internal business affairs of American distributors and not to comply with laws demanding Jews be fired from local boards of trade.[24] Domestically, however, the general response from those marked as Jews was tepid. A few, including István Székely and his wife, the actress Íren Ágai, read the tea leaves and emigrated.[25] Gyula Kabos fled in 1939, followed closely by the dramatist Ferenc Molnár. Others fell back on rehearsed lines and empty platitudes. The Jewish director Béla Pásztor, for example, restated his belief that Hungarian film was based in the 'Hungarian atmosphere' and the 'Hungarian soil', not the religion or race of its maker.[26] There was no such thing as Jewish film, he protested. *Egyenlőség,* one of Hungary's most important Jewish dailies, was an exception. Its editors contradicted the racist logic of 'Aryan film production' and the politicization of culture through 'fake national catchwords'. *Egyenlőség* warned that:

> If we understand 'Aryan production' to mean that which has already been tried in other countries and from the standpoints of politics or ideology is that which they [the Germans] want to smuggle on to the film screen, then in principle they will guarantee themselves defeat….Film must instruct us in accordance with noble morality; [it must] cause us to weep and to laugh. It must not, however, agitate, organize, and recruit. *And above all, it must not propagate division and hatred.*[27]

Figure 9: Celebrating exile. Hungarian 'film Jews', led by István Szekély and Irén Ágai (both second from left), stage their happy departures from Budapest. Source: MaNDA, Budapest.

The editors of independent film trade journals spoke out even more forcefully against Hungarian antisemitism in general and film industry racism in particular. *Magyar Filmkurír,* edited by Andor Garami, and *Filmújság,* edited by László Zsolnai, published articles condemning the 'Aryanization' of Hungarian film with increasing frequency between July 1937 and the end of 1938, when the government shut down their publications. Garami and Zsolnai, both of Jewish origin, exhorted Jewish audiences to embargo Aryan films and Jewish producers to stand up to the radical Turuls and fascist Arrow Cross.[28] Theirs were certainly the most courageous voices in the Hungarian film establishment.

Zsolnai was the first to respond to the drive to de-Jewify the Hungarian film world. In a July 1937 rejoinder to the Turul Society's Végváry memorandum calling for a *numerus clausus* in the industry, Zsolnai systematically challenged each charge of Jewish dominion.[29] Of the hundred or so new theaters that opened since Hungarian sound film production began, remarked Zsolnai, over 80 percent were Christian owned and run. Outside of Budapest, most of those who owned, operated, and profited from cinemas were Christian. Zsolnai ridiculed Végváry's assertion that 'those who really control film are always Jews, [Jews] who are not familiar with the Hungarian people and indeed do not know the Hungarian language' by citing a litany of 1930s movies based on the novels, stories, and plays penned by the most famous Christian Hungarians, including Zsigmond Móricz, Ferenc Herczeg, Kálmán Csathó, and Mór Jókai. He refuted Végváry's charge that Hungarian films were infused with a 'ghetto-flavored immorality', insisting that nearly every Hungarian film contained majority Christian casts and crews. Further, if films possessed some 'ghetto mentality', ultimate responsibility belonged to the censors who approved them.[30]

Andor Garami, the editor of *Magyar Filmkurír*, likewise scolded those who used race and faith as measures of a film's Hungarianness. 'Those who make these provocations deeply wound Hungarian culture and the interests of the Hungarian economy'.[31] The problem, Zsolnai and Garami concurred, was not that a particular religious denomination or 'race' held sway over film, rather that so-called 'Christian capital' had failed to participate in cinema. The solution was simple. Nothing prevented Christian capital from investing in film production or distribution, wrote Garami. Considering also that the government controlled Hungary's studios, these so-called Christian production companies had all the advantages they needed.[32] Zsolnai concurred. All the Turul Society had to do was to mobilize a fraction of the capital held by the anti-Jewish aristocracy of Hungary, he wrote, and Hungarian film would no longer be reliant on Jewish money for its security. Zsolnai pledged to be the first to celebrate this infusion, just as soon as it occurred.[33]

Zsolnai never held that party. There was no sudden outpouring of Christian money. Through 1937 and into 1938, as the debate over Jews and culture proceeded, Hungarian film production remained funded almost solely by domestic Jewish investors, made by Jewish professionals, and supported by credit from the state Film Industry Fund. 'If [the threatened Turul] boycott [of Jewish films] begins', wrote *Cinema and Film World* [*Mozi- és Filmvilág*] editor György Guthy, '...oh God, who would there be who...could produce?'[34] Guthy saw the crux of the production problem in film's commercial side. As long as producers handled the manufacture of film in the same manner as stock certificates, the majority of Hungarian film would continue to be 'tasteless, stupid, incomprehensible, spiritless, lacking in good dialogue...'[35] In Guthy's prose we see an echo of part of the populist mantra: modern capitalism and mass media eroded Hungary's ancient culture and materialist filmmaking would do further damage. Not just the Jews, but all parties, including the Hungarian state, were guilty in this conspiracy. Echoing Gyula Szekfű's critique of Hungary's postwar plight, just as Zsolnai and Garami had, Guthy singled out Hungary's Christians who had forfeited the movie business to the Jews. They had no one to blame but themselves. Clearly, Guthy's brand of populist nationalism was not the antisemitic variety. Like Zsolnai and Garami, he defended the Jews of the film profession. 'They say' there is 'Jewish predomination' in film industry, wrote Guthy, and 'it's true'. However, 'it doesn't matter to me that [the actor Gyula] Kabos is a Jew, only that he be talented. It doesn't matter that the non-Jew [Pál] Jávor is slightly less riotous, only that he is talented'. Good Hungarian films should be well-written, well-acted, well-directed, and well-funded. Boycotts and a *numerus clausus,* he countered, would merely exacerbate the difficulties faced by the motion picture industry.[36]

What Zsolnai, Garami, Guthy, and others in the film world agreed upon was that even if one consented to the argument that non-Jews should assume a greater role in Hungarian cultural production, there was not an elemental difference between Hungarian culture and Jewish culture. Nor could an alternative 'Aryan film production' be created by legislative fiat. No pool of experienced movie professionals or investors ready to jump into the mix existed. In the months preceding Hungary's adoption of the First Jewish Law in May 1938, during the parliamentary debate over the law, a number of well-known Hungarian intellectuals, including Béla Bártok and Zoltán Kodály, came out in support of Hungary's Jewish-origin culture producers. These artists and intellectuals signed a declaration denouncing any attempt to create unequal strata of the citizenry, an unprecedented statement against discrimination. Published in the daily *Pesti Napló*, the petition quite adamantly rejected the notion that the Christian professional middle-class could only flourish if others lost their positions. It concluded that patriotic Hungarians should oppose the law.[37] However, rather than persuading the majority of the intellectual elite to make a stand, the declaration provoked a bitter backlash. The petition's signatories were heavily criticized by the right, and this was the last substantive public pronouncement in opposition to antisemitic legislation made by non-Jews in the professions and culture industries.

Practicalities, in general, did not dissuade zealous antisemites and their cronies in government, whose push for numerical limits on Jews accelerated. By late-1937, schemes and proposals for anti-Jewish legislation clouded the political horizon, spreading fear, insecurity, and instability throughout Hungarian Jewish society. In the movie profession specifically, early 1938 rumors of an impending anti-Jewish law brought paralysis, as it made the jobs of thousands of cinema specialists tenuous and provoked Jewish communities to consider retaliatory boycotts of Hungarian films made by producers who excluded Jews from their casts.[38]

The Jewish question, however, was not the only factor edging the Hungarian film industry toward ruin. Hungarian filmmaking, particularly that done in the Hunnia studios, was perched perilously on bad credit, a house of cards that was to come tumbling down in 1938 and 1939. The huge and rapid increase in the quantity of Hungarian features made between 1934 (13) and 1937 (37), according to some observers, had caused Hungarian film to lose its novelty. Inferior films with stale, repetitive themes, plots, and characters produced disinterest, especially among those ever so essential rural audiences.[39] 'Rapturous interest' in Hungarian film has declined, remarked Károly Freidrich, a cinema manager in Sopron.[40] Contemporaries surmised that the right's branding of film as a 'Jewish' cultural product helped to dissuade rural Hungarians from frequenting movie theaters. Women within cities and without were more interested in hoarding their 'household money' than spending it on movie tickets, claimed János Magyar, a theater owner in Nagykanisza. Commentators repeated practised fears that Hungarian audiences were being 'weaned from the habit of movie-going'.[41] By early 1938, an industry that had experienced blockbuster growth appeared to be in freefall. The principle audiences for Hungarian film appeared to be increasingly alienated from their country's national product.

Gleichschaltung, Hungarian Style: The Institutionalization of the Christian National System

Part I: The Theater and Film Arts Chamber

On 15 March 1938, only two days after the German annexation of Austria, Prime Minister Darányi announced a massive plan to modernize and re-equip Hungary's armed forces. The proposal, known as the Győr plan, also included legislation restricting the involvement of Jews in the Hungarian economy. Darányi's anti-Jewish efforts were meant to appease Germany and Hungary's growing far right by giving them some of what they wanted. Instead of being sated both Germany and the Hungarian right saw Darányi's plan as an appetizer. Worried about his government's failure to halt the extremist turn, Regent Horthy dismissed Darányi just before Parliament adopted his new anti-Jewish law. Shepherding through the legislation was Darányi's last act as prime minister. The First Jewish Law, Act XV/1938, went into effect under the stewardship of a new Prime Minister, Béla Imrédy, in mid-May 1938.

The battle over the meaning and shape of the 'Christian' national film industry entered a new phase with the First Jewish Law. The law's avowed aim was to bring about 'more effective safeguards' to create 'balance' in Hungarian economic life and 'to combat unemployment among the intelligentsia'.[42] Balance meant the eventual reduction in the number of Jews in

Hungarian universities or employed in professions to a maximum of twenty percent by 30 June 1943 in order to propagate Christian Hungarian capitalists and culture-producers.[43] Over the course of the next year, the government issued order upon order mandating how the principles outlined in the First Jewish Law were to be applied within specific industries and trades. On 28 August 1938, on behalf of Prime Minister Imrédy, Interior Minister Ferenc Keresztes-Fischer signed the critical decrees which created the Hungarian Theatrical Arts and Film Arts Chamber [*Szinművészeti és Filmművészeti Kamara,* hereafter the Theater and Film Chamber].[44] With great fanfare and high expectations, this umbrella organization set out to recreate Hungarian culture along ethno-nationalist lines.

Similar to most developments in the motion picture industry, the film portion of the joint Chamber had both domestic and international antecedents. The domestic roots of the Film Chamber can be found in the 1935 origins of the National Hungarian Film Union [*Országos Magyar Filmegyesület* or OMF], a short-lived organization designed to better coordinate motion picture interests by bringing together leading film industry figures. Its members included the corporate directors of the Hungarian Film Office and Hunnia, as well as most of Hungary's top film professionals, numerous high ranking bureaucrats and members of parliament. As a result, the Film Union immediately assumed its place alongside the MMOE, OMME, and IMB as one of the most powerful associations in the industry.[45] Through 1937, the directors István Székely and Béla Gaál served on the OMF board and the Union included other important 'film Jews'. However, in 1937 and 1938, as the historian Tibor Sándor points out, the OMF began to marginalize and exclude these individuals while becoming as an incubator for many of the centralizing and Christianizing ideas which later gained sway in the film industry.[46] The very ideas discussed by the OMF inspired the creation of the Film Chamber, whose existence in turn rendered the OMF redundant. The latter disbanded in 1939.

Between May and August of 1938, the Inter-Ministerial Committee began to discuss implementation of the First Jewish Law in the movie business. Its members reached a consensus on several issues. First, there must be a chamber to centralize and unite, to bring all branches of the profession together so that they could make decisions collectively and harmonize all movie matters. Second, the Chamber must provide some 'production direction' or instruction to filmmakers regarding the content and casting of their films. Third, all production professionals must be members of the Chamber, and all necessary steps must be taken to prevent the proliferation of the 'strawman' system. The strawman system was a well-known means of circumventing anti-Jewish regulations. In brief, it meant that while on paper a company was owned and directed by Christians, in reality it was Jewish managers who ran the business and Jewish money which funded it. The IMB discussions sketched the ideal vision many of its members shared, one which would bind together all segments of the cinema enterprise, encouraging them to put aside their biases and self-interests in order to work for the well-being of the nation and its Christian culture.

From the very beginning, however, there was dissension among those shaping the Chamber. While the Chamber, as part of Hungary's creative cultural apparatus, was to be part of the Minister of Religion and Education's portfolio, movie theaters and censorship had always been the domain of the Interior Ministry and the Ministry of Trade historically had supervised film distribution. These jurisdictional questions portended all sorts of future problems. In part because of this inter-departmental struggle, the IMB could not agree whether cinema license holders must be members of the Chamber nor whether the major film industry associations – namely the MMOE and the OMME – should be required to transfer their duties and powers to the Chamber.[47] In mid-August 1938, the Ministerial Council, Hungary's equivalent to the Prime Minister's Cabinet, took up the issue of the Film Chamber on short notice. Despite widespread industry opposition to a joint film and theater chamber, Ferenc Zsindely, the Ministry of Religion and Education's delegate to the Council, pressed for precisely such a chamber, perhaps feeling that it would better cement his boss Bálint Hóman's control over Hungarian mass culture.[48] For several days in late August, the future Film Chamber's contours and reach were the subjects of intensive negotiations involving the Ministry of Religion and Education, the Interior Ministry, the Justice Ministry, the Trade Ministry, and additional members of the IMB. In its initial configuratioin, the Chamber amounted to Hungary's most revolutionary attempt to bring the film arts under state control since the Republic of Councils nationalized the industry in 1919. Its actual compromise form, however, disappointed all involved. The late-August decrees that constructed it were a patchwork of concessions that, perhaps purposely, failed to meet the expectations of radical elements in the film industry and their allies in government.[49]

In a broad sense the Film Chamber was a component of a corporatist wave in Hungary, one which led to the establishment of professional chambers for nearly all the liberal professions. While these guild-like chambers incorporated certain Italian influences, the more direct comparison can be made with the Nazi professional chambers.[50] Like its German equivalent, the *Reichsfilmkammer*, the Hungarian Film Chamber design seemed to allow political authorities to control all aspects of the cinema, to solve chronic credit and finance problems, and to purge most if not all Jews from the motion picture industry. It was clear to most contemporaries that the Film Chamber was based on the Nazi example and was symbolically constituted to demonstrate that elements in the upper echelon of Hungarian government were willing to change Hungarian institutional structures to better conform to the ideology of the continent's dominant power.[51]

Of course, these positions were only rarely stated so clearly, and were secondary to domestic concerns. The Chamber's purpose was, in accordance with the First Jewish Law, to 'effectively secure equilibrium in [Hungarian] social and economic life'. The Chamber's specific tasks, as outlined by Prime Ministerial decree, were:

- To assure and enforce the requirements of the national spirit and Christian morals in film and theater circles
- To represent the social interests of the [professional] bodies belonging to the Chamber
- To safeguard the moral quality and prestige of their vocations
- To protect the rights involved with the practice of their vocations and to supervise the fulfillment of their obligations
- To impose disciplinary authority on members
- … to offer opinions and recommendations in questions related to film and theater.[52]

Conspicuously absent from the Chamber's tasks were the promotion and protection of the economic interests of its members. There was good reason. What made money for distributors was not always in the best economic interests of theater owners, for example. Nor could financial success always be reconciled with Christian morals and the national spirit. By defining its tasks vaguely, yet first and foremost as social and ideological, the Film Chamber's organizers allowed seeds of professional discord to germinate. Division spread because of the second major flaw in the Chamber blueprint: its lack of enforcement powers. Deprived of a real mechanism to insure that its members abided by its decisions, the Chamber was handicapped from the very beginning.

The Prime Minister's Council awarded the job of organizing the Chamber and its operations to the Minister of Religion and Education, who assembled a Theater and Film Chamber Planning Committee.[53] Regarding film issues, the primary and most difficult role of this multi-ministry planning group was to determine who could and should join the Film Chamber. The committee decided to select Chamber members on the basis of two factors: the individual's role in the movie business and the individual's origin. Film professionals possessing foreign citizenship were to be excluded from the Chamber, and those of Jewish origin were to be granted membership in limited numbers. All film professionals were required to submit applications, and the Planning Committee's decisions, made by 30 November 1938, were to be final. Once the Film and Theater Chambers had individually determined their memberships, they were to convene a general meeting to elect delegates to attend a representatives' meeting. At this second meeting, Theater and Film Chamber delegates would go through a complicated series of joint and separate elections. When they occurred, these elections determined the president of the combined Theater and Film Chamber, the Film and Theater constituencies' leaders, the vice-presidents of the combined Theater and Film Chamber, the members of professional and technical committees, and the remainder of the upper leadership of the Chamber, including its secretary general, attorney, treasurer, auditor, and secretaries.[54] Once this surprisingly democratic process played out and selection of the entire managing structure was complete, the joint Chamber was to become the single governing body for the theater and film professions, effective 1 January 1939.

While the Theater and Film Chamber constructed itself in late 1938, the government took action to further alter Hungarian culture. In a December 1938 proposal that would eventually lead to the passing of the Second Jewish Law in May 1939, Prime Minister Imrédy and Minister of

Justice András Tasnádi-Nagy recommended additional drastic restrictions on Jews.[55] Surprising political pundits by acting more rightist and more pro-German than his predecessor, Prime Minister Imrédy had multiple objectives in mind as he mulled supplementary antisemitic legislation. In November 1938, Hungary won its most stunning diplomatic victory in the post-Great War period when Germany and Italy awarded it a large chunk of the Slovak section of the rump of Czechoslovakia known as the Uplands. This territorial revision, called the First Vienna Award, fortified Imrédy. The Prime Minister viewed additional anti-Jewish laws as a method of appealing to Nazi authorities and perhaps achieving further alteration of the Treaty of Trianon.[56] Second, the radicals whom Hungary's ruling elite had hoped would be appeased by the First Jewish Law instead were chomping at the bit, anxious to further restrict Jewish involvement in Hungarian society.[57] Imrédy and his partners posited that only a draconian Jewish law would allow the government to outflank these extremists. Third, Imrédy was mired in a political crisis that threatened (and ultimately ended) his tenure as prime minister. To create a sustainable cabinet, he drifted right, and his antisemitic proposals became more stringent in an attempt to hold together his fragile coalition.

The resulting Imrédy/Tasnádi-Nagy proposal argued the First Jewish Law did not go far enough to curtail the 'exaggerated amount of influence exercised by Jewry by means of the press, [and] the theaters and cinema enterprises'. This prevailing influence so affected Hungary's intellectual life that it altered 'national individuality'.[58] In order to end this catastrophic destruction of Hungarian identity, the Prime Minister and his Justice Minister proposed legislation that was qualitatively and quantitatively different from the First Jewish Law. First, it defined Jews as constituting a distinct 'racial, physiological, mental, and sentimental unit'.[59] Second, it not only placed a percentage limit on Jewish participation in Hungarian cultural life but also banned Jews from taking leadership positions in Hungarian culture all together. Jews were now legally defined as more than a threat to the national culture and economy. They were alien and outcast, something other than Hungarian.

On 5 May 1939, Hungary adopted Act IV/1939, known as the Second Jewish Law. Hungarian law now considered as Jewish anyone with one parent or two grandparents who were born Jewish. The law also mandated that Jewish participation in all professions and higher education drop from the twenty percent required by the First Jewish Law to six percent. The new legislation had a massive impact on the movie industry. It forbade Jews from being publishers or journal editors, although most Jewish film journal editors had already had their ability to publish circumscribed by the end of 1938. Paragraph 11 of the new law forbade Jews from being artistic directors in plays and films, screenwriters, or any employee credited during a film showing. Jews were no longer allowed to serve as chief executives, managing directors, or administrators for companies that exhibited, produced, or traded motion pictures. Further, paragraph 30 of the law required that all chief executives of film companies be members of the Film Chamber. These two requirements made it all but impossible for Jews to remain in business without ceding their companies to non-Jews, at least on paper. After the passage of the Second Jewish Law, lines of cinema managers and film distributors formed before the Chamber, hoping to gain entry into the organization that now controlled their destiny. Naturally, most top Jewish actors, actresses, directors, writers, movie theater owners, film executives, and others were denied membership, stripping them of the ability to legally participate in filmmaking. Because the language of the Second Jewish Law did not completely ban all Jewish participation in the world of film, there were some loopholes for desperate Jewish film professionals and their allies to exploit. Despite these, the situation was bleak for Hungary's Jewish film professionals, not to mention for the rest of the film industry, in mid-1939.

In certain segments of the profession, such as film journalism, the First Jewish Law and the creation and empowerment of the Film Chamber had already had a devastating effect. On 20 November 1938, the government banned all of the most influential cinema journals, including the Andor Lajta-edited *Filmkultúra,* Andor Garami's *Magyar Filmkurír,* László Zsolnai's *Filmújság,* Ferenc Endrei's *Mozivilág,* Imre Somló's *A Film,* and Pál Krisztinkovich's *Movie Life* [*Mozgó Élet*]. Almost every other film journalist lost the right to own or operate a periodical, eviscerating independent film journalism. The single privately-funded publication devoted entirely to film which continued to publish, György Guthy's *Mozi- és Filmvilág,* did not last long. Authorities curtailed its run in July 1939. During the tempestuous last months of 1938 and early 1939, precisely when the Film

Chamber's planning committees were selectively bestowing memberships, Hungary's film industry was left essentially without autonomous monitoring. Rumors were ubiquitous, exacerbating an already uncertain situation.

To compensate for this total demolition of film literature, the Film Chamber made *Magyar Film* its official publication, replacing *Filmkultúra* as the industry's internal informer.[60] By anointing this periodical as its mouthpiece, the Film Chamber announced the triumph of the principles of the Turul movement, to which *Magyar Film* traced its origins.[61] In its 18 February 1939 inaugural issue, *Magyar Film* announced that its task and the task of the Film Chamber were to Christianize and nationalize the Hungarian film industry. This was to be accomplished by infusing the film business with a 'national spirit and Christian morals' and by insuring that 'national goals' were the industry's number one priority.[62] This, of course, would be achieved by expelling Hungary's movie-Jews and putting Christian Hungarians in their places.

In the initial issue of *Magyar Film,* the leadership of the Theater and Film Arts Chamber officially introduced itself. The actor Ferenc Kiss, the Chamber's first president, made the opening statement, a relatively benign exhortation to work together and make Hungarian film 'useful'.[63] These sparse, innocuous words belied Kiss' true intentions. A popular screen and stage actor, Kiss was also an outspoken antisemite. As President, Kiss exercised great influence during the formative stage of the Theater and Film Chamber.[64] A long-time Turul activist, he was a mobilizing force behind the Christianization of Hungarian culture. Well aware of the Imrédy government's intention to introduce a stricter anti-Jewish law in late-1938, Kiss did all he could to prevent Jews from playing a role in the future of Hungarian film. He allegedly advocated a *numerus nullus,* a complete ban on Jewish employment in the theater and film arts, although this proved untenable.[65] Under his direction, the Film Chamber did the next best thing. It filled its quota of Jews by purposely hiring Jewish clerks and stenographers in order to then deny important Jewish film company directors, cinema owners, and film stars entry to the Chamber.[66]

The second film personality whose words made the pages of *Magyar Film* was Zoltán Tőrey, the director of the Hungarian Film Office.[67] Tőrey was one of the Theater and Film Arts Chamber's two vice-presidents. His words illuminated the huge gulfs that separated leading figures in the Chamber from the outset. While Kiss was ideologically motivated, Tőrey was a pragmatist. Unlike Kiss, Tőrey was not affiliated with the Turul Society, nor was he known to have antisemitic proclivities. In fact, he later claimed he only declared as a candidate for the vice-presidency at the behest of Jewish members of the film profession, who viewed him as an ally and protector.[68] After all, it was during his tenure that the MFI entered into numerous contractual and cooperative ventures with István Gerő, the so-called Jewish 'film dictator'. Tőrey's introductory words could not have been more different than those of his colleagues in the upper echelon of the Film Chamber. In the course of four paragraphs he proceeded to denounce government intervention in film, a barely disguised attack on government support for the Hunnia studios. He also pointed out the already apparent faults of the Film Chamber, especially the absence of most movie theater management from the Chamber. Finally, and most glaringly, Tőrey failed to say a word about the necessity for a 'Christian national' orientation for the Hungarian film industry.[69]

Géza Ágotai, the editor-in-chief of *Magyar Film*, and Dezső Vaczi, the magazine's senior editor, deserve kudos for not covering up divisions in the industry and for permitting a good deal of conflicting opinion to appear on their magazine's pages. A handful of other leading Christian film professionals published in the first issue of *Magyar Film*, several of whom echoed Tőrey's criticism of how the Chamber was constituted. Béla Rennard, the managing director of Budapest's massive Uránia theater, agreed that the Chamber was flawed, but for wholly different reasons. Rennard criticized the government for 'truncating' the Chamber and limiting its power, for hindering the progress of Hungarian national film.[70] Still others avoided conflict altogether and instead envisaged the dawning of a new golden age of Hungarian cinema. Béla Mihályfi, who along with Tőrey served as one of the combined chamber's vice-presidents, predicted a flood of new Hungarian films. 'In these new Hungarian films, Hungarian actors, stripping [themselves] of foreign influenced mentalities, will find, through their sublime artistic works, the ancient Hungarian sentiment, the noble Hungarian thought, the true Hungarian heart…'[71] From February through July 1939, as production dropped precipitously, this language of resurrection, of a vanguard struggling to bring about the rebirth and purification of all Hungarian culture, was common on the pages of *Magyar*

Film.[72] In journals, newspapers, and in the halls of Parliament, Hungary's self-appointed Christian nationalist film professionals indulged in self-promotion by arguing that only their profession could and should create a new (or reinvigorated ancient) Hungarian national culture, spirit, and style.[73] They would forge the national consensus and clarify the Hungarian essence that their Jewish predecessors had deliberately distorted.

The alteration of Hungary's borders, particularly the return of the Slovak Uplands as a result of the First Vienna Award, further convinced the movie profession of the urgency of its calling. Hungarian film, wrote the editors of *Magyar Film,* now had a special obligation 'to promote national unity' in order to encourage the smooth and rapid reintegration of the regions that had been under foreign occupation for the last twenty years.[74] Cheering the recent border revisions, Alfréd Szöllőssy, the head of the Censorship Committee, announced that the 'rigid isolation' which had confined the distribution of Hungarian culture to within 'our vilely restricted borders' had ended. He boasted that film would stimulate precisely the cultural rebirth that residents of the returned lands had long desired, since 'in Hungarian hearts it is not possible to diminish the Hungarian feeling – ever, in any place, by any means'. But the revival Szöllőssy foresaw would not be limited just to the returned lands. The border changes would catalyze a cultural transformation based 'once and for all…[on a] Christian and national spirit', which, in turn, would result in a new patriotism, a new way of life, and a better future for Hungary.[75]

This dream was shared widely by the Christian Hungarian film elite. After Hungary annexed the Carpatho-Ukraine (Ruthenian) region of Slovakia in March 1939, MMOE Vice President Pál Morvay suggested that Hungarian film professionals had a 'patriotic duty' to affect the area's rapid cultural re-incorporation into Hungary, 'to again acquaint our Ruthene brothers – who for twenty years were separated from Hungarian culture under the Czechs – with our national culture'.[76] Likewise, noted Zoltán Tőrey, only film could insure that '[audiences] abroad are correctly acquainted with our nation and our reality'.[77] This goal is within our grasp, remarked Hunnia Chief János Bingert, if our government only had the gumption to create a cultural plan, just as it has created an agricultural plan. Money and ideas would beget a new national style, a durable cinematic vision of a national ideal.[78] Achievement of this ideal must be paramount, *Magyar Film* told its readers in July 1939. The Jewish Laws and the formation of the Chamber were not planks of a simple jobs program. 'It was not the highest goal of the reorganization that Christians leap into places [occupied] until this time by Jewish actors, directors, producers, etc. Rather, [the aim was] to insure the rebirth of the film arts and the widest distribution of one of our culture's most powerful tools. We want a Hungarian national culture, spirit, and style to come out of the film production of Christian Hungary'.[79]

Part II: The National Film Committee

1939 was a watershed year for the film industry as the government reorganized it along Christian national lines. On 1 January, the Film Chamber officially assumed its role as the primary organ supervising the industry's activities. On 24 February 1939, the government took another major step in the restructuring of the profession, a step which leading movie men hoped would allow their new visions of Hungary to flourish and project on to the silver screen. Prime Minister Pál Teleki abolished the Gyula Gömbös-created Inter-Ministerial Committee, replacing it with a national film committee [*Országos Nemzeti Filmbizottság*, hereafter ONFB]. This new committee oversaw the Film Industry Fund and became the core of a new pre-production censorship system.[80]

On the surface, it appeared that the ONFB was merely a more powerful version of the Censorship Committee. However, the reach of the ONFB extended far beyond censorship, making it a substantively distinct organization. It did not replace the Censorship Committee, which continued to review films post-production under the leadership of Alfréd Szöllőssy, now a ministerial advisor in the Interior Ministry. Rather, it regulated all production. No film could be made without the pre-production approval of the National Film Committee nor could any movie be screened by the Censorship Committee without an ONFB certificate. Until 1939, private producers could film without the support of the Film Industry Fund and without the approval of the IMB, though this was a rare occurrence. As of March 1939, however, film production was impossible without official state sanction. As Ferenc Zsindely, the Committee's first president, explained, the creation of the ONFB marked a major change in the business of Hungarian movie production. The ONFB was designed 'under the direct control of the Prime Minister' as the high authority in motion pictures, with responsibility for determining what films would be made and who would receive permission to make them.[81]

To produce a film, companies now had to submit a script and, more importantly, a list of stars and directors, a budget, and an estimate for the number of studio days. The National Film Committee then audited the financial records, read and edited cast lists and screenplays, and evaluated the feasibility of the plan before allowing the producer permission to contract with a studio. Even movies to be made at the supposedly private Hungarian Film Office studios were subject to this review, and the total number of films produced there was no longer determined by the MFI itself, but by the National Film Committee. Moreover, the ONFB's charge included enforcing the Jewish Laws by vetting the creditors and production personnel of proposed films.[82] This authority gave ONFB members the power to shape a distinct national concept and film culture by determining what was to be made, who would make it, and in which studio. The National Film Committee could encourage the production of certain types of films that dealt with specific ideas, themes, characters, and histories. It could prohibit not just the screening but the making of films of unwanted style or content and it could prevent the emergence of themes that did not correspond to its narrow notion of nation. In short, the formation of the ONFB represented a major step toward centralizing, nationalizing, and de-Jewifying Hungarian film production.

While the ONFB may have technically been part of the Prime Minister's portfolio, the Minister of Religion and Education, Bálint Hóman, was the true authority behind it.[83] His control over the ONFB created a shift, displacing the Interior Ministry as the predominant power governing film matters. Four of ten ONFB members were Religion and Education appointees, including the Committee's two leading figures, Baron Gyula Wlassics, Jr. and László Balogh.[84] Both Wlassics, who was to become President of the ONFB in early 1940 and thus assume the mantle of the most influential individual in the film world, and his right-hand man Balogh, who won appointment as ONFB Secretary in 1939, were veteran members of the now defunct IMB. In 1938, Wlassics had chaired the preparatory committee of the Film Chamber. He remained a power-wielding vice-president of the Censorship Committee through 1944.[85] In the convoluted logic of the 1930s Hungarian oligarchy, Wlassics also was a member of the Hunnia Directorate and a major stockholder to boot. Even better, Balogh held shares of Hunnia and the Hungarian Film Office while simultaneously occupying seats on the boards of directors of both firms, thereby spreading his conflict of interest as evenly as possible.[86] Because they had stakes in the companies' profitability, Wlassics and Balogh may have allowed movies that did not quite meet the letter of the law because they looked like hits. Conversely, they may have weeded out films they believed abided by the appropriate laws but were likely to bust or whose themes or content might run them afoul of the Censorship Committee. This constrained those few Hungarian filmmakers who may have wanted to push the artistic envelope by engaging democratic themes, applying experimental techniques, or utilizing larger budgets.

Paradoxically, the aforementioned conflicts of interest sometimes compelled the ONFB to shelter Jewish film professionals. When needed, its members simply chose to ignore continued Jewish participation in film production. It must be recognized that the ONFB was entirely separate from the Film Chamber, symbolic of the government's inability to trust the 'unruly masses' of the Chamber and its preference for an elite cadre of leaders who ruled the film world more dispassionately. At first, the Chamber was not overly concerned with the separate sphere occupied by the ONFB. It appeared that the formation and operation of the ONFB would blaze a path for the forces which favored a more racist, antisemitic point of view to impose their designs upon the Hungarian film world. Minister of Religion and Education Hóman favored a relatively strict application of the Jewish Laws. However, the first ONFB chief, Zsindely, waffled and Hóman's underlings, Wlassics and Balogh, actually did much to mediate the laws' effects in the realm of filmmaking. I do not wish to promote the impression that the ONFB welcomed Jewish involvement in film creation. It did not. Yet as the following chapter will show, neither did it harshly crack down, a development which irritated radical 'Christian nationalists' who soon recognized ONFB independence as one of the most significant institutional shortcomings hampering the function of the Film Chamber.

Creating a Christian National Culture through Destruction: The Production Crises of 1938 and 1939

Even before implementation, fear of the impending Jewish Laws had a seismic impact. First, Hungarian movie making lost its top two directors, Béla Gaál and István Székely, and a bevy of talented actors, actresses, and production personnel.[87] This brain drain visibly changed the face of Hungarian filmmaking, and frightened those thinking of investing in these endeavors. Second, and of even greater significance, was the impact the Jewish Laws had on the mechanisms of picture production, particularly on the funding of features and the credit-worthiness of the studios that made them, Hunnia and the MFI. The production crises were thus a 'ceasura', a transition period when Hungary's production structure splintered.[88] The laws and trepidation about them produced a palpable loss of desire on the part of Jews to put money into production. This caused a domino effect, shattering the traditional methods of funding film and causing a near production stoppage in late 1938 that lasted through mid-1939. Hungary's response to the crisis was shaped, once again, by its 'backwardness', its aversion to capitalism, its penchant for government-dictated change, and also by its disinclination for radical solutions.

The first production stall occurred in May 1938, a direct result of the lingering anxiety engendered by Germany's annexation of Austria and by Hungary's adoption of the First Jewish Law. For several months after the Jewish Law's approval, the studios of Hunnia and the MFI were entirely empty, palpable evidence of Jewish producers' unwillingness to invest.[89] This especially hurt Hunnia, whose empty studios generated almost no income, and whose creditors briefly froze the company's assets. As the world discovered that Germany was not ready to go to war and the Hungarian film industry realized that the Jewish Law would likely have a limited initial impact, production hesitantly restarted. Jewish capital was coaxed back, and production reached a fevered pitch in the last third of 1938. Of the 33 films made in Hungary that year, only seven premiered between February and August, meaning that the large majority were made between July and December. Incongruously, the announcement in August that the Film Chamber would be created seemed to have had little effect on scheduled projects.

Once the Film Chamber was actually convened in January 1939 and rumors of a second round of antisemitic legislation became pervasive, production again screeched to a halt. Hunnia and the bulk of Hungarian filmmaking now teetered on the edge of collapse. There were four primary causes for this crisis, three of which were long-term developments. First, in its bid to insure that its studios were always busy, Hunnia had granted credit to filmmakers too freely. This resulted in the production of low quality films that did not yield returns high enough to allow the production companies to pay their debts, leaving Hunnia with unrecoverable losses, particularly in the calamitous 1938 year.[90] In addition, because the Hunnia studios were state-owned, the firm often found itself obliged to make unprofitable pictures because some government bureaucrat deemed the films as necessary in the interests of the nation.[91] Second, in 1935, officials including Kozma and Bingert had negotiated a 'gentlemen's agreement' permitting the Hungarian Film Office to begin feature film production. MFI studios were limited to a maximum of seven features annually, ostensibly to prevent the MFI from taking business away from Hunnia.[92] However, Hunnia had lost its monopoly on filmmaking, a problem with huge ramifications in 1938. Third, Hunnia had experienced a considerable decline in foreign income. The disappearance of Austria meant Austrian companies no longer made films at the Hunnia atelier. Germany showed little interest in compensating Hungary for its lost production, and France and the United States had bid the age of international co-productions adieu. No foreign movies were made in Hungary between the middle of 1938 and the beginning of 1941, which severely affected Hunnia's bottom line. The fourth cause of Hunnia's credit crunch was the advent of the Film Chamber. After the Chamber began functioning in January 1939 and once it became clear that Jews would, in great numbers, be excluded from the profession, Jewish creditors again stopped funding films. Hunnia's credit sources, already minimal, shriveled. The only legitimate creditors remaining were the Hungarian state and the country's largest private banks. 'On principle', László Zsolnai had pointed out months earlier, 'large Christian banks will only invest if they are guaranteed a profit…[and in general] most

Christians will only invest in such companies that come with low risk and are secure'. Film, he acknowledged, was hardly a sure bet, meaning that options for Hunnia were few.[93]

Hunnia's credit crisis became the film industry's black hole, a discursive place where film establishment arguments over the role of the state in film production, the role of Jews in the industry, and the role of capital in Hungarian society swirled, collided, and mixed. Discussion of the nature of Hunnia's problems and the varied solutions recommended, illustrate that within the movie realm, there was no consensus concerning the efficacy of the Jewish Laws, nor was there uniform support for state-led centralization of the film business. Put plainly, few of Hungary's leading film figures had any sense of what the future Christian national film production system should be, other than not Jewish. One who ventured a guess was Count Géza Lipót Zichy. In late January 1939, Zichy wrote that as Hungary stood on the threshold of Christian national production, it must steel itself for a total stoppage in filmmaking. To compensate for the retreat of Jewish capital, the Count suggested changes in the structures of film finance and the methods of organizing the film industry. Some of his ideas were quite practical, such as increasing ticket prices and requiring Hunnia to assume that every third film it made would not produce profits. He also recommended that a semi-private Film Bank be created, capitalized mostly by banks but also by film producers.[94] This new bank would take over the function of a pre-existing film crediting firm owned by two Jews, Sándor Faludi and Endre Somló, a company known as the Film Trade Company.[95] In later correspondence, perhaps in response to the Zichy letter, Hunnia Director János Bingert proposed cost cutting and financing reforms, namely the elimination Jewish 'private usury capital' in the funding of films, replaced by government-backed Christian capital.[96]

The problem with all new financing proposals was that few wealthy Christians stepped up to fill the void left by the fleeing Jews. In the first half of 1939, Hunnia made no films, and the Hungarian Film Office made only three. This prompted the Minister of Manufacturing to name a special delegate to supervise János Bingert's management of the state-run studio. Insulted by this, Bingert offered to resign, escalating the industry's sense of imminent implosion. His resignation was rejected and although several new appointees took positions on the Hunnia Directorate to monitor him, Bingert remained a steady force in motion pictures through 1944. Official lack of faith in him, however, further destabilized the industry.

As the crisis remained unsolved, new explanations for its persistence emerged. Whereas Bingert and Hunnia repeatedly called for more direct state investment in film production, others disagreed, believing further government intrusion was unhealthy. In fact, the editor of the only remaining independent film trade magazine, György Guthy, blamed the production crisis on excessive government incursions. The government-created Film Chamber refused to admit most Jewish film professionals with good reputations and influence. The emergence of neophyte, underfunded Christian movie companies, also encouraged by the government, rattled the business. Experienced producers and authoritative cliques of distributors and theater owners lost faith in the integrity of their industry. If the business were to survive, Guthy wrote, the Chamber needed to provide moral, social, and artistic vision, and exercise economic leadership. Government operatives should rein in radical elements, which would restore confidence and ultimately place the industry on a proper, sound, Christian footing.

For Guthy, the concept of a Christian national film industry was an appealing one, but it needed to be founded with long-term sustainability in mind. The withdrawal of 'private [Jewish] capital' from film production left the industry standing on a 'bum leg', a 'catastrophe' in terms of Hungary's national culture and its economy. Since experience was difficult to gain without actual jobs, no training of new Christian film industry employees could occur unless conditions changed. In the long term, Christian capital must fund film production and theater licensing procedures must change. In the short term, wrote Guthy, the survival of Hungary's Christian national movie business and the integrity of its culture would only be assured if two conditions were met. First, the Chamber must encourage the re-entry of private investment, even Jewish capital, into film production. Specifically, it must allow Jewish producers to invest and work. Second, Jewish theater license holders should retain their legal rights to their licenses. In the end, argued Guthy, it matters less who holds the license and who pays for the movie than whom one hires as employees.[97]

The irony of Guthy's main point, that the future of a Hungarian Christian national movie business was dependent on the participation of

Jewish professionals and especially Jewish capital, does not seem to have been lost on Hungary's authorities. The issue of *Mozi- és Filmvilág* in which this article appeared was the last Guthy penned. The magazine either failed or, more likely, was banned in July 1939, just as public debate over the film production crisis rose to a crescendo. That same month, Hunnia requested and received an emergency infusion of HP 80,000 from the state Film Production Fund, barely escaping financial disaster. Not only did Hunnia feel it rightfully deserved greater support from the government, it felt that the government now favored its semi-private competitor, the Hungarian Film Office. The MFI had recently received licenses to three theaters as a result of the Interior Ministry's slow unwinding of the Gerő Trust. These theaters included the Gerő empire's crowned jewel, the Royal Apolló premier theater. Jealous of this sop, which they correctly attributed to behind-the-scenes maneuvering by Miklós Kozma, Hunnia officials feared they would lose even more ground to the MFI. They fretted that the MFI might deny Hunnia-made films access to its theaters, perhaps in a power play to force the alteration of the 1935 'gentlemen's agreement' that capped the annual total of feature films the MFI could produce.

Competition and envy now ignited simmering Hunnia-MFI disputes just as the film production crisis reached its zenith. A summer 1939 'news war' broke out between the two sides as they wrestled with each other and sought scapegoats for the industry's demise.[98] Rejecting Hunnia's calls for a state-controlled monopoly over film production, Miklós Kozma, representing the MFI, claimed that what prevented Hunnia from being a successful, nimble, money-making enterprise were in fact the shackles of being a state institution.[99] Filmmaking, Kozma concluded, relied on free market capitalism. For Hungary to emerge from its filmmaking funk, it needed to resist, not give in to the 'tendency to want to put the state in charge of everything'. Was it logical or even possible, asked this deeply ensconced government insider, 'to choose such a path where in the place of Christian employees and good businessmen we rear [government] bureaucrats?'[100]

In the final analysis, the production crisis of 1939 came down to matters of risk, entrepreneurship, and interwar Hungary's tormented relationship with the capitalist system. In response to a proposal from Ferenc Zsindely describing the function of the National Film Committee and its plans for reorganizing Hungarian film, the head of the Pesti Magyar Kereskedelmi

Bank, Károly Lamotte, was quite frank in his reading of the situation. He wrote that he did not believe the problems afflicting the movie business concerned access to capital, nor did he feel that the extension of greater public credit which Zsindely requested would be a viable solution. The primary problem, argued Lamotte, was that filmmakers were not contributing enough to their own projects. If film producers were not asked to increase their shares and they failed to attract credit from private banks, why should government guaranteed funds be extended? Banks, with no expertise in film should hardly be compelled by politicians to bear more risk.[101]

On the other side of the coin were those Christian producers who professed a desire to make 'Christian national' film but refused, as Lamotte noted, to tolerate the risk inherent in filmmaking. They made arguments highlighting the internal contradictions of the 1938–9 era or by resorting to racist obfuscation to justify their failure to invest in their own films. Like Lamotte, the Christian producer Péter Bajusz denied that the lack of capital was the cause of Hungary's crisis. The root issues, Bajusz declared, were the industry's and the government's failures to take bold action by expelling the Jews and making state-backed production subsidies readily available. Claiming to speak for the majority of filmmakers in the Chamber, Bajusz swore that while government prescribed half-baked measures, such as an enfeebled Chamber, and did not enforce existing ones, such as the Jewish Laws, production would cease. The new generation of Christian filmmakers, he menacingly wrote, had

> no desire to make films using those whom the law requires be kicked out of the [filmmaking] realm… no desire to cover up those strawmen, who make filmmaking impossible…no desire to make films based on the old usurious construction, which guaranteed bankruptcy…or while any sort of trust arrangement still existed… no desire to make films until high authorities made the proper arrangements vis-à-vis [the] Jewish Laws, quotas, licenses, etc…[and] no desire to make films until capital was organized on a moral basis.[102]

This socialist-populist rationalization, from a man who claimed not to see the question of funding at the bottom of the film crisis and who asserted that the ideology of Hungarian Christian nationalism came before the desire for material success, is revealing. According to Bajusz, unless 'high authorities' essentially guaranteed Christian filmmakers a

Jew-free, risk-free filmmaking environment, they would not build a Christian national film industry. Denunciations of the 'Jew', 'usury' and the government's inability or unwillingness to crack down on them were often rhetorical bargaining chips used by people like Bajusz to extort better terms before they committed to making a movie. By mid-1939, the atmosphere in Hungary had become so rancid that the mere mention of 'money' convinced antisemites that Jews were involved and the film was somehow tainted. 'Faces would sour and the film, its story, its cast, its mode of production, [and] its mentality...[would be] assumed to be Jewish'.[103] This observation, combined with the comments of Bingert, Guthy, Zichy, Kozma, and Bajusz, demonstrates how the murky matters of Jews, money, and film coalesced during the 1939 crisis period to expose Hungary's inability to come to grips with capitalist modernity and racial nationalism.

As it did in many other areas, Hungary emerged from this crisis through happenstance, calculated violation of its own laws, and corporatist compromise. What evolved in the fall of 1939 was a hybrid system, one which partially sated rightist desires for centralization, increased government direction, and minimized risk. But this transformation also required the willful ignorance of and sometimes purposeful countenance of the reintegration of Jewish capital and talent into film production. The post-1939 system was one divided against itself. Torn between ideological goals and financial bottom lines, film figures relied upon plastic phrases such as 'Christian national' and 'Jew' to paper over vast differences. This semantic dance was a pragmatic negotiation, one based on a recognition of the fact that were Jewish investors and industry professionals suddenly and without exception thrown out of the film business, the entire edifice would crumble.

Conclusion

If self-evident ideals, great Hungarian men, and the popular spirit were to be the basis of a new national identity or national culture, they certainly were not the basis of the Christian national film industry. No great men of pure intention and irrepressible Hungarian racial character came to the rescue of the industry as it purged its Jewish members. Rather, a collection of lesser men fomented chaos between 1938–9. Their profession was fraught with contradictory bureaucratic, financial, and moral imperatives. Bankruptcy threatened Hunnia and other smaller film companies and cinemas throughout Hungary. The crisis came to a head in matters concerning the Hungarian Film Chamber, whose first year of operation was, if one wishes to be kind, tumultuous, fractious, and ineffective.

There were a multitude of reasons for the Chamber's disarray: poor organization, divisions among its leaders, jurisdictional disputes among governmental ministries which affected the Chamber's operation, the re-surfacing of old divisions in the film industry, and the lack of legal clarity concerning the Chamber's power and reach. The primary reason for the crises of 1938–9, however, was that the government had not delineated the power and purview of the Film Chamber with precision. It will soon become clear that certain segments of Hungary's political leadership desired this ambiguity. They purposely diluted the Chamber's power to restrain the more radical elements of the film establishment. This prevented unity from being imposed upon the film industry and created obstacles and uncertainty, which in turn exacerbated Hunnia's credit calamity and the industry-wide production collapse. It also created space, giving government bureaucrats the opportunities to stall, thwart, or promote film industry initiatives and competing visions of nation as they saw fit. For most Jewish Hungarians, 1938–9 was the beginning of a catastrophic period of oppression. For some movie industry Jews, however, the period's turmoil unexpectedly extended to their professional lives.

As 1939 drew to a close, few of the film realm's endemic problems had been resolved. In the face of all these difficulties, Hungary's movie enterprise still managed to emerge stronger in 1940 than it had been at any time since 1937. Credit should not, however, go to the film elite. Rather, a number of fateful events changed the course of world history, most notably the outbreak of World War II. These events, described in chapter six, created opportunities for Hungarian film and helped distract members of the industry from their day-to-day disorder, leaving the cinema world happier and healthier, although still burdened by unresolved dilemmas about who should make, exhibit, and regulate Hungarian film.

Notes

1. The most important contemporary contributor to the debate was the historian Gyula Szekfű, who initially argued that in the absence of a Hungarian middle-class, Jews first had taken over Hungary's commercial, financial, and industrial enterprises, and then moved to culture, which they used to advance their own causes, particularly liberalism. Hungary's hope, he argued, was to return to its pre-1848, pre-liberal 'Christian-Germanic' traditions and to reject the antinational Jewish/liberal spirit. Gyula Szekfű, *Három nemzedék és ami utána következik* (Budapest, 1935), esp. 240–67, 332–48.
2. Victor Karady, 'Different Experiences of Modernization and the Rise of Anti-Semitism. Social-political Foundations of the *numerus clausus* (1920) and the "Christian Course" in Post World War I Hungary', *Transversal* 4/2 (2003), 13.
3. János Smolka, *Mesegép a valóságban* (Budapest, n.d.[1938?]), 16; György Guthy, 'Árjafilm', *Mozi- és Filmvilág* II/1 (15 January 1938), 6. Guthy claimed that through 1937, 100 percent of the producers and 80 percent of the stars of Hungary's first 100 films were Jewish. Smolka believed there was perhaps one viable non-Jewish production company.
4. László Zsolnai, 'A 20%', *Filmújság. Zsolnai László Hétilapja* (14 May 1938). Reprinted in Sándor, *Őrségváltás,* 112–19. The last official statistics concerning numbers of actors and actresses [in film and theater] indicate that as of 1930, approximately 24 percent of all Hungarian professional players were Jewish. In Budapest, the numbers showed Jews were better represented, constituting nearly one-third of the capital's performers. See 'VI. Táblázat–A zsidók száma és aránya az egyes fontosabb értelmiségi pályák keresői közt Magyarországon a főváros kiemelésével (1890–1930)', in F. L. Lendvai, A. Sohár, & P. Horváth (eds), *Hét évtized a hazai zsidóság életében I.rész* (Budapest, 1990), 193.
5. Zsolnai, 'A 20%'.
6. *AAA* - GsB, Fach 27, Akt. Kult 12, Nr 4a; Filmtheater in Budapest; Filmwesen, 1939–41. RMVP (Fischer) to GsB, V 5513/Ung/10.1.39/959-1,7, 10 January 1939. This source asserts that István Gerő controlled all but three of Budapest's premier theaters, and the Jews 'Lederer' and 'Zuckermann' controlled two of the remaining three. See also 'Konzept!' 7 February 1939, AA. Berlin B. Nr. 123. These estimates matched earlier Hungarian estimates.
7. 'Hites moziengedélyeseket', *MF* I/16 (3 June 1939), 1–2. The article explained that an additional 31 theaters were licensed to Christians, but were run by Jews; 23 had licenses held by a Christian/Jewish partnership (sometimes silent partnerships) with Jewish employees running the business; four Budapest-area cinemas had Christian licensees, Jewish partners, and Christian business managers; and three theaters had Jewish licensees but were run by Christians. The remaining 12 were directly licensed to and run by Jewish Hungarians.

8. *MOL* - KMI, K429-cs59-t3. Vegyes MTI levelezés 1940, 150. 'Statisztika', 26 May 1939. In Sándor, *Örségváltás után*, 9–10.
9. Quoted in Carl Schorske, *Fin-de-siècle Vienna: politics and culture* (New York, 1980), 145.
10. E.g. Tim Cole, 'Constructing the "Jew", Writing the Holocaust: Hungary 1920–45', *Patterns of Prejudice* 33/3 (1999), 19–27.
11. Manchin, 'Fables of Modernity', 65.
12. Samu Stern, *A zsidókérdés Magyarországon* (Budapest, 1938), esp. 21; and 'Zsidók a magyar kulturában', *Esti Kurir* (17 March 1939), 4. See also Guy Miron, 'History, Remembrance, and a "Useful Past" in the Public Thought of Hungarian Jewry, 1938–1939', *Yad Vashem Studies* 32 (2004), 131–70 and Anna Szalai, 'Will the Past Protect Hungarian Jewry? The Response of Jewish Intellectuals to Anti-Jewish Legislation', *Yad Vashem Studies* 32 (2004), 171–208.
13. See, for example, the anonymously authored pamphlet *A második zsidótörvény célja, indokolása, következményei. Néhány szó a magyar zsidó értelmiség nevében a magyar közvéleményhez* (Budapest, 1939), 8–11.
14. László Szűcs, 'Kormányzói audiencia az első Zsidótörvény előtt (1937. November 11)', *Levéltári Közlemények* 1–2 (1993), 146. A life-long antisemite, Horthy did want to remove Jews from Hungary, but at a slower, staged pace.
15. János Pelle, *Sowing the Seeds of Hatred* (Boulder, CO, 2004), 16–68.
16. István Milotay, *Új világ felé*, vol. II, 80, quoted in A. Sipos, 'Who is a "True Hungarian"?' 122.
17. 'Pángermán agitáció a magyar filmek ellen', *Magyar Filmkurír* XI/29-32 (20 September 1937), 8. While anti-Germanism was strong among some segments of the populist movement, this sentiment was not powerful among film industry populists. When anti-German sentiment was expressed, it was frequently conflated with attacks on Jews.
18. 'Művészi és írói kamarákat sürgetett a Turul Szövetség Szépmíves Bajtársi Törzsének Nagytábora', *Bajtárs* (April 1938), cited in Sándor, *Őrségváltás*, 107–12, quotes from 109, 111.
19. Macartney, *October Fifteenth*, vol. 1, 213; I. Deák, 'Hungary', in H. Rogger and E. Weber (eds), *The European Radical Right: A Historical Profile* (Berkeley, CA, 1966), 393–4; Hanebrink, *In Defense*, 141.
20. Balázs Ablonczy, *Pál Teleki: The Life of a Controversial Hungarian Politician*, trans. T & H. DeKornfeld (Boulder, CO, 2006), 182.
21. Németh, quoted in Pelle, *Sowing the Seeds*, 37. Gyula Illyés, *Magyarok-Naplójegyzetek* (Budapest, 1938), 276–7.
22. Manchin, 'Fables of Modernity', 72–5.
23. Székely, *A Hyppolíttól a Lila Akácig*, 21–2.
24. Herrick Library - MPPAA Archive, MPAA General Correspondence Files, Reel 5 [1938–41], 'General Letter', from MPPDA Vice President to all foreign branches of American film companies, 24 January 1938.

25. Endre Tóth (Andre de Toth), László Vajda, and others also left Hungary in the late 1930s.
26. Imre Somló interview with Béla Pásztor, in 'Meg kell teremtenünk a tökéletesen magyar filmstilust', *A Film. A Magyar Mozi és Filmszakma Lapja* IV (October 1937), 3–4.
27. 'Árja filmek kellenek?' *Egyenlőség* (9 September 1937), 5.
28. Born Jewish, Zsolnai converted. His choice of Christianity, however, became immaterial with the 2nd Jewish Law.
29. See Chapter 2.
30. László Zsolnai, '"Sürgősen állítsák be a filmgyártásba a keresztény erőket" Hozzászólás a Turul Szövetség gettófilm memorandumához', *Filmújság. Zsolnai László Hétilapja* (17 July 1937), 1–4.
31. Andor Garami, 'Magyar film, vagy zsidó film', *Magyar Filmkurír* XI/23–6 (26 August 1937), 4–5.
32. Ibid., 4.
33. Zsolnai, 'Sürgősen állítsák be…', 3.
34. György Guthy, 'Árjafilm', *Mozi- és Filmvilág* II/1 (15 January 1938), 6.
35. György Guthy, 'Sok a túlprodukció', *Mozi- és Filmvilág* II/1 (15 January 1938), 1. The stock certificate comment refers to István Gerő, whom Guthy saw as the plague upon Hungarian film.
36. György Guthy, 'Hagyjuk a numerus clausust!' *Mozi- és Filmvilág* II/2 (20 February 1938), 8.
37. The petition is discussed by Nathaniel Katzburg, *Hungary and the Jews* (Ramat-Gan, Israel, 1981), 111 and by Vera Ranki, *The Politics of Inclusion and Exclusion*.
38. In an introduction to the published 26 April 1938 MMOE meeting minutes, the editors of *Filmkultúra* wrote that 'the peace of the movie profession was destroyed…and the ability of several thousand to earn their keep was made insecure' by the possible anti-Jewish legislation. 'Megkezdődött a munka az ujjászervezett Moziegyesületben', *FK* XI/5 (1 May 1938), 2.
39. 'Válságban a vidéki mozgóképszakma', *FK* XI/6 (1 June 1938), 2–13. This issue included a series of articles, many written by countryside small theater owners. Repeating the refrain that the Hungarian public still loves Hungarian film, but only good Hungarian ones, many complained of overproduction harming the quality of Hungarian films. They worried about their publics' loss of enthusiasm for Hungarian film and a shift in preferences toward foreign film.
40. Károly Freidrich, 'Visszapillantás a vége felé közeledő 1937–38. évi soproni moziidényre', *FK* XI/6 (1 June 1938), 3.
41. János Magyar, 'A rossz magyar filmek elszoktatták a közönséget a moziból', *FK* XI/6 (1 June 1938), 7. Magyar suggests that women made up the majority of film viewers. Research on interwar Germany has shown this to be the case,

but I have located few theater attendance statistics for Hungary, and those that exist do not disaggregate by gender.

42. XV/1938, quoted in Yehuda Don, 'Anti-Semitic legislations in Hungary and their implementation in Budapest – an economic analysis', in R.L. Braham (ed), *The Tragedy of Hungarian Jewry: Essays, Documents, Depositions* (Boulder, CO, 1986), 49.
43. Gyurgyák, *A Zsidókérdés Magyarországon,* 138.
44. The most important decrees were 6090/1938 M.E. sz. and 6095/1938 M.E. sz.
45. The Film Union lacked a mandate from any segment of the film industry and thus only operated as an advisory organ and a forum where film elites exchanged ideas.
46. Sándor, *Őrségváltás után,* 13–14, 16–17, 21–2. The OMF shift to the right and toward Christianization began in mid-1937. Of the four voting board members elected in June and July 1937, three were already or later became exponents of radical right, anti-Jewish politics (the directors Viktor Bánky and István György, and the Hunnia production manager Sándor Nagy). The Film Union's voting membership, which in 1936 was at least one-third Jewish, was diminished and the numbers of Jewish participants reduced. Without expelling all of its Jewish members, the Union, in the summer of 1937, adopted a rather anti-Jewish point of view, formulating plans for the creation of an independent film chamber complete with a *numerus clausus,* plans which were later realized in the genesis of the Film Chamber.
47. *MOL-Ó* – Filmipari Alapot Kezelő Miniszterközi Bizottság [FAKMB] 1926–40, Z1129-r1-d1. Ülési jgyzk. 1938. See Jgyvk, May & June 1938.
48. *MOL* – Minisztertanácsi jgyvk xeroxmásolatai, K27, 12 August 1938 Minisztertanácsi jgyzk, 28–9.
49. Sándor, *Őrségváltás után,* 26.
50. Yehuda Don, 'The Economic Dimensions of Antisemitism: Anti-Jewish Legislation in Hungary, 1938–44', *East European Quarterly* 20/4 (January 1987), 448. Filmmakers, through the Chamber, sought to remove risk and competition, and guarantee themselves profit. In that way, the Chamber resembled a medieval guild. In comparison, Italy did not fully centralize and corporatize its film industry until 1937. Ricci, *Cinema and Fascism,* 67.
51. For example, the short-lived *Film Officials' Journal,* representing upper management in Hungarian film production, distribution, and premier theater ownership, wrote that the Chamber's role was to bring Hungary's film industry more in line with the 'totalitarian, military, radical right nationalist Zeitgeist' and 'corporate order' that now presented itself in Central Europe. József Pethö, 'Koreszme és Kamara', *Filmtisztviselők Lapja* (November 1938), 1.
52. A m. kir. minsztériumnak 6.090/1938 M.E. sz rendelete. 'A szinművészeti és filmművészeti kamara felállítása tárgyában', paragraph 2.

53. Of the committee's nine members, the president and four others came from Religion and Education. The Prime Minister, the Minister of the Interior, the Justice Minister, and the Trade Minister each named one delegate.
54. The film professional and technical committees were: an artistic committee for directors and their immediate assistants; a committee for artistic aides involved in film music and dance; a performance committee for all actors, singers and dancers involved in film; a business committee constituted by those involved in the trade portion of the film industry; and a production committee for personnel involved with the technical aspects of filmmaking. Other leaders picked included members of a joint disciplinary court and delegates to a National Theater and Film Council.
55. Tasnádi-Nagy in particular saw the Jewish laws as a means of fulfilling the obligation of Hungarians to protect their 'national character', which he saw in racial terms. Gyurgyák, *A Zsidókérdés,* 140.
56. Imrédy denied German pressure convinced him to enact the second law, but the incentive of further revision certainly swayed him. Péter Sipos, *Imrédy Béla a vádlottak padján* (Budapest, 1999), 125.
57. The elections of May 1939 demonstrate the rightward trend in Hungarian politics. The fascist Arrow Cross Party received the second largest chunk of the vote, some 750,000 out of the approximately 2 million votes cast. Some argue the Arrow Cross in fact received the largest vote, but the government party, the official first place finisher, fixed the overall result. Other sources say the total number of Arrow Cross votes was closer to 674,000. The Arrow Cross and other national socialist parties became the second largest bloc in the Lower House of Parliament, with 49 seats (out of a total of 260). For conflicting figures, see Katzburg, *Hungary and the Jews,* 158–9; István Deák, *Hungary from 1918 to 1945* (New York, 1989), 28; and Gy. Ránki, (ed), *Magyarország története,* 993.
58. *NARA* – RG59 [State], M-1206, Roll 6, 864.4016/Jews, 1, 21–2. Translation – 'Indication of the Bill Limiting Jewish Participation in the Cultural and Economic Life', submitted to Parliament by Prime Minister Béla Imrédy and Minister of Justice András Tasnádi-Nagy, 23 December 1938.
59. Ibid.
60. *Magyar Film* remained the only film trade journal until 1942, when a variety of private magazines began to reappear.
61. The Turuls originally conceived of *Magyar Film* in 1938 as an answer to Jewish film press hegemony. They selected Dezső Vaczi, an industry veteran who headed the MFI screenwriting division, to be the managing editor. Publication began in February 1939.
62. 'Feladatunk', *MF* I/1 (18 February 1939), 1.
63. 'Útravaló', *MF* I/1 (18 February 1939), 2.

64. *BFL* – Robert Bánky [Bánki] (a.k.a. Robert von Bánky), Nb. 17670/1949. See also *BFL* – Victor Bánky [Bánki], Nb. 2540/1945. 'Bánki Gyula Viktor, Filmrendező', Nü 4959/1945, 8 August 1945', 7.
65. *BFL* – Ferenc Kiss, Nb. 4077/45. In particular, see the testimony of Klári Tolnay, Géza Staud, Tamás Major, and Anna Tőkés. At his trial in November 1945, Ferenc Kiss denied it was his choice to reduce Jewish representation in the Theater and Film Chamber to below six percent. That decision, he claimed, was a grassroots decision reached by the Chamber members themselves. See Kiss statement in *A budapesti népbiróság* Nb.IX.4077/1945-2.sz. Other key original members of the admissions committee, such as the producer József Daróczy, were also alleged to have supported a *numerus nullus*. *BFL* – Documents related to MFSzSz XVII 647/6 – 287 B. István Kertész testimony, Jgykv felvétetett a MFSzSz-éhez a Nb. által kiküldött IB III.ülésről, 25 June 1945.
66. In addition to material in the above footnote attesting to these actions, US sources also confirm the Film Chamber founders' antisemitic intent. See *NARA* – RG59 [State], M-1206, Roll 6, 864.4061/Motion Pictures 68. F.L. Herron, Foreign Manager of the MPPDA, to P.T. Culberson, Dept of State, 3 February 1939.
67. In May 1939, Taubinger Hungarianized his name to Tőrey. 'The Germans', he said, 'always thought I belonged to them' and I wanted to demonstrate in clear terms that I did not. *BFL* – Budapest Nb. docs. re: Magyar Hiradó, MFI Rt. XVII 789/6 – 399 A, file on Zoltán Tőrey [Taubinger].
68. *BFL* – Nb. docs. re: Magyar Hiradó, MFI Rt. XVII 789/6 – 399 A, Zoltán Tőrey [Taubinger] file. Dr. Zoltán Tőrey testimony, Nb.ig. XIV 991/1947-5, 12. While one could dismiss this claim as a defendant's exaggeration, several Jewish film figures who survived the war, including Ernő Gál and László Huszár, confirmed Tőrey's assertion.
69. 'Útravaló', op. cit., 2.
70. Ibid., 4.
71. Ibid., 3.
72. Through mid-July, Hungarian studios produced only eight films. Five were made at the MFI and a stunning three at Hunnia.
73. 'Nemzeti magyar filmművészetet!' *MF* I/24 (29 July 1939), 1–2.
74. 'Új területek, új feladatok', *MF* I/5 (18 March 1939), 1.
75. Alfréd Szöllőssy, 'Magyar feltámadás, magyar film', *MF* I/8 (8 April 1939), 1.
76. Pál Morvay, 'Kárpátalja. Kedvesményekkel alapozzuk meg Ruszinföldön a magyar filmkultúrát', *MF* I/6 (25 March 1939), 2. Morvay recommended Carpathian theaters receive free newsreels, films, projectors, chairs, still photos, etc. and be exempted from musicians fees and transport charges so that they might re-establish Hungarian culture more rapidly.

77. Zoltán Tőrey, 'Tőrey Zoltán dr., a M.F.I. ügyv. igazgatója, a Filmművészeti Kamara alelnöke eladása', *MF* I/17 (10 June 1939), 5–6.
78. 'Bingert János dr., a Hunnia Filmgyár rt. igazgatója előadása', *MF* I/17 (10 June 1939), 6–9.
79. 'Nemzeti magyar filmművészetet!' *MF* I/24 (29 July 1939), 1–2.
80. Decree 2240/1939 M.E. sz., 24 February 1939.
81. *MOL-Ó* – PM, Z40-cs55-Ikt18: Magyar játékfilmgyártás pénzügyi alapjainak biztositása. Zsindely to Lamotte, 5 May 1939.
82. Ibid.
83. Márk Záhonyi-Ábel, 'A magyar filmes intézményrendszer 1938–1944', *Metropolis* 17/2 (2013), Filmcenzura. http://metropolis.org.hu/?pid=16&aid=503
84. Prior to 1938, the Interior Ministry controlled the two major film regulatory agencies, the OMB and the FIF. From 1939 on, after it gained control of the ONFB, the Ministry of Religion and Education was ascendant. In addition to the four VKM officials, the ONFB consisted of appointees from the Interior Ministry (1), Army (1), Ministry of Manufacturing (1), Ministry of Transport and Trade (1), Prime Minister's Office (1) and Foreign Ministry (1). After December 1942, a representative of the National Security and Propaganda Minister joined the ONFB, as did, on occasion, an agricultural ministry official. See Sándor, *Őrségváltás után,* 189, ft. 48; Langer, 'Fejezetek', 168; Nemeskürty, *A Magyar Film 1939–1944: Egész müsort betöltö játékfilmek* (Budapest, 1980), 11–12.
85. At his 1945 trial, Ferenc Kiss testified that the true power in Hungarian film, the one person who decided which films would reach the screen and which films would not, was Baron Gyula Wlassics, Jr. *BFL* – Ferenc Kiss, Nb.IX.4077/1945-2.sz. Testimony at the Zenemüvészeti Főiskola on 26–7 November, 1945, 24–5.
86. Both held board positions through the late 1930s and until the mid-1940s. Balogh became a Hunnia stockholder in 1941 at the latest but was a board member at least as far back as 1939. See *MOL-Ó* – Hunnia, Z1123-r1-d1. Igazg. jgyzk. 1936–45; Hunnia, Z869-r1-d1. Közgyülési jgyzk., 1941–5.
87. By 1940, along with Székely, the writer László Vajda, actors Szőke Szakáll, Gyula Kabos, Irén Ágai, and Zita Perczel had emigrated. Directors Viktor Gertler and Márton Keleti; stars Imre Ráday, Kálmán Rózsahegyi, Gyula Gózon; and one of Hungary's top female comics, Ella Gombaszögi, all lost their jobs. Vajdovich, 'A magyar film 1939 és 1945 között', *Metropolis* 17/2 (2013), http://metropolis.org.hu/?pid=16&aid=480
88. Sándor, *Őrségváltás után,* 44.
89. Ibid., 38, 41–2.
90. Hunnia's income in 1938 was approximately 35 percent of the 1937 figure. *BFL* – Hunnia Filmgyár Rt. – Cg29830-3636. 'Hunnia Filmgyár Rt. zárszámadási 1938.év', *A Budapesti Közlöny Hivatalos Értesitője* 157 sz. (16 July

1940), 4; 'Hunnia Filmgyár Rt, Mérlegszámla 1937. év', December hó 31-én', *A Budapesti Közlöny Hivatalos Értesitője* 202 sz. (11 September 1938), 6.

91. Langer, 'Fejezetek', 165. Langer cites an 11 July 1938 meeting of the IMB requiring Hunnia to make the films *Pagans* [*Pogányok*], *Storm on the Puszta* [*Vihar Kemenespusztán*] and others.
92. *MOL* - KM, K66-cs470, 1940, III-6: Kulturális ügyek. Kozma to Ullein-Reviczky, 19 May 1940. In this letter Kozma references the 1935 MFI/Hunnia concord and proposes changing it. The 1935 agreement was re-confirmed in a July 1938 meeting of the committee in charge of the FIF, as Hunnia and FIF bosses became more concerned with MFI competition. *MOL-Ó* - Filmipari Alapot Kezelő Miniszterközi Bizottság 1926–40, Z1129-r1-d1. Ülési jgyzk., 1938. FAKMB –1938. július jgykv-ei.
93. Zsolnai, 'A 20%'.
94. This had been tried in 1936, but the bank's capital apparently was too Jewish for Zichy. Despite his misgivings, Zichy engaged in numerous projects with Jewish partners.
95. Zichy proposal sent to János Bingert, 26 January 1939, cited in Langer, 'Fejezetek', 188–9.
96. *MOL-Ó* - Hunnia, Z1123-r1-d3. János Bingert, 'Tervezet a folyamatos magyar játékfilmgyártás pénzügyi lebonyolitásának biztositására', 6 February 1939. In this same proposal, Bingert admitted that most production and distribution firms, as of the time he wrote the plan, remained Jewish-owned.
97. György Guthy, 'Miért szünetel a magyar filmgyártás?' *Mozi- és Filmvilág* III/6 (June 1939), 1–2.
98. The 'news war' is described in Sándor, *Őrségváltás után*, 85–7.
99. Miklós Kozma, 'Hozzászólás a játékfilmgyártás válságához', *MF* I/22 (15 July 1939), 8–13. See also *MOL* - KMI, K429-cs59-t2. Vegyes MTI levelezés 1938–9, particularly Kozma's rejection of the suggestion that Hungary establish a state run monopoly made by the Italian head of Luce, in the 28 May 1939 document 'Feljegyzés filmügyben Paolucci olasz rendkivüli követ és meghatalmazott miniszterrel folytatott megbeszélésről'.
100. Kozma, 'Hozzászólás a játékfilmgyártás válságához', 12.
101. *MOL-Ó* - PM, Z40-cs55-Ikt180: Magyar játékfilmgyártás pénzügyi alapjainak biztositása. Lamotte to Zsindely, 12 May 1939.
102. Dr. Péter Bajusz, 'Ceterum censeo', *MF* I/13 (13 May 1939), 3–4.
103. Sándor, *Őrségváltás után*, 41.

5

The Flawed Christian National System

Introduction

By 1939, the pathogens of antisemitism had sickened and split the entire motion picture business. The creators of the Film Chamber envisioned their organization as the integrative palliative to the industry-wide fractures. When conceptualizing the Chamber in 1938, they foresaw a healthy, centralized, and unified film world. In this utopia, all film professionals organizations, including the OMME and the MMOE, would dissolve into a representative, all-encompassing structure. Each organizations would cede its provenance to a High Office for Film [*Filmfőosztályának választmánya*], a group of experts within the Film Chamber who would weigh the concerns of all sides – producers, distributors, exhibitors, actors, and regulators – and dispassionately tease harmony from cacophony, acting in the national interest. This victory of social engineering premised on Christian nationalism promised radical change. The result would be film production which was a genuine expression of a biologically distinct Hungary, a people unique among all others.

Throughout the world of culture, this racist vision denied Jews a role in the production of the nation. Merely enunciating this vision, however, did not make it so. Culture was not created by the flick of a wand. It required institutions, which in turn became sites of contestation. Between

1938 and early 1942, Hungary's film powers – its new radical right, entrenched government bureaucracies, and professional organizations – engaged in a rough-and-tumble struggle over the future of the profession. Their battles exposed numerous flaws and inconsistencies inherent to the idea of a Christian national reconstruction. They also revealed real diffusions of power which further complicated the culture creation process. This chapter examines the rhetoric and actions of the critical players and the institutions over which they fought. It demonstrates the existence of power centers resistant to the racial, antisemitic strains of nationalism that became increasingly mainstream during this period, and explains how some Hungarian Jews maintained a presence in the film world through the wartime era.

In 1942, Andor Lajta, the Jewish former editor of *Filmkultúra* and ex-Agfa company representative in Hungary, tried to convince his Christian colleagues that harmony had been achieved and a Christian national industry realized. Ironically, his presence was symptomatic of Hungary's twisted road to Christianization. In recognition of his longtime service and his unparalleled insider's knowledge, Lajta received a special government dispensation to edit the *Film Arts Almanac.*[1] In the 1942 edition of this annual industry informer, he wrote that the aims of the Second Jewish Law had been substantially achieved, marked by a 'totally new vanguard' of film distributorships and theater managers, particularly in Budapest movie houses.[2]

Other accounts, provided in this chapter, indicate that the transition was not as seamless or complete as Lajta presented it. On the contrary, Jews continued to influence Hungarian film and the Jewish question remained the fault line from which all other fractures emanated. The range of the discussion was defined by multiple centers: the Film Chamber leadership, most of whom advocated a strict interpretation of the Jewish Laws; the Interior Ministry, which supported a more inclusive, long-term vision of how to Christianize Hungarian culture; and segments of the film industry, which flouted all restrictions placed upon them. Within this admixture were opportunists, flame-throwing ideologists, sympathetic Ministers of State, protective philo-semites, turf-defending professional organizations, and even the Regent's son. Most parties came to two related insights by the early 1940s. First, Hungary was so dependent on its cultural producers of

Jewish origin that without their help, a Christian national culture would miscarry. Second, professionals and bureaucrats alike recognized how little they enjoyed giving up power. These realizations grievously injured the prospects for a recognizably Christianized and nationalized Hungarian culture.

Undermining the Jewish Laws: Miklós Kozma

From its initial conceptualization, the central component of the Christian national project, the Film Chamber, was grossly flawed and frequently criticized. It never came close to rivaling its Nazi counterpart, the *Reichsfilmkammer,* in terms of administrative control over the industry. Neither did it inspire the professional unity its advocates desired. There was a simple reason for this. Not all of Hungary's cultural and political elites believed the concept of a Christian renaissance, led by a state-directed Film Chamber focused on the complete elimination of Jewish men, women, ideas, and money, to be wholly desirable. One of the most intriguing and influential movie business powerbrokers to share this opinion was Miklós Kozma. Known for his ardent public antisemitism, his radical revisionist politics, and his alliance with Gyula Gömbös, Kozma was also a member of Regent Horthy's inner circle. Hints of his cultivated ambivalence in defense of 'film Jews' were revealed during his tenure as Interior Minister and in his negotiations with Germany. Prior to an assignment as the Plenipotentiary of Carpathia in 1939, he used rhetoric, illusion, and contradiction to assuage the far right while enacting a less extreme agenda.[3] As a member of the Upper House of the Hungarian Parliament, Kozma made practiced statements about the need to reduce Jewish influence in Hungarian society and in the closed film community. He warned of the threats posed by Jewish cultural domination. He helped dethrone István Gerő, Hungary's so-called film czar, by fracturing Gerő's theater trust.[4] Kozma never, however, called for or supported the complete Christianization of Hungarian film. In fact, while disgorging pedantic anti-Jewish rhetoric in public pronouncements, he remained an astute businessman whose actions often protected Jewish film interests. Kozma regularly walked the line between generic denunciations and the defense of specific Jews, both in public and in private.

Unless one wanted to obliterate Hungarian film, Kozma concluded, it made no sense to rapidly replace Hungary's 'film Jews' with Christians. Since there were few trained Christians, doing so would mean replacing businessmen with bureaucrats, a formula the bureaucrat-businessman Kozma ironically considered anathema.[5] In January 1939, he wrote a piercing letter denouncing the impact of the Film Chamber legislation to his compatriot, Prime Minister Béla Imrédy, only days after the Chamber began functioning:

> Hungarian film production is indeed a question of great propaganda, cultural, and economic significance. By all means, the regression of the Hungarian film industry must be stopped. Unfortunately, the industry is in great danger and the Jewish Laws only make conditions worse…If the Hungarian film industry fails or regresses, Hungarian propaganda, culture, and economic life would experience serious misfortune.
>
> It can be established that due to the Jewish Laws, attendance at the movie theaters has significantly declined (on average by 30 percent) and it is foreseeable that this will remain the case for a long, albeit temporary, period. This means that Hungarian films will not be able to earn as much as they had and film businessmen, 90 percent of whom were until now Jews, will not dare to take the risks which today are part of the preparation of a Hungarian film. Hungarian film companies … cannot do without that first payment of HP 15–20,000, which presently is the amount the theaters regularly pay to have a film put at their disposal. These Jewish theaters, in any case I could say [those] in Gerő's hand, precisely because of the Jewish Laws, already cannot pay out these premier advances.[6]

Kozma saw a clear connection between the 1938–9 crisis and the cascading effects of anti-Jewish legislation and rhetoric. If Jewish theater owners experienced declines in box office revenue, they would not have the funds to pay distributors. Distributors would not have the cash to underwrite production, and producers (whether Christian or Jewish), already immobilized by the threat that they might lose their investments, would abstain from filmmaking. Thus, the creation of the Chamber risked destroying the entire Hungarian film industry. For Kozma, there had to be a less radical excision of Jewish film capital in order to save the motion picture industry and, by implication, Hungarian culture. His prescription was a combination of state loans and credits for production and mandated increases in exhibition quotas. In Kozma's estimation, the increase in income produced by these requirements would more-or-less offset the loans offered by the state. Taken in total, Kozma believed his proposal to transfer state monies to private production companies – both Jewish and Christian – in order to maintain national production would result in a more 'natural' cultural distribution, one in which Christian-made films would eventually outstrip those made by Jews.

Further, Kozma proposed not to disband the Film Chamber but to make it more inclusive by inviting select Jewish cinema and film company owners to be members. In rationalizing this stance, Kozma took aim at the hypocrisy of Christian nationalism, arguing its advocates shared responsibility for making the film industry as Jewish as it was. In a March 1939 letter to Interior Minister Ferenc Keresztes-Fischer, Kozma reminded his comrade that it was the Christian National League [*Keresztény Nemzeti Liga*], with Imrédy as its president, which had leased its theater licenses in Budapest, Debrecen, and Szeged to one István Gerő. It was 'through the good offices of the League [that] Gerő had become the uncrowned king of Hungarian film'. Jews had not taken the film business from Hungary's Christians; Christians had handed it to them. 'In nearly all cases', claimed Kozma,

> ...film companies would hire a skilled Jew, [and] pay him a little more than unemployment insurance would. In [the public] forum, out flowed great anti-Jewish screeds; in practice the Jew took the business, the power, and a good part of the financial position as well. In the headlines, the Christian denounced the Jew, but in the financial columns he protected the Jew more. If the person is truthful, the blame must be assigned not to the working, business-smart, risk-taking, entrepreneurial Jew, rather the lazy, or comfortable and non-working Christian.[7]

Kozma, like the story of Hungarian cinema, was complex. A realist and an opportunist, he was not morally opposed to the Jewish Laws. His self-serving disagreements with them emerged from pragmatic objections. Rather than facilitate the creation of a Christian Hungary, he told Regent Miklós Horthy, an ideologically inflexible assault on Jews would destroy its economic and national foundations.[8] In a different setting Kozma remarked that his Hungarian Film Office must be managed 'free of party politics', meaning insulated from antisemitic absolutism.[9] Kozma relied on these rationalizations to justify his company's continued cooperation with Hungary's top 'film Jews', including István Gerő, Ernő Gál, and others.[10] He was not, however, above taking advantage of the Christianizing process to enrich himself and the MFI. In return for not blocking the dissolution of the Gerő Trust, the MFI received the licenses to three of Hungary's top movie houses. All three, Budapest's Royal Apolló, Szeged's Corsó, and Debrecen's Apolló theaters, were formerly part of Gerő's stable.[11]

By adopting positions such as these, Kozma, like the Regent himself, appeared to espouse a more 'gentlemanly' or 'selective' strain of antisemitism, one that saw no contradiction between spewing antisemitic venom while at the same time benefitting from, doing business with, and even befriending Jews.[12] According to this view, limited Jewish 'assimilation' was possible and Jews could become good Hungarians, perhaps even without abandoning Judaism. This was in contrast to the intolerant, race-based strain championed by important Film Chamber leaders and their supporters, who stuck to this belief even if it risked destabilizing the very culture and nation they held so dear. At first glance, the latter position carried the day among many segments of Hungarian society. Prime Minister Imrédy and his successor Pál Teleki, the main author of Hungary's Second Jewish Law, largely ignored Kozma's proposals for increased state aid to prop up the movie business.[13] Journalists, politicians, and Christian nationalist film professionals all regularly called for immediate de-Jewification. In practice, however, Kozma's notion of a film industry tolerating a background Jewish presence remained the operational norm into 1941 and beyond.

Undermining to the Jewish Laws: Movie Theaters, the Interior Minister, and Chamber Membership

In the midst of the early 1939 industry crisis, *Magyar Film* published a spate of articles bemoaning the constraints on the Film Chamber, calling for action to revitalize the film industry and to promote professional unity. Most of the articles blamed the Film Chamber's stunted development on the issue of membership. There was no requirement that cinema owners, license holders, or regular cinema employees be members of the Chamber.

Only senior theater management were so obliged. At first glance, cinema owner/license holder reluctance to join the Chamber and its Christian-national project seems unexpected. Antisemitism had spread through the MMOE in the mid-1930s. Of Hungary's major professional film organizations it was the most committed to Christianization. Logic dictated that the MMOE would embrace participation in the Film Chamber. Remarkably, the MMOE opposed mandatory membership for its constituents, and the organization's resistance gravely damaged the Film Chamber in its early stages.

Why was this organization, still led by the antisemite Gábor Bornemisza in early 1939, so resistant to the extension of the Chamber's powers and its de-Jewification of the film business? Professional self-interest was certainly a factor. While the MMOE had proudly carried the flag of antisemitism through 1938, it was one tactic among several in a broader assault that was also anti-big capital and anti-urban in character. MMOE antisemitism had a very particular target: big city premier theaters and the Gerő Trust. Many theater owners, some of them Jewish, joined the fight against the trusts because they believed it promoted their own economic well-being. As the trust issue seemed to move toward resolution in 1939 and István Gerő was forced to renounce his contracts and sell many of his theaters to Christians, calls for a broader antisemitic push from forces within the MMOE abated.[14] Many urban theater owners were themselves Jews, and Christian owners throughout Hungary had Jewish employees. Independence from the Film Chamber made hiding this reality easier.

Because the status of theaters was not specifically delineated in the First Jewish Law, debates about a Second Jewish Law included specific proposals obliging all theater license holders to join the Chamber. A leading voice in these discussions was MMOE Vice-President Pál Morvay, supervisor of the MMOE's day-to-day operations. His dissent against the excesses of discrimination was unexpected. Long an opponent of trusts and a man who until 1938 had held István Gerő in contempt, he directly challenged the wisdom of the First Jewish Law and the Film Chamber it spawned.[15] In March 1939 correspondence with government officials and articles in *Magyar Film,* Morvay and his co-MMOE Vice-President Ödön Ruttkay spoke out against mandatory Chamber membership, and also urged the authorities to reject any language which would require the Interior

Minister to reappraise all cinema licenses.[16] Christian MMOE members were anxious that their links with Jews – either directly through Jewish employees or indirectly through secret contracts with Jewish distributors or shadowy arrangements with Jewish managers – might be discovered in the course of a full-scale audit that would be part of any license reevaluation. Morvay and Ruttkay also worried that movie house owners would be paralyzed by widespread 'legal insecurity' stemming from more stringent anti-Jewish laws.[17] What would happen, they asked, if tomorrow the government enacted even harsher antisemitic legislation, dropping the six percent quota of Jews to two percent or to nothing at all? Morvay and Ruttkay argued that no theater owner would invest in normal upkeep or future contracting, not to mention renovation or new theater construction, if the threat of losing one's license was ever-present.[18] Strict application of the Jewish Laws would jeopardize the very existence of the cinema sector.

In spite of these fears, Morvay suggested that theater operators would endorse the overhaul of the entire licensing system with the caveat that it occur in the distant future. His organization also consented to language that was ultimately included in the Second Jewish Law mandating that all top tier employees of theater companies, including the company owner and the managing director be Christians and members of the Film Chamber. What Morvay extracted, with the aid of Miklós Kozma and an Interior Ministry determined not to forfeit its control over theater licensing, were two important concessions. First, the MMOE and its allies won a blanket exemption for theater license holders, even those who were also active managers, from required Chamber membership. This meant that temporarily, their licenses were protected and they were insulated from the political extremism of the Chamber. Second, drafts of the Second Jewish Law dropped all language compelling the Interior Ministry to review theater licenses. In essence, the alliance of the MMOE, Kozma, and the Interior Ministry carved out space for a functioning strawman system, making it a legitimate form of business for cinemas.

This compromise drew fire from many quarters. *Magyar Film* denounced MMOE and Interior Ministry collaboration as half-measures that would destroy the Film Chamber's mission of professional unification.[19] The radical right condemned the MMOE as callously materialist, a frequent criticism it leveled against any organization it deemed to be protecting Jewish interests. They must have been disappointed when one of their own, MMOE President Gábor Bornemisza, publicly thanked Miklós Kozma for overseeing the revision of the paragraph concerning theaters and insuring that the new Jewish Law would apply only to those it was meant to encompass.[20] They were likely even more disappointed with Pál Morvay's reply to their attacks. Morvay stated boldly that the MMOE was a 'non-political organization' with 'a large Christian majority and a pronounced Christian orientation' but in no case would it 'countenance bloodthirsty excess[es]' such as those demanded by its critics. The only way to insure the proper functioning of the film industry was to prevent the 'sword of Damocles' – meaning uncompromising antisemitism – from falling on the MMOE.[21]

In making these statements, Morvay spoke for the majority of theater owners, including those in the countryside. His constituents argued, as they had the preceding decade, that the health of Hungary's film culture depended primarily on the economic well-being of their theaters. Making something Christian and national, most theater owners understood, was parochial and personal, not a matter requiring the ideological centralization of the entire film world. As a result, large numbers of cinema owners and their employees ignored their obligations to join the Chamber, something which inflamed the editors of *Magyar Film* and other rightist periodicals. MMOE members, eager to preserve their independence, justified their stance by maintaining that the Chamber represented neither the interests of the theaters nor the Christian national project. By remaining outside of the Chamber, theaters retained veto power within the profession. They could refuse to adopt measures approved by the Chamber merely by claiming the Chamber lacked the jurisdiction over the license holders who would have to implement them. Once inside the Chamber, they would lose that power.

In the late summer of 1939, Morvay publicly clarified the MMOE's position on the Jewish Laws. Once more, his view was symptomatic of the instrumental variant of anti-Jewish thought, one which validated antisemitism and opposed big capital but was not absolutely racist in form. The Jewish Laws, he believed, had one major purpose in regards to movies: to eliminate the dominance of Jewish 'moguls', not to eliminate Jewish involvement in the film profession *in toto*. Naturally, the laws had gaps.

These were by design, loopholes meant to allow for the employment of low and mid-level Jewish technicians and managers, those who 'for two to three decades have done admirable work in the [film] profession'.[22] The future film industry ought to be 'mogul-free' but the complete expulsion of all other Jews from the Hungarian film world was something Morvay and his organization had no wish to encourage.

If professional self-interest, the exercise of power, and divergent opinions about the nature of Hungary's new Christian movie culture led movie houses to distance themselves from the Film Chamber, then the vagaries of Hungarian politics also divided the industry. Scholars have suggested that Regent Miklós Horthy purposely promoted a degree of political diversity by placing individuals with conflicting beliefs in high offices to allow himself room to maneuver.[23] The film industry, a bureaucratic morass predisposed to jurisdictional battles, was a perfect example. Passage of the First Jewish Law and the reorganization of the film world which it spawned gave the Minister of Religion and Education jurisdiction over actors, film production (aspects of which he shared with the Minister of Manufacturing), and the future Film Chamber and National Film Committee. The Minister of Trade controlled film export and import. The Interior Minister supervised post-production censorship and theater licensing. The main clash over regulation of the film industry occurred between the Ministry of Religion and Education, led by Bálint Hóman, and the Interior Ministry, led by Ferenc Keresztes-Fischer. A long-time member of Horthy's inner circle, Keresztes-Fischer replaced József Széll in May 1938 and immediately became one of the most powerful and respected figures in government. An ally of István Bethlen, the 'moderate' former prime minister who had presided over the watering down of Hungary's first anti-Jewish legislation, the *numerus clausus* of 1920, Keresztes-Fischer was also an opponent of the increasingly severe Jewish Laws passed in 1938 and 1939.[24] He quietly worked to weaken them, albeit without publicly announcing his opposition. A key figure in government until the German occupation, Keresztes-Fischer was one of the pivotal forces guiding the transformation of Hungarian society in the late 1930s and early 1940s.[25]

This influence had a substantial impact on the film industry. Keresztes-Fischer and his proxies wielded formidable authority in determining how the Film Chamber would be constituted and how the majority of the Jewish

Laws would be enforced through 1944. The primary reason the theater firms were able to maintain their independence from the Film Chamber as long as they did was because of the tacit and sometimes blatant support of Keresztes-Fischer's ministry.[26] In 1939–40, Keresztes-Fischer parried calls from the editors of *Magyar Film,* many of the Film Chamber's leaders, and other high officials that he force all cinema employees to join the Chamber and that he carry out a comprehensive review of cinema licenses. Wanting to avoid the mistakes caused by the first cinema licensing revision of the early 1920s, he merely tweaked the status quo, issuing a 14 December 1939 order revoking some Jewish licenses while allowing most Jews to retain their licenses and continue to engage in partnerships.[27] Later that month, political powerbrokers, likely including Keresztes-Fischer, further mitigated the extremist wave by supporting a palace coup within the MMOE which resulted in the replacement of the radical Gábor Bornemisza by the more moderate Zoltán Bencs as the organization's president.[28] The chess game of Hungarian étatist politics again played out in cinema.

Keresztes-Fischer's deliberate yet decisive actions may have prevented panic from erupting among theater owners and spreading throughout the cinema business. Certainly, they crippled the Chamber and thwarted Hungarian radicals' dreams of a unified Christian national cinema industry. By 1940, the majority of MMOE members still had not joined the Chamber, insuring that it would never become the film industry's policymaking center. The absence of theater licensees virtually guaranteed a divided industry consumed with matters of jurisdiction rather than the substance of cultural 'renewal'. Keresztes-Fischer's actions served as an intra-governmental check, reducing the clout of the more antisemitic Religion and Education Ministry and Trade Ministry. While these and other ministries continued to work towards a racist Christian national vision, the limits of that vision were becoming clearer.

Undermining the Jewish Laws: Distributors

Efforts to construct a Chamber-centered Christian national film system were damaged, but not destroyed, by the fact that exhibitors and license holders remained outside the Chamber. The mortal blow came from within, dealt by the Association of Producers and Distributors, the OMME.

Producers and distributors in interwar and wartime Hungary were closely linked, often in the form of the same person or company, as distributors frequently were the only individuals with adequate cash and the desire to risk it in the precarious enterprise of filmmaking. Unlike theater license holders, film production personnel and top employees of all production and distribution companies were specifically enumerated in the First or Second Jewish Law and required to be members of the Film Chamber.[29] Plus, many of these individuals had great incentives to join, whether ideological or financial. The newly empowered 'third generation' of filmmakers was committed to realizing the ideal of a Christian national film industry. The affairs of distributors were primarily the domain of the Trade Ministry, run by Géza Bornemisza, an avowed antisemite who had actively labored to Christianize Hungarian businesses since the Gömbös period. In sum, in early 1939 it appeared that the stars were aligning in such a way that producers and distributors would become the core support for the Chamber.

That expectation was entirely incorrect. Tensions between the Chamber and the OMME surfaced almost immediately after the Chamber's inception. On 22 April 1939, in the same issue in which its editors incorrectly claimed that cinema concerns would join the Film Chamber, *Magyar Film* ran a banner headline announcing that the Film Chamber would take over the OMME sphere of competence.[30] Instantaneously, the OMME leadership convened and angrily scolded the Chamber leadership, prompting *Magyar Film* to write a follow-up describing the OMME's refusal to give up its authority just one week later.[31]

The OMME's pronouncement was noteworthy for several reasons. First, it was crafted and signed by the organization's secretary István Erdélyi. Erdélyi, the owner and chief of the Kárpát Film Trade Company, was one of the most prominent and experienced Christian movie producers and distributors in Hungary. He was also an early opponent of the Jewish Laws, openly declaring that individuals should be evaluated by merit, not background. Second, Erdélyi argued that the OMME was a commercial organization, overseen by the Ministry of Trade. The Film Chamber, on the other hand, had no legal authority to grant the certificates, permits, and identifications necessary to engage in the film business. Its primary function was social, cultural, and moral: to create an amorphous Christian national spirit and unity and to insure the welfare of those in the profession.

The OMME's more concrete purpose, to protect the financial interests of the film profession, meant that it and the Chamber had entirely distinct goals and spheres of influence. Thus, the pronouncement denied having ceded any authority to the Chamber and revealed that the OMME's executive board had voted to proclaim that the association would not cease operations.

This declaration of independence was not a minor matter. It was a clear demarcation of difference and a terrible blow to the prospects of the Chamber. Erdélyi and Frigyes Pogány, the President of the OMME, regularly lambasted the Chamber and the Jewish Laws, hammering away at the economic damage they did to the cinema world. At his association's grand parliament in June 1939, Pogány blamed his Christian colleagues' fear of investment for Hungary's failure to create a Christian national industry. Erdélyi followed and declared that the Second Jewish Law had transformed 'what earlier appeared to be an unsolvable situation [into one that was] totally hopeless' by initiating a purge of Jews from the distribution business without any thought of how to replace their connections, experience, and money. Both asserted that the Chamber, as constituted, could not represent the business interests of film distributors.[32] Six months later the two men reiterated this stance, leaving no doubt where the OMME cadre stood. In his opening remarks at the December 1939 OMME general assembly, Pogány warned that until the government radically reorganized the Chamber, the OMME had the 'obligation to protect itself', a position strongly seconded by Erdélyi.[33] Erdélyi again dealt with the Jewish question, issuing what contemporaries considered a forceful condemnation of the Jewish Laws:

> Concerning this changing of the guard I feel this way: It is our [the OMME's] obligation, within a short period of time, to bring it to a halt, and, concerning the laws, declare with the greatest respect and by means of a totally clear pronouncement, that we regret the departure of so many old and valuable colleagues among those representatives and employees expelled from the profession.[34]

The true test of the continued independence of the OMME came in October 1939, when it clashed with the Chamber Directorate over film imports. It was hardly a contest as Erdélyi and his OMME men adroitly sidestepped the Chamber, negotiating the procedures for import and the quotas on foreign films with Ministry of Trade officials. The Chamber leadership, bypassed entirely by a ministry it assumed to be friendly, was rather peeved. *Magyar Film* demanded the government grant the Chamber greater powers and the ability to unite the profession. Its editors accused bureaucrats of embarking on a divisive and self-defeating course. By ignoring the Chamber, the government yet again allowed Jewish distributors access to import permits, 'while worthy Christian companies [received] absolutely nothing'.[35]

A month later, a special committee assembled by the Chamber made a new bid for supremacy. In an attempt to outflank the OMME, it proposed that all three branches of the movie business, as well as the Foreign Trade office, the National Bank, and any number of other interested parties, be involved in import negotiations. In the interests of the whole, the committee claimed, an industry-wide agglomeration capable of weighing all sides and correctly enforcing the Jewish Laws should determine who received import permits.[36] The Chamber's initiative went nowhere. The OMME, already engaged in negotiations, reached an agreement with the Trade Ministry days later. The concord confirmed that officials placed greater trust in a private organization to balance the conflicting ideological, political, and commercial agendas of the industry than they did in an organization the government itself had created expressly for that purpose. OMME independence, and government preference for it, all but exploded the dream that the Film Chamber would serve as the keystone of the Christian national system.

Although able to defend its own autonomy, the OMME was not insulated from Christianizing winds howling through Hungary. In fact, the OMME's sovereignty may have been preserved by its willingness to take actions illustrating it was capable of Christianizing on its own. By the fall of 1939, the association's upper leadership was entirely different than it had been in 1938. When Erdélyi asserted in a November speech that every member of the OMME leadership was Christian and a member of the Chamber, his claim was by-and-large true, although there were notable exceptions.[37] Several Jews did remain as active voting members of the OMME, including Richard Horovitz, owner of the storied Müvészfilm production and distribution company, and Emil Kovács, a former OMME vice-president whose production and distribution company had funded Hungary's first smash feature, *Hyppolit*. Horovitz, however, was stripped of his co-president title and demoted to a mere voting member of the OMME board. Jews who had held top positions in the OMME in 1938, such as Fox distributor Károly Matzner; MGM branch chief Miklós Salamon; the country's leading producer/distributors Adolf Fodor, Dezső Frankl, Imre Tsuk, Ferenc Pless, and Rezső Faragó; all found themselves purged from the association's official decision-making strata in 1939. Even Manó Guttmann, the Honorary Eternal Co-President of the OMME, lost his position in late 1938.

The requirement that all upper managers and owners of distribution companies be members of the Film Chamber hit Jewish distributors hard, as all but a select handful were denied entrance into the Chamber. In 1938 and 1939, tens of Jewish-owned businesses were re-constituted with Christian upper management and ownership or were forced into bankruptcy. The film production crisis of 1939 also devastated Jewish distribution companies, especially the smaller ones that dealt primarily in Hungarian film. Christian-owned companies such as the Atelier Feature Film Production and Trade Company, the Mária Hausz Film Distribution and Production Company, the Palatinus Film Distribution Company, and Mester Film Company, arose from this carnage. All were founded in 1938 or 1939 in the aftermath of the Jewish laws. Their ranks were augmented by new distribution branches established by Hunnia and the MFI, and a host of other fly-by-night production/distribution ventures.[38] 1939 marked the year that Hungarian film distribution, at least in terms of numbers, became Christian.[39]

Much to the chagrin of those who advocated this change, the roots of this Christianization were quite shallow. The fragility and the difficulties faced by many of the new Christian distribution companies in the early stages of their development help explain why the OMME turned against the Film Chamber and its fanatical Christian national project. With so few films produced in late 1938 and the first half of 1939, there was little to trade, and many of these distribution companies failed within months. Without experience and foreign contacts, few novice firms could hope to break into the import market. Because of the insolvency and insecurity it created, it quickly became apparent that Christianization would hinder the objective of forging a durable national industry.

Hunnia's attempt to build a distribution department was illustrative of this clash between ideological and commercial goals. Realizing that it must reduce the numbers of Jews in every department in order to abide by the First Jewish Law, Hunnia's board actively sought qualified Christian professionals. Sándor Nagy, the director of Hunnia's neophyte distribution section, noted how difficult a task this was. Claiming he came to Hunnia because he 'was a lover of Hungarian film and a proponent of the creation of a Christian film industry', Nagy rose through the upper echelon and during the Nazi occupation became Hunnia's managing director. An outspoken antisemite and a member of a known circle of Christian racialists, he sought out Christian distribution agents but discovered there were only three or four in all of Hungary in September 1938.[40] This paltry selection forced a reluctant Nagy to hire several Jews as key members of Hunnia's film placement office and to recognize the fiscal imperative of short-term reliance on Jewish expertise. Most senior among his hires were Sándor Balla, the director of the distribution office, and his assistant, Sándor Gárdonyi. Both were of Jewish origin. When government officials demanded the removal of Balla and Gárdonyi in late 1938, Nagy replied that 'the new generation must study from the old', because there was not a supply of 'those with appropriate experience, knowledge, and skill'.[41] Until Christians were properly trained, Nagy asserted, his Jewish employees would stay, which they did through 1941. The Hunnia Directorate came under increasing pressure from István Kultsár, the Commissioner for Unemployed Intellectuals [*Értelmiségi Munkanélküliség Kormánybiztos*, hereafter EMK], in 1940. Appointed in December 1938 by Pál Teleki and Prime Minister Imrédy as a special Commissioner and as the Ministry of Religion and Education official designated to enforce the Jewish Law, Kultsár soon became a driving force in extripating Jews from Hungary's culture industries.[42] While the Hunnia Directorate eventually succumbed and agreed to fire the two men effective of 1 January 1941, it later reneged. The Directorate informed Kultsár that until he purged Jewish employees of other firms, businesses which would likely snatch up Balla and Gárdonyi, Hunnia would retain them.[43] This game of 'you first' was repeatedly played out, resulting in the continued presence of Jewish film professionals in a variety of vocations.

Other companies mimicked Hunnia, although the smaller the company, the greater the need for discretion. Some distribution companies in existence prior to 1938 and supported by Jewish monies simply amended their bankrolls and management hierarchies. One example among many was Palló Film, a distribution firm owned prior to 1938 by the retired Habsburg general János Palló and the Jewish banker Imre Lewin. After 1938, it was listed as owned by Palló and Mrs. Sándor Molnár. Mrs. Molnár was likely a 'strawwoman' who held Lewin's interest for him.[44] Many of the large American distribution companies substituted Christian manager for Jew, at least on paper, although most, including Fox, Paramount, and Universal, kept Jewish employees in upper level positions. These types of responses to the Jewish Laws enabled some Jews to remain active and in fact thrive. Manó Guttmann, Béla Weissburg, Sándor Ungár, Ferenc Pless, Károly Matzner, and others continued to openly participate in the distribution business through 1941, and several managed to survive in diminished positions after that. Thus, despite the sudden increase in Christian-owned and Christian run firms, Hungary remained reliant on its Jewish film professionals into the early 1940s.[45]

Perhaps one of the more convoluted ironies of the period was that several of the most successful Christian distribution firms were those which circumvented the Jewish Laws most blatantly. Géza Kormos, a Jewish film distributor who worked for István Erdélyi throughout the 1940s, labeled Erdélyi's office a 'yellow casino, because it swarmed with yellow-starred co-religionists whom [Erdélyi] supported and protected'.[46] Other top production/distribution companies, such as Mária Hausz's company, the Nazi-funded Mester Film, Modern Film, and others involved silent Jewish partnerships and even publicly known Jewish management through 1941 and possibly well beyond.[47]

Overall, however, despite important exceptions, the distribution wing of the film industry experienced a fairly meaningful overhaul beginning in late 1938. Christian managers gained crucial experience, often, incongruously, in failed ventures or under the stewardship of Jews, and some made more successful forays into distribution in the later war years. Although the radical right protested that the professional home of the distributors, the OMME, remained a Jewish redoubt and that Jews continued to dictate distribution terms, the claim was only partially true. Jewish predominance in film distribution began to decline in 1939, a trend which accelerated quite rapidly in 1941, then stabilized until the German occupation of Hungary in March 1944. Much of this occurred not because of the Film Chamber, but in spite of it.

The Demise of the Film Chamber

The combined efforts of Hungary's two top trade organizations, the OMME and the MMOE, with help from key political figures, including Ferenc Keresztes-Fischer and Miklós Kozma, rendered the Film Chamber impotent, a place where radical Christian nationalists had a voice but only a minor role in policy. As early as autumn 1939, the powerlessness of the Film Chamber was fully exposed, when 14 of the 16 members of the High Film Committee, the equivalent of the Chamber's board, resigned after less than eleven months in office. The positive interpretation was that the industry-wide production crisis had been surmounted, the trusts beaten down, and order restored to the film world, courtesy of the Film Chamber. The Chamber had advanced the cause of Christian national Hungary as best it could, under the circumstances, and it was time for new leadership. The more likely reason, published in an open letter to the President of the Theater and Film Chamber, was that this was a mutiny against the dictatorial rule of the Chamber's President, Ferenc Kiss. Those who quit protested that Kiss

> has not observed the Chamber's statutory decrees, pays no attention to the decisions of its select committee, does not enforce them, takes contrary actions, in general operates in such a way as to create dissension in the profession and illegally place people in disadvantageous positions.
>
> After struggling against this, we have come to the decision…since the majority of your [Kiss'] actions are contrary to the committee's and the committee has no way of enforcing decisions…to resign.[48]

Implicit in this criticism is that Kiss pursued an antisemitic agenda too energetically, even for those who might have agreed with the need to eventually purge all Jews from the film world.[49] This criticism was confirmed in print in 1942, when Géza Staud indicted Kiss for his over-zealous application of the Jewish Laws and his contribution to the atomization of the film industry.[50] In later testimony at an assortment of postwar trials, numerous witnesses, including one of the two vice-presidents of the Film Chamber, Zoltán Tőrey, and one of the High Film Committee members, Viktor Bánky, attested to the fact that the resignations were due to Kiss' fundamentalist antisemitic fervor.[51]

Kiss was not without allies. Days after the *en masse* resignations, Minister of Religion and Education Hóman promised to solve the Chamber's root problems and made Kiss the Government Commissioner [*Kormánybiztos*] in charge of film, effective 1 March 1940. Essentially, he gave Kiss authority over the Film Chamber and its bullhorn, *Magyar Film,* allowing Kiss to act on behalf of the Film Chamber until such point as a suitable High Film Committee re-formed. Yet even this step contributed to the increasing powerlessness of the Chamber. After Hóman took action, the Chamber's High Film Committee, realizing the threat of an unchecked Kiss, reconstituted itself, bringing back many of the same figures who had resigned in December 1939. They quickly set out to reorganize the Chamber in order to resolve jurisdictional disputes, and produced the following language:

> The task of the Film Chamber in the fields of film production, film trade, and film exhibition are: to secure and enforce the requirements of the national spirit and Christian morals and the protection and advancement of the moral, commercial, organizational, and social interests that bind the Chamber together. Its job includes legal protection and the fulfillment of obligatory supervision. Its disciplinary committee is charged with engaging and making suggestions concerning all questions regarding film production, trade, and exhibition.
>
> The Film Chamber looks after the advance of national and artistic quality in film arts, organizes the film labor market, and arranges peaceful cooperation between members of the film profession with conflicting interests.[52]

This purposely obtuse statement, acknowledging a role in 'enforcing' the national spirit and Christian morals but explicitly noting that the disciplinary committee had only suggestive power, did little to mollify Chamber radicals hoping for authority to fundamentally reconstruct the movie business.

Competing government bureaucracies continued to clash over the Film Chamber's scope, meaning that throughout 1941, the Chamber served as little more than a rhetorical forum whose purpose and reach remained unresolved, and whose powers to enforce law were intentionally checked. Frustrated by these limits, Kiss frequently threatened to quit his multiple positions. He finally did resign as Chamber President on 30 December 1941, prompting his closest ally, the Minister of Religion and Education, to act yet again. On 1 January 1942, Hóman extended Kiss' appointment as Film Commissioner and 'temporarily' absorbed the High Film Committee into his wing of the government until such time as the Film Chamber was independently reconstructed. While it continued to exist, the Theater and Film Chamber was now literally and officially only a discussion group, stripped of all power to determine its own direction.

The short, divisive life of the institution birthed as a unifier reveals how difficult the process and practice of nationalizing and Christianizing a culture truly was. The Chamber became an impediment to, or at least an excuse for maintaining existing divisions within a highly fractured industry. This did not, however, mean that the Christianization project stalled completely. Despite the Chamber's implosion and the professional organizations' protection of Jews, by the end of 1939, the listed leadership of the OMME and MMOE were virtually *Judenrein*. But beneath the registers of associational titles, Jews retained membership and some degree of influence, shielded from the Chamber's more radical initiatives through 1941 and beyond.[53]

Undermining the Jewish Laws: The ONFB and the Studios

Prior to the summer of 1941, Jewish involvement in film production followed a trajectory similar to distribution and exhibition. In production, however, the direct government role enabling the persistence of Jewish activity was even more pronounced. As discussed earlier, the government had empowered the National Film Committee [ONFB] to review film scripts, casts, and funding matters pre-production. Once Hungarian filmmaking revived in September 1939, the ONFB and its *de facto* leaders, Baron Gyula Wlassics, Jr. and László Balogh, became the prime movers in, and definers of, Hungary's Christian national system.

Soon after its creation, the ONFB took steps to extend its authority. To insure that all studios came under its purview, the ONFB unilaterally 're-authorized' the gentlemen's agreement, the half-secret arrangement between Hunnia and the Hungarian Film Office regulating the maximum number of feature films the MFI could produce annually.[54] Because the MFI was semi-private, it had the right to negotiate contracts with companies with which it wished to do business. However, the studio was limited to making seven feature films in 1940, and the ONFB more-or-less determined which companies would produce on its premises. It mandated that two of the 1940 features be István Erdélyi-backed projects and gave permission to a select group of other firms to take their scripts to the MFI.[55] Thus, by installing itself as the arbiter of production ratios and by making the gentlemen's agreement official film industry policy, the ONFB tightened its grip on deciding who would make movies and where they would work, further centralizing Hungarian film production.

As its power grew, the ONFB created the impression of government support for filmmakers. Its authority allowed for the creation of a new system of film finance, whereby movies could be made with little real seed money. Proposals need only demonstrate minimal financial backing to receive ONFB approval. Then the producer could seek other investors, as the film was assured studio space and virtually, though not always, censors' approval. While this system allowed more films to be made in the course of the early 1940s, it also provided the surreptitious means for Jewish capital to enter into the picture. The film need only be certified as Christian-funded when the proposal was submitted to the ONFB. The OMB censors, who reviewed the film post-production, concerned themselves only with the content of the film, not its origins nor its financial backers.

The ONFB, then, was the primary organization that dealt directly with workaday matters involving Jews and film production. Its imprimatur was required for any changes made to the film script or casting and production decisions. On numerous occasions its members did meddle during filming, writing dialogue and altering the portrayals of individual characters, but only rarely did it stop the making of a movie mid-production. If it discovered Jews too conspicuously engaged, it could refer the case to the EMK, István Kultsár, whose office would then investigate. Emblematic of this procedure was a Művelődés Film case from August 1940. The company submitted a plan to film in the Hunnia studios and hand-picked the Jewish director Béla Gaál. The ONFB chiefs, Wlassics and Balogh, both of whom also sat on the Hunnia board, found this to be too excessive. Wlassics recommended that Gaál neither direct nor influence the artistic development of the film. Balogh, however, took a harsher line, suggesting the EMK investigate whether Művelődés Film was actually an Aryan company.[56]

In light of such minimal adherence to the Jewish Laws on the part of production companies in particular, Kultsár began to regularly attend ONFB meetings in 1940. When Kultsár was present, Jewish issues appear to have been dealt with more severely, at least in terms of promises of remedies. When Kultsár was absent, however, the Committee behaved less rigidly, favoring continuous production over the disruption of ideological purification. In numerous cases, when important producers presented the ONFB with plans that explicitly featured Jews, the Committee tended to recommend that Jews merely occupy less prominent or unofficial positions for which they would not be credited in the movie. For example, in a September 1940 meeting of the ONFB, Lóránd Szöts informed the committee that Béla Gaál had, once again, been contracted by Művelődés Film, this time to cast a film to be directed by Endre Rodriguez. Since the Ministry of Trade and Public Transport believed this to be a 'suitable' arrangement, the ONFB likewise raised no objection. At the same meeting, László Balogh reported that István Erdélyi wished to hire the Jewish director/composer, Sándor Szlatinay [Szlatinai], to direct the film *Property for Sale* [*Eladó birtok*]. Erdélyi indicated that should this impede his receiving production approval, he was willing to hire Viktor Bánky in Szlatinay's stead. The ONFB concluded that it was undesirable for Szlatinay to direct the film, 'but it had no comment against involving [Szlatinay] in the production as a composer'.[57] Flexibility and discretion seemed to go a long way toward smoothing the way for continued Jewish participation in movie making, so long as Kultsár was not looking.

That questions concerning Jews and production came up so regularly at ONFB and Hunnia meetings indicates that small but prominent numbers of Jews remained active in Hungarian film production. Screenwriters of Jewish origin, in particular Károly Nóti, István Békeffy, and Jenő Szatmári, appear to have written tens, if not scores of films between 1939 and 1944, many of which were simply credited to another writer. Nóti, the single most prolific and successful screenwriter in Hungary in the 1930s, secretly wrote the films *Hello Peter* [*Szervusz Péter*], *The Bercsényi Huszars* [*Bercsényi huszárok*], *Property for Sale* [*Eladó birtok*], *Entry Forbidden* [*Behajtani tilos*], *The Talking Robe* [*Beszelő köntös*], and *African Bride* [*Afrikai vőlegény*] for István Erdélyi's companies alone.[58] Working illicitly with the director Viktor Bánky, the man who eventually directed the most antisemitic film made in Hungary, *Changing of the Guard* [*Őrségváltás*], and who at best was a reformed antisemite or victim of blackmail, Nóti also wrote, co-wrote, or sketched the films *István Bors* [*Bors István*], *The Ball is On* [*All a bál*], *Yes or No?* [*Igen, vagy nem?*], and *Andrew* [*András*].[59] Among this list were the most commercially successful films of the period. István Békeffy wrote, co-wrote, or acted in films directed by nearly all of the prominent Christian directors of the wartime period, including works made by Baron Félix Podmaniczky, Jenő Csepreghy, D. Ákos Hamza, Ákos Ráthonyi, Kálmán Nádasdy, and Viktor Bánky.[60] On occasion through 1943, he was acknowledged overtly, on advertising posters and in film credits.[61] Jenő Szatmári was publicly credited for writing two films released in 1942, *Jelmezbál* [*Costume Ball*] and *Kisértés* [*Temptation*], and likely wrote the scripts for several more.[62]

Jews also continued to provide the money needed to make films and the contacts necessary to assure these films the widest showings. The rightist newspaper *A Nép* announced in May 1940 that Jewish interests still controlled at least 67 percent of Hungary's distribution firms, and these distributors frequently doubled as producers, underwriting Hungarian production.[63] Numerous antisemitic journals continued to suspect or to charge that István Gerő remained the puppeteer behind Hungarian film, either making decisions himself or through his surrogate, Henrik Castiglione, the owner and manager of the Corso Theater. They also alleged that Jewish interests continued to 'ooze' back into the film profession in the course of 1940, frequently by means of confidential partnership arrangements with cinemas, production companies, and distribution firms.[64]

In the summer of 1940, *Magyar Film* resumed its fusillade against Jews and strawmen. In a 1 July 1940 article, the editors berated their readers for becoming 'Jewish stooges'. It was 'never the primary goal that Christian gentlemen come into well-paying positions, rather that in these positions they would actually work for and achieve a national Hungarian film and cinema profession....All members of the vanguard of the new order have a spiritual and moral obligation not to act as paid puppets [*fizetett bábként*]....' We've had enough 'transition time and the unwanted solutions it has created'. Now was the time to enforce the laws and establish a healthy Christian national film profession.[65]

The Jewish Laws were repeatedly ignored or arbitrarily enforced in numerous professions.[66] Yet in late 1940, enough Jews had overtly returned to the filmmaking fold that even ONFB members, who sometimes facilitated this reemergence, grew ever more frustrated with the industry's failure, Hunnia's especially, to police itself. Ignoring his own inconsistent actions, László Balogh complained that whether contracted by movie producers or as advisors, many previously expelled Jews had resumed work at Hunnia. Balogh named the well-known set designer József Pán and the producer János Smolka, whom he claimed 'walks around the studio unhindered'. When confronted with evidence of this sort of blatant contempt for the Jewish Laws, the Hunnia Directorate consented to an entrance permit system and urged its own managing director, with whom the ultimate responsibility for policing the studio lay, to be more vigilant.[67]

This, however, was not the only pressure weighing on Hungary's studio leaders and its filmmaking elites. The gatekeepers of Hungary's studios found themselves facing yet another challenge in 1940 and 1941. The Vienna Awards and the expansion of Hungary's borders meant that millions of Hungarian speakers, long deprived of Hungarian culture, now clamored for Hungarian movies. Plus, the government had adopted increasingly high quota requirements for domestic theaters, mandating that 20 percent of all feature films be of Hungarian origin in July 1939, hiking that number to 25 percent in July 1940, and raising the bar again to 33 percent in July 1941.[68] In addition to the domestic need, Hungarian films were enjoying extraordinary success throughout Europe, meaning that pressure to quickly manufacture features multiplied. If they were to crack down on Jews in the movie business, film bigwigs might sacrifice a once-in-a-lifetime chance to appreciably expand the export possibilities for Hungarian film. Beyond forfeiting tremendous cultural-political capital, strict enforcement would jeopardize a domestic income windfall, the very same profits which could be used to cement the ground beneath the fledgling Christian filmmaking vanguard. Hence the dilemma: abide by the letter of the Jewish Laws and permanently damage the long-term prospects of the industry, or remain lax and establish Hungary as a continental film power.

Much to the chagrin of the pro-German, Christian nationalist segments of the film industry and antisemites in general, Hungary's studio heads chose the latter path, ignoring or even accepting Jewish participation in production. Hunnia especially seemed to be guided by the premise that to claw back from the brink of bankruptcy, it should make as many films as physically possible, willfully ignoring Film Chamber credentials and papers certifying religious origin. On occasion, enraged by the empty promises made by the upper management of Hunnia and the MFI regarding enforcement of the Jewish Laws, antisemitic film personnel would express dissatisfaction by personalizing the struggle. The mid-1941 filming of *Girls' Fair* [*Leányvásár*] gave rise to what rightist circles viewed as a *cause célèbre*. Noticing the producer Ferenc Pless at Hunnia during filming,

Figure 10: Zita Szeleczky, still from *One Night in Transylvania*. Szeleczky was one of the leading advocates of the anti-Jewish purge of Hungarian film. Source: MaNDA, Budapest.

the actress Zita Szeleczky, one of the starlets of the late 1930s/early 1940s and a reputed member of a ring of radical actors associated with Ferenc Kiss, allegedly threatened that 'as long as there is a Jew in the studio, we will not proceed'.[69] Pless, who through 1941 worked semi-secretly for the Imágo film company and who may have been the actual producer of *Girls' Fair*, denied the confrontation ever occurred.[70] Whether imagined or not, this vignette illustrates the degree of frustration the right felt well into the early 1940s.

Undermining the Jewish Laws: The Return of István Gerő

In matters of film and culture, nothing irked Christian nationalists and the radical right more than the continued presence of the supposedly deposed Hungarian film mogul, István Gerő. For antisemites, Gerő was an icon, the burning symbol of the parasitic semitic capitalist scourge Hungary could not shed. Since 1938, Gerő's empire had been besieged. According to film magazine editor György Guthy, the profession had made great progress building a Christian national foundation simply by heavily restricting Gerő's activities.[71] Forced to sell his theaters and most of the stock in the film businesses he himself had founded, Gerő could have retired and lived off his wealth. Yet he chose not to, continuing to find ways to exercise influence over Hungarian film both within Hungary and without. *Magyar Film* protested Gerő's attempts to organize the theaters from the Uplands region returned to Hungary with the 1938 Vienna Award.[72] Gerő was one of the leading figures who raised cash for the new 'Christian' filmmaking company Films of Hungarian Writers [*Magyar Irók Filmje, Rt*].[73] The Hunnia Directorate discussed Gerő's activities at its January 1940 meeting, acknowledging that although Gerő's participation was not overt, he 'arranges all sorts of things' behind the scenes.[74] *Nemzetőr*, an antisemitic and anti-liberal paper, asserted in July 1940 that Gerő had secretly reassembled the movie empire he had been forced to surrender in mid-1939, making a mockery of the Jewish Laws. The article alleged that the 'cunning Jew' in the background won permission to show films using Christian workers and intermediaries, and thus maintained covert contracts to supply 11–12 theaters. *Nemzetőr* charged that Gerő's power to control film production and exhibition was little diminished from 1938, and menacingly concluded that public opinion demanded Gerő once-and-for-all vanish from the Hungarian film business.[75]

Whether or not the allegations were true, it is undeniable that Gerő retained a hand in the Hungarian film trade well into 1941. Gerő was protected from on high, not only by Interior Minister Ferenc Keresztes-Fischer and MFI chief Miklós Kozma, but also by his friendship with and business connections to some of the most important political figures in Hungary, including the son of the Regent, Miklós Horthy Jr. The latter was, in fact, president of the board of directors of Gerő's Moving Picture Company [*Mozgóképüzemi Rt.*] and maintained contact with Gerő long after the mogul was forced to divest himself of the company.[76]

In late 1940, while Gerő's position within Hungary remained tenuous, he discovered a new way to make himself an irreplaceable cog in the movie machine, through distribution of Hungarian works outside of Europe. The onset of war caused massive transportation problems, declines in film production, and a general shortage of film throughout the world. American film officials, impressed with the quality of Hungarian film and the success Hungarian films were having in the Balkans, began expressing interest in Hungarian products. Miklós Horthy Jr., then ambassador to Brazil, returned to Budapest for the holidays and commenced discussions with industry leaders concerning the export of Hungarian film beyond Europe. Because the global structures of film distribution were, in the words of Horthy Jr., 'unfortunately' Jewish, he turned to the well-connected Gerő. Gerő devised a scheme whereby United Artists would distribute Hungarian films in the 43 countries in which it currently operated. If the plan succeeded, wrote an anonymous author who drafted a memorandum detailing the discussions, the best Hungarian movies would earn 'many hundreds of thousands of *pengő* abroad…'[77] Gerő's involvement, however, prevented the export proposal from garnering universal support. The State Secretary from the Ministry of Manufacturing objected, bemoaning the excessive role Jews played in film export. These objections aside, the memorandum's author claimed that there was no film person, 'neither old nor new', who had the network Gerő did or who could carry out such a project.[78] Even Bálint Hóman offered reluctant support for the Gerő design, remarking that he would not stand in the way of the endeavor but that the concept should remain confidential.[79]

Other National Film Committee documents indicate the original plan called for Gerő to contact a Mr. Kastner, the United Artists manager in Lisbon, bringing with him five of the best Hungarian films.[80] In mid-January, Gerő actually signed a contract with both Hunnia and the MFI, with the blessing of the National Film Committee and the Ministry of Religion and Education, to expand export possibilities for Hungarian film abroad.

Soon after, Gerő requested and received permission to travel abroad. On 22 January 1941, the ONFB informed the EMK's office about the contract and about Gerő's request for selections of the best Hungarian films to carry to Lisbon.[81]

Whether Gerő and Kastner met is uncertain, but Gerő took advantage of his permission to travel, opportunistically hawking Hungarian wares in several countries. In mid-February he set out for Lisbon, choosing the circuitous route of traveling via Belgrade. Once in a then peaceful Belgrade, Gerő optimistically arranged for continued travel through Western Europe and America. He also considered buying a theater and setting up a central distribution point for Hungarian film in Yugoslavia.[82] A month later, conditions had changed. Stranded in Belgrade with visa difficulties, Gerő wrote to László Balogh, the second-in-command on the National Film Committee. In desperation he begged for Balogh's aid, volunteering to make additional stops in Bulgaria, Turkey, Syria, Iraq, Greece, and North Africa on his way to Lisbon and eventually to North and South America.[83]

Only a day before receiving this letter from Gerő, Balogh had written to István Kultsár, detailing the worldwide Hungarian film distribution plan hatched by Horthy Jr., Gerő, and, according to Balogh, the Jewish former top representative of MGM in Hungary, Károly Guttmann. Balogh confirmed that Hunnia and the MFI had agreed to participate in a cooperative export pact and that the head of the National Film Committee, Baron Gyula Wlassics Jr., supported the proposal. Neither Károly Kádas of the Minister of Manufacturing, nor Bálint Hóman raised official objections. Several of Kultsár's underlings, personally informed of the arrangements that had been made with Gerő, remained mum. In other words, men high in the government bureaucracy were not only informed, but backed the idea of a Jewish-led effort to increase the global distribution of Hungarian film. Preempting the Commissioner's opposition, Balogh argued that the Gerő machinations were beyond Kultsár's jurisdiction. He also reminded Kultsár that Kultsár himself had admitted 'this man [Gerő] knew the rules of trade in Hungary better than anyone else, and it was [unfortunately] a tragedy for him that he was Jewish.'[84]

Neither the right-wing press nor Kultsár would be dismissed that easily. Word of Gerő's plan leaked in mid-March, prompting demands that Kultsár respond.[85] Kultsár bided his time and when he replied in mid-April, stated his opposition in stark terms. Rejecting any hint of official government sanction of Gerő's cross-continent jaunt, Kultsár tagged Gerő as the primary obstacle to the Christianization of the film profession. He catalogued the numerous on-going legal proceedings against the dethroned czar of the movie business for flaunting the Jewish Laws. Kultsár concluded that 'he believed there to be no reason whatsoever to have sent Gerő abroad and from the perspective of the Jewish Law it was undesirable.'[86]

While Kultsár's staunch opposition of Gerő's activities undoubtedly contributed to the general failure of the Gerő mission, what truly thwarted Gerő was a combination of geopolitical and domestic circumstances. Stranded in Yugoslavia because he was unable to procure the appropriate travel visas, Gerő was forced to abandon plans for his worldwide expedition when Yugoslavia's military launched a coup in late March. He somehow returned to Budapest before 2 April 1941, when Hungary joined the German invasion of Yugoslavia and entered World War II. The destruction of Yugoslavia was the beginning of a downward spiral for Gerő, who by mid-1941 saw his influence drastically curtailed and his businesses looted. First, the government forced him to sell the majority of his Moving Picture Company to those with certified right-wing credentials, including Miklós Mester.[87] Second, authorities began to enforce Prime Ministerial decree 5.660 of August 1940, which declared that partnership agreements with Jews must cease by 31 July 1941, closing the loophole which had allowed wealthy, influential 'film Jews', albeit in limited numbers, to maintain their involvement in movies. The decree's activation coerced Gerő to officially withdraw from all cooperative agreements with theaters he had previously operated or for which he had done programming.[88] Many more Jews also suffered great losses, as partnership arrangements involving cinema ownership, management of distribution firms, and financial support of movie projects were now prohibited.

The Commissioner for Unemployed Intellectuals and the Antisemitic Crackdown

The July 1941 enforcement of Decree 5.600 and additional legislation requiring Jewish cinema license holders to turn over their licenses to Christians constituted a direct assault on the foundations of the 'strawman' system, the means by which Jews remained active participants in many Hungarian businesses.[89] Whether these anti-Jewish measures had ephemeral or long-term impacts depended upon the degree to which the Commissioner for Unemployed Intellectuals, István Kultsár, chose to enforce them. Pressure on Kultsár to launch a final attack on Jewish interests had been building throughout 1940 and the first half of 1941. During that time, which presaged promulgation of the draconian Third Jewish Law on 2 August 1941,[90] frustration with the ineffective implementation of the Second Jewish Law smoldered on the right and antisemitic sentiment spread.[91] Within the government, Bálint Hóman called for a 'more radical line regarding the Jewish question'.[92] The Party of Hungarian Renewal [*Maygar Megújulás Párt*], headed by the powerful former Prime Minister Béla Imrédy, demanded that 'the Jewish question be solved in such a way that Hungary is completely liberated from the Jews and the Jewish spirit [*szellem*]'. Point 14 of the party's program stated specifically that 'Jews must be expelled from Hungarian cultural life'.[93] German pressure to act more forcefully against the Jews likewise increased, particularly after Hungary joined the Tripartite Pact in November 1940.

The rhetorical bombast relating to the Jewish film question reached new heights in early February 1941. Hundreds of Hungarian theaters played *Jud Süß* to packed houses,[94] and the movie's popular reception emboldened *Magyar Film* and the papers to more assertively push the EMK to act decisively and strictly enforce the Jewish Laws in the film world.[95] Also encouraged was the flame-throwing fascist Lower House parliamentarian and future Arrow Cross Minister of Culture, Ferenc Rajniss. In a parliamentary hearing, Rajniss savaged the MMOE for protecting Jewish interests and for exercising impermissible influence on Hungarian cultural politics. He directed similar venom toward the OMME, accusing the producers and distributors of thwarting the development of a state-directed Christian national film industry. Rajniss then condemned the Minister of the Interior, Ferenc Keresztes-Fischer, for not acting more vigorously to strip all Jews of licenses to operate cinemas. He completed his tirade by demanding that all film matters be transferred from the Interior Minister's portfolio to that of the more antisemitic Minister of Religion and Education.[96]

Interior Minister Keresztes-Fischer responded forcefully to Rajniss' provocations on 12 February 1941. He claimed that of the 84 cinemas in Budapest, only 13 remained officially in Jewish hands. Covert partnerships may exist, he admitted, but of the 13 theaters run by or licensed to Jews, all had been legally exempted from the Jewish Laws, mostly for distinguished service or wounds received in the Great War. Refusing to punish either the MMOE or the OMME, Keresztes-Fischer again demonstrated his distaste for legal discrimination against Jews. He told the Lower House it was his obligation to enforce the law 'honorably and enthusiastically' while ensuring his actions did not 'transcend the law', eliciting rounds of applause from the government party.[97]

Keresztes-Fischer and other moderates like him, however, were swimming against the tide. When war with the Soviet Union broke out in June 1941, as Hungary considered its Third Jewish Law, antisemitic feeling crested. László Bárdossy, who had replaced Pál Teleki after the Prime Minister committed suicide in April, favored a more stringent application of the Jewish Laws. Therefore, in early August 1941, just days after the partnership and license forfeiture decrees came into force, István Kultsár recognized he finally had the tools to ratchet up his anti-Jewish crusade, including growing popular support and strengthened strategic allies inside the government. He commenced an offensive which would rattle the industry and mark the nadir of antisemitic excess in film prior to the Nazi occupation of Hungary in 1944.

Since early 1940, Kultsár had employed a force of detectives to investigate whether companies were adhering to applicable Jewish regulations. His men appeared in offices and homes, seized financial papers and contracts, and sorted through sources of capital and actual business hierarchies. They collected files on numerous film companies and dissolved several firms that ran afoul of the Jewish laws. In June 1941 his office began compiling statistics on Jews in all industries, and in August, Kultsár decided to go after the big fish in the culture industries. Regarding film, he set his sights on the main professional organizations. The OMME and the MMOE had been cooperating ever more closely in order to resist the radical pressures emanating from the Film Chamber and Kultsár's office. They deflected Kultsár's questions and impeded his efforts to collect data on Jews in film. The frustrated Commissioner then dispatched his men directly to production companies and the MFI and Hunnia studios, intent on finding those producers who collaborated with Jewish screenwriters.

Unable to win full cooperation 'to liberate the activities of the MMOE or the OMME from the destructive influence of the Jews', Kultsár took unprecedented steps.[98] In early August he assembled the industry's leadership. Accompanied by his assistant László Répássy and Ferenc Kiss, Kultsár addressed MMOE Vice-President Pál Morvay; OMME chiefs Frigyes Pogány and István Erdélyi; the editors of *Magyar Film* Géza Ágotai and Dezső Váczi; László Balogh; János Bingert; and a handful of other powerful film figures. He issued this ultimatum: Hungary's 35 production and distribution firms had 30 days to sever their contacts with Jews who directed or assisted in the direction of distribution departments.[99] Companies which failed to comply would be sanctioned, broken up, or dissolved. The allegations made by Kiss, Balogh, and Kultsár were quite specific. They charged that 'Jews write nearly every film', and Kiss fingered Erdélyi individually and the OMME in general as protectors of these screenwriters. Erdélyi claims he mustered the courage to issue a rejoinder, warning that the film industry could not be transformed overnight and that Jewish colleagues were necessary as a 'crutch' [*mankó*] to support the profession.[100] Still, Kultsár, Kiss, and their coterie, buttressed by Arrow Cross support in Parliament, bullied the OMME into taking a more antisemitic public position and adopting the following statement:

> Any Jewish employee taking a leading role in the production, sale, and distribution of film (as film salesperson, head of distribution department, film buyer, and employees immediately below them [agents or emissaries]) as well as dramaturges and officials operating as in the service of publicity chiefs shall be immediately dismissed and the Prime Minister shall be informed by writing of these actions.[101]

In addition, Kultsár demanded the OMME and MMOE immediately survey their members regarding relations with Jewish companies. OMME members, for example, were asked to reveal whether they used news or advertising services linked with Jews. They were compelled to name the Hungarian Jews engaged in export and import. Finally, they were questioned about which production firms were actually owned by Jews and in what capacity these Jews worked.[102] Likewise, Kultsár issued a stern warning to the MMOE. Those recognized as Jews, he reminded the theater owners and license holders, could not be involved in determining programming or directing film provision for cinemas, nor could Jews be active advertisers. In no cases, he asserted, could Jews have an advisory role, as this would allow them to exercise influence over Hungarian culture. Kultsár intimated that he had 'knowledge that Jews, [under the guise of being] lawyers, coatroom clerks or buffet attendants, bookkeepers, private secretaries or other employees, have illegal influence in the preparation of cinema programs for certain companies', and that this must cease at once.[103]

Kultsár's detectives visited numerous firms between mid-August and November 1941. In the course of these 'professional patrols' Kultsár himself dropped in on film companies to personally determine if they were abiding by the Jewish laws, checking names and baptismal records.[104] It soon became clear to Kultsár that the film industry continued to play fast and loose with anti-Jewish regulations. Dissatisfaction inspired him to enact a purge campaign that went well beyond the letter of the Jewish laws and sent shock waves through the industry. Between October and December 1941, Kultsár's office began blackmailing American film companies with Budapest offices.[105] Warner Brothers/First National, for example, was told that if it did not immediately dismiss its Jewish bookkeeper Gyula Kastner, it would be refused licenses to import its films into Hungary and denied approval for those films presently before the Censorship Committee.[106] In November, the Commissioner made his most dramatic move, ordering the arrest of seven leading Jewish figures: dramaturge, advertiser, and postwar director Béla Pásztor; scriptwriter Andor Klein-Kolozsváry; Lena Film Company's distribution head Károly Öhler; chief film distributor for the Mária Hausz Film Company, Mrs. Tódor Várdai; dramaturge, advertiser, and former film journal editor István Radó; advertiser András Brody; and former film journal editor and Olympia theater director István Borhegyi.[107]

A few days later, the producer Géza Bolgár-Csöröghy was also interned. These arrests, Kultsár told the National Film Committee, were carried out with the approval of Prime Minister László Bárdossy.[108] Another round-up soon followed, one which targeted other prominent and respected Jewish professionals. On 15 December, police interrogated and detained world-renowned screenwriter Károly Nóti; producer and agent János Smolka; former Fox Films representative Sándor Ungár; former Fox Budapest branch chief Károly Matzner; and Andor Verő, an Atelier Film Company partner. Officers also issued a warrant for producer Ferenc Pless, who turned himself in on 19 December.[109] In the span of a month, authorities jailed 15 of Hungary's most admired, longest-serving Jewish movie professionals, sending a sharp, clear message to the industry. 'Business as usual' was no longer acceptable.

Magyar Film sheepishly applauded the EMK's resolute actions, embarrassed that the industry it represented had not, on its own, replaced its Jews. 'Unfortunately, we lied' when we claimed we could solve the Jewish problem without the interventions of Kultsár, wrote the editors. As a result, this 'radical solution', though distasteful, was necessary.[110] Invigorated and determined to put teeth into his anti-Jewish program, Kultsár took his campaign to government offices. No one in the film industry, not the Hunnia Directorate, not even the *éminence grise* of the ONFB, Baron Gyula Wlassics Jr., escaped Kultsár's wrath. The Commissioner scolded Wlassics for his committee's feeble enforcement of anti-Jewish measures.[111] Kultsár's subsequent actions indicated that he and his staff did not trust Wlassics to follow through. Thus, Kultsár demanded the ONFB provide him earlier notification about filmmaking permits it granted in order to allow him more time to investigate production company compliance with the Jewish Laws. Although he did not occupy an official position on the ONFB, Kultsár and/or his deputies started to regularly attend ONFB meetings, squashing proposals which dared to incorporate Jews in any creative capacity.[112]

After jailing Jews and pressuring the ONFB, Kultsár's assertion of authority continued at the December meeting of the Hunnia Directorate. He muscularly rephrased the message he delivered to Wlassics, telling his colleagues that 'the most recently completed investigations reveal that production and distribution firms have not enforced the firing of banned Jews, indeed in this area they are guilty of negligence and violations of the law'.[113] Kultsár's solution was a Nazi-style centralization and nationalization of the Hungarian film industry, a suggestion which provoked strenuous objections. Undeterred, the Commissioner warned:

> As the great significance of film production and film trade day-to-day becomes more manifest, the government cannot close itself off from consideration of this question [of government directed and controlled film production]...There is no argument that prohibited Jewish elements want to take part in secret, or that they wish to damage and endanger [the film industry] by permeating it with their pliant, non-Christian based business intellect. The cleansing along so-called Christian lines must go forward.[114]

Hunnia's János Bingert replied that in principle, his firm had no issue with central control over production and distribution, shrewdly suggesting it would be possible if Hunnia possessed a monopoly over the Hungarian film market. Creating such conditions, however, would have enormous costs, such as putting 'a multitude' of companies out of business and many people on the streets. László Balogh, surprisingly, unconditionally opposed total centralization, supporting the less drastic step of culling the number of production firms while keeping those 'worthy Christian producers', who 'have demonstrated that they know how to create good and valuable [films] with a suitable mentality'.[115] The discussion closed with Kultsár's exhortation that Hunnia be more industrious in its enforcement of the Jewish Laws, but with no firm policy changes.[116]

Whether Kultsár's onslaught was prompted by a nudge from the Prime Minister's office or a nod from the Ministry of Religion and Education, or whether Kultsár acted on his own volition is not known. Whatever the impetus, in late 1941, the EMK became a steady and somewhat unwanted presence alongside the National Film Committee, objecting to any whiff of Jewish involvement in the manufacture or sale of film.[117] Kultsár's actions and the responses to them among various industry circles point toward two broad conclusions. First, although Kultsár's assault did not mean the absolute end of Jewish involvement in Hungarian filmmaking, it marked a new stage. Jewish participation in filmmaking was now severely curtailed. Second, reactions among Hungary's film elites to Kultsár's strenuous efforts to de-Jewify and potentially nationalize the industry demonstrate most recognized the dangers of carrying such actions to extremes. Even the Hunnia Directorate, composed as it was of a horde of government bureaucrats masquerading as movie-makers, realized that complete nationalization would be a disaster. A system in which the state decided which movies to make, whom to cast, and how they would be distributed and exhibited, declared most of the nation's leading film figures, was wrong for Hungary. Backward or not, the gentry capitalists who ran Hunnia and who constituted the ONFB's majority made an important stand in December 1941. Even while promising to more effectively weed out Jews, they blocked the more immoderate position held by Kultsár and his cronies. In doing so, they preserved a degree of autonomy for themselves and preserved a role for private capital in the business of making films.

There is delicious irony in the fact that Bingert, Balogh, and Wlassics, the triumvirate of government operatives whose task was to shape and control the images Hungarians consumed of themselves and of others, were the steadiest voices against a complete government takeover. In speaking out against Nazi-like nationalization, they recognized that private capital was necessary to keep Hungary's film culture vibrant. This entailed reluctant acceptance that for the foreseeable future, Jewish capital would continue to fund Christian Hungarian film. Surely, they found this situation 'incorrect' and its cessation desirable, but they acknowledged in December 1941 that 'to this day, Jewish money takes part in paying the costs of Hungarian film production.'[118] On a regular basis, ONFB members spoke of the pressing need to 'rid Jewish capital from film production' yet passed the buck to other bureaucrats.[119] They routinely requested that the EMK or the Treasury Minister establish a film bank so that Jewish capital would not be required, but in wartime and cash-poor Hungary, such proposals never yielded fruit.[120] Into 1942, the key film industry policymakers committed to the rhetoric of antisemitism and greater state involvement in film, yet not to zealous antisemitic practices. The Christian national film industry envisioned by Bingert, Balogh, and Wlassics was neither completely Jew-free nor completely dictated by the state.

What developed in the aftermath of the Kultsár offensive was a more discreet business, one chilled by the arrests of its members but not entirely frozen. The Hungarian film world remained a safe haven for an important, but dwindling group of Jews. Protected by their connections on high, these individuals stayed active in the background as silent production partners, as ghost-writers of scripts, as distributors, as film advertisers, and occasionally as members of a film's cast or production personnel. János Smolka, for example, was one of the Kultsár detainees to win freedom, released in February 1942 with help from Ferenc Herczeg and Ferenc Keresztes-Fischer. Herczeg, then perhaps Hungary's most famous playwright, an occasional film writer, and a member of the Upper House of Parliament, had worked with Smolka and urged the Interior Minister to free him.[121] Keresztes-Fischer did so, despite the extreme right having labeled Smolka the film profession's 'third-ranked public enemy'.[122] Both András Brody and Károly Nóti stayed employed even while in jail, both working for István Erdélyi. Brody arranged film deals, while Nóti, the writer *extraordinaire*, conceptualized movies and wrote scripts from his cell.[123]

But by 1942, these were the rare exceptions. The other interned film professionals languished in jail between six weeks and three years.[124] Hundreds of other Jewish film professionals were now on the streets or forced into labor service, their businesses abandoned and their personal safety at risk.[125] The prevalence of antisemitism and simmering political pressure to de-Jewifiy meant that while there was still room for public objection to antisemitic measures, the space was shrinking. This atmosphere forced OMME Secretary István Erdélyi to resort to backhanded opposition to Kultsár. At the 21 December 1941 annual meeting of the OMME, Erdélyi promised his association would pay more attention to enforcement of the Jewish Laws, yet he assailed Kultsár's 'excessively strict' steps as counterproductive. In Erdélyi's estimation, the forced dismissal of Jews from distribution companies, rather than ending the strawman system, perpetuated it, as many of those fired were illegally rehired due to an industry-wide need for expertise.[126] Critiques such as Erdélyi's, who paid lip service to Christianization while continuing to pay several Jews Kultsár had interned, still had a place in Hungary in 1942. Unabashed public defense of Jewish culture producers, like that heard in some quarters in 1938, no longer did.

Conclusion

From crisis came talk of the rebirth of the Hungarian film industry, minus the albatross of the Jews. Yet the reality was that, as Europe descended into all-out war in 1942, none of the industry's problems were resolved and divisions remained as wide as ever. The industry had failed to train replacements for the Jews it cast out. Film production still rested on a very fragile financial base. Neither of the primary agents of the Christian national renaissance, the Film Chamber and the ONFB, proved to be the vehicle for harmonizing the film world. Fatally handicapped by ideological blinders and blocked by the continued independence of the profession's two major associations, the MMOE and OMME, the Film Chamber collapsed upon itself, serving as little more than a gathering place for industry firebrands. The ONFB, perhaps the one organization which could have transformed the film industry, instead became a brake on radicalism.

By the end of 1941, these state institutions provided homes for two distinct mentalities. On the one hand, centered in the Film Chamber and allied with Kultsár, figures in the Imrédy party, and the fanatical right were those who accepted the essentialist myth of Jewish difference. They uncompromisingly believed that the only way to insure the future viability of Hungarian film and to build an authentic Christian Hungarian culture was to excise 'film Jews' and the strawmen who hid them. In their way stood a circle of mostly long-term government officials and industry insiders, many of whom held financial stakes in Hungary's film businesses, especially its studios. These men, often found on the ONFB and the Censorship Committee, and on the boards of Hunnia and the MFI, were generally content with the system that evolved in 1941 and were unwilling to permit drastic alterations to it. While few were vocal opponents of antisemitism, some did view racial antisemitism as repugnant, while others were inconsistent or outright hypocrites. If they deserve any credit at all, it is that they prevented Hungary's film world from experiencing a total, Nazi-style *Gleichschaltung*, or synchronization of ideology and practice, which would have meant the end for all Jewish film interests.

Despite Hungary's inability to *gleichschalten*, the mere creation and presence of structures such as the Film Chamber, the ONFB, and the EMK did shift the paradigm for Hungary's Jewish film professionals. Unlike the late-1800s, when antisemitism was not tolerated by Hungarian authorities, in the 1920s and especially the late-1930s, the Hungarian government itself took steps which insured that antisemitism would became a constituent part of all state-supported concepts of Hungarian national identity. For the film profession, the founding of the Theater and Film Chamber and the

ONFB established institutional bases for exclusion. Through these institutions, the state not only sanctioned antisemitism, but legitimized its escalation. The Film Chamber and its sister chambers deprived thousands of Jews of their livelihoods, made it nearly impossible for Jews to overtly contribute to Hungarian culture and society, and reaffirmed the Jew's pariah status. This became especially apparent in late 1941, when Hungary expanded its racial definition of who was a Jew with the passage of the Third Jewish Law and unleashed the Commissioner for Unemployed Intellectuals' purge of Hungarian culture. Once state structures, legislation, and actions confirmed Jews to be outsiders with little social utility, they ultimately, even if unwittingly, helped to smooth the path toward elimination.

Yet those same state structures could provide protection and provide a check when conditions seemed ripe for radicalization. Just as the Chamber's efforts to exclude Jews belittled them and Kultsár's treatment of some of the industry's most respected Jewish figures dehumanized them, the National Film Committee's episodic willingness to accept Jewish participation in filmmaking affirmed the value of Jews to Hungary and its culture. A modicum of moderation could also be seen in the Interior Minister's efforts to restrict the impact of the anti-Jewish laws and to occasionally intervene on behalf of Jewish film figures. The private professional organizations were even more active and vigorous in the defense of the human and economic value of Jews in film, even if their motives were primarily material. The continued existence of professional organizations, and the devolution of power they represented, brought a degree of rationality and stability to the industry. In resisting the zealots, government and industry leaders grudgingly accepted the economic realities of the late 1930s and early 1940s. They realized it was impossible to remove risk and speculation, the so-called 'Jewishness' of capitalism, from a creative economy. This meant that their *pengő*-poor and bureaucracy-rich state had to compromise its Christian national dogma in its attempt to establish a financially viable Christian national system.

Notes

1. Pál Morvay interceded on behalf of the MMOE to retain Lajta as *Filmművészeti Évkönyv* editor. *MOL* – BM, K158-cs10, OMB, 1940. Alapszám 375/1940. MMOE [Morvay] to OMB [Szöllőssy], 14 October 1940.
2. Andor Lajta, 'Az év története', *Filmművészeti Évkönyv* (1942), 177, 179.

3. As late as 1941, Kozma argued that radical solutions to the 'Jewish Question' needed to be delayed until the war's end. There is, however, a massive exception to this generalization about Kozma's stance as an obstacle to radical action toward Jews: his involvement in deportations which resulted in the mass murder of over 19,000 Jews when he was Plenipotentiary of occupied Carpathia in late summer 1941. That said, Kozma's defense of 'film Jews' was not a complete shock, as rumors about his family origins and disappointment regarding his antisemitic actions led to his being 'detested' by the radical right. See Macartney, *October Fifteenth,* vol. I, 105.
4. Kozma, as previously indicated, opposed 'deposing' Gerő. His role in trust busting was to insure Gerő's soft landing and prevent Gerő's complete expulsion from the film business. Langer, 'Fejezetek', 202–3.
5. Ormos, *Egy magyar mediavezér,* 585.
6. *MOL* - KMI, K429-cs59-t2. Vegyes MTI levelezés, 1938–9, 31. Kozma to Imrédy, 9 January 1939. Enclosure 'Intézkedések a magyar filmgyártás érdekében', Budapest, 4 January 1939.
7. *MOL* - KMI, K429-cs59-t2. Vegyes MTI levelezés, 1938–9, 42–9. Kozma to Keresztes-Fischer, 2 March 1939.
8. *MOL* - KMI, K429-cs45-Napló II, Bejegyzés. Conversation with Miklós Horthy, 17 January 1939.
9. Kozma quote recounted by Zoltán Tőrey. *BFL* - Magyar Hiradó, Magyar Filmiroda Rt., XVII 789/6 - 399 A, Tőrey [Taubinger] Zoltán, 'Tárgyalási Jgykv' Felvéve 1945 November 14, A Magyar Központi Hiradó Rt. /ezelőtt Rádió és társvállalatai/ -hoz kiküldott IB.
10. *BFL* - Tőrey Zoltán, 8222/1945 *(found within Bánki Viktor file, coded Nb.2540/1945).* Tőrey and the MFI were criticized in a 14 May 1944 *Magyarság* article for working with Jews such as Gál, the director Márton Keleti, and the actress Ella Gombaszögi; for keeping on Jewish secretaries; for working with Jewish advertisers; and for mocking the Szeged idea while 'becoming filled with Jewish film and Jewish business sense'. All occurred under Kozma's watch.
11. Langer, 'Fejezetek', 202–3. The MFI also received 20-year extensions on the terms of the licenses. These were the same theaters originally leased to Gerő by the Christian National League.
12. Ormos, *Egy magyar médiavezér,* 587. This essentially racist position allowed Kozma and Horthy to be party to the murders of thousands of Galician Jews in Carpathia, where Kozma served until his death in 1941. Horthy's stance permitted Hungarians and Nazis to combine forces and eliminate over 450,000 Jews in 1944.
13. Ablonczy, *Pál Teleki,* 181–2. Imrédy, ironically, resigned in disgrace when opponents discovered 'Jewish blood' in his family tree.
14. The Royal Trust was first divided in July 1938, and further reduced in 1939. György Guthy, 'Miért szünetel a magyar filmgyártás?' *Mozi- és Filmvilág* III/6 (1939 June), 1–2.

15. It is worth noting that in July 1938, when Gerő's Royal Trust was first broken into three parts, Morvay benefited substantially. He was placed in charge of providing the programming for Budapest's top second week theaters: the Lloyd, Simplon, Corvin, and Élit. Most likely, this required Morvay to work with Gerő. It is possible this new arrangement softened Morvay's stance regarding Jews in the profession. He warned of the dire effects far-right radicalism and the lack of Christian talent needed to replace Jews in September 1938. Pál Morvay, 'Az uj évad küszöbén', *FK* XI/9 (1 September 1938), 12–13.
16. Ödön Ruttkay and Pál Morvay, 'MMOE hivatalos közlemények', *MF* I/4 (11 March 1939), 5.
17. Morvay's concerns were well-founded. He himself took over control of the previously Jewish-owned Hungarian Film Contracting Company, and would have been intimately aware of the threat of legal uncertainty. *BFL* - Magyar Filmbeszerző Rt - File Cg15639.
18. Not only would Hungary's exhibition system suffer from a fundamental reexamination of all licenses, claimed MMOE vice-presidents Ruttkay and Morvay, it would kill the industry. If frightened theater owners would not invest anew, they would not conclude deals with distributors. If distributors could not distribute, they would not fund films, and filmmakers, it followed, would be unable to produce. Freefall would ensue. See *MOL* - VKM, K507-cs91-t12, 1937–44: Filmügyek. Ruttkay and Morvay, to Hóman; and 'MMOE hivatalos közlemények', *MF* I/4 (11 March 1939), 5. Miklós Kozma made a similar argument in the Upper House of Parliament during April 1939 discussions concerning the text of the Second Jewish Law. Concerning paragraph 12, which would have mandated all cinema personnel be members of the Film Chamber, Kozma explained that such a mandate would harm 'Christian interests'. The constant threat of license review and possible revocation would turn the spigot of investment in theaters, barely a drip at the time, off entirely. Kozma quoted in 'Módosította a felsőház a 12. Paragrafust', *MF* I/10 (22 April 1939), 1–2.
19. See numerous early March 1939 articles on theater membership in the Chamber in *Magyar Film*.
20. 'Választmányi ülés', *MF* I/10 (22 April 1939), 7. What Bornemisza seemed to suggest was that his organization saw a comprehensive licensing examination as bureaucratic overreach, and that the Jewish Laws were designed to eliminate large societal imbalances, not every exchange between Jew and non-Jew.
21. Pál Morvay, 'A moziegyesület és a zsidótörvény', *MF* I/5 (18 March 1939), 5.
22. Pál Morvay, 'A zsidótörvény végrehajtása és kijátszása', *MF* I/25 (5 August 1939), 6–7.
23. Thomas Sakmyster, *Hungary's Admiral on Horseback. Miklós Horthy, 1918–1944* (Boulder, CO, 1994), 207–36.

24. Even antisemites who chided the Interior Ministry as the 'helping hand' that had allowed trusts to persist, feared that an ill-prepared plan to review and re-issue cinema licenses could result, as it had in the early 1920s, in disaster. Cash-poor and ill-trained cinema owners might be forced into strategic partnerships with Jews and/or trusts or into bankruptcy. See 'A Filmkamara a trösztkérdés megoldásáért', *MF* I/9 (15 April 1939), 1–3.
25. C.A. Macartney labels Keresztes-Fischer, Interior Minister from May 1938 through March 1944, the 'bulwark of the "Liberal-Conservative" tendency' in the upper echelon of ruling circles, the 'defender in chief of the Jews' and the 'best-hated figure in Hungary to the Germans and the Arrow Cross'. He was arrested and interned by the Germans when the Nazis occupied Hungary. Macartney, *October Fifteenth*, vol. I, 104–5.
26. *MOL-Ó* – Filmipari Alapot Kezelő Miniszterközi Bizottság, 1926–40. Z1129-r1-d1. Ülési jgyek., 1938. Jgyek. a FAKMB-nak a Hunnia Filmgyár tanácstermében, 13 June 1938.
27. 347.347–1939 B.M. sz. required that licensees involved in partnerships concluded before 1936 have their contracts examined. Those partnerships involving Jews and not exempted by other portions of the decree would lose their licenses. In addition, as of 14 December, no Jew could receive a new theater license. However, partnerships involving Jews or Jewish capital, which were scheduled to be in effect beyond November 1940, were not subject to review and were permitted to continue to exist. As such, those partnerships were able to retain and even extend the length of their licenses, at least through July 1941.
28. Sándor, *Őrségváltás után*, 105–6. Bencs was an advisor to the Prime Minister on social and cultural politics. Under Prime Minister Teleki he headed the office of social politics.
29. This included all producers, screenwriters, production personnel, actors, and cinema managers.
30. 'A Filmkamara átvette az OMME munkakörét', *MF* I/10 (22 April 1939), 2–3.
31. 'Az OMME beolvadása a Kamarába. A filmkölcsönzők egyesületének nyilatkozata', *MF* I/11 (29 April 1939), 4.
32. See Erdélyi's speech, quoted in 'Az OMME közgyűlése', *MF* I/20 (1 July 1939), 8.
33. 'Az OMME közgyűlése', *MF* I/45 (23 December 1939), 3, 6. Speeches of Pogány and Erdélyi. Pogány chose to 'protect himself' by keeping his Jewish partner, Dr. Richard Horovitz, as the business head of Pogány's company, Müvészfilm.
34. 'Az OMME közgyűlése', *MF* I/45 (23 December 1939), 3ff. Throughout his speech, Erdélyi references the Jewish question and emphatically states that the excessive enforcement of the Jewish Laws as advocated by the Film Chamber and *Magyar Film* would undermine the goal of building a strong Christian national industry.
35. 'Egységes szakmai irányítást kérünk', *MF* I/35 (14 October 1939), 2.

36. 'A Filmkamara állásfoglalása a külföldi filmbehozatal kérdésében', *MF* I/38 (4 November 1939), 2–3.
37. István Erdélyi, 'Az OMME, a Filmkamara és a külföldi filmbehozatal', *MF* I/39 (11 November 1939), 4.
38. Atelier, run and likely owned by MMOE president and prominent rightist Gábor Bornemisza and István Szentpály, a founder of the radical right newspaper *Virradat*, was for a short time one of Hungary's largest distributors. It provides an example of how radicals did attempt to reshape the film industry. However, once there was competition from other Christian firms, Atelier faded.
39. For individual companies, see Ábel, 'Magyar filmgyártó…' and Lajta, 'A magyar film története', 176–96.
40. *BFL* –Hunnia Filmgyár Rt, XVII 781–2, 395 B, file L-Z, Nagy Sándor documents. Nagy to I. Antal, 20 May 1942. Nagy estimated there were three or four Christian distribution agents and three Christian distribution department heads in September 1938. Other *BFL* documents condemn Nagy as a fascist, pro-German, blackmailing, lying, Jew-hater who was Kultsár's stooge at Hunnia. *BFL* – Dr. Nagy Sándor – Nb 3482/1946.
41. *MOL-Ó* – Hunnia, Z1123-r1-d1. 1938. 6 September 1938 Hunnia Igazg. jgyek, minutes 2. Nagy allegedly paid his Jewish employees 'starvation wages'. *BFL* – Dr. Nagy Sándor – Nb 3482/1946. Sándor Gárdonyi testimony, 'Tanúvallomási jgykv' 4 May 1946, 14356/1945.7/3–34.
42. *BFL* – István Kulcsár [Kultsár], Aü.83377/1949. Kultsár to Pál Teleki, 23 December 1938, Miniszterelnökség 9651/M.E.I. Teleki was then a privy councilor in the Ministry of Religion and Education.
43. Balla was still a Hunnia employee as of 3 September 1941, when the EMK demanded his dismissal. On Balla and Gárdonyi, see *MOL-Ó* – Hunnia, Z1123-r1-d2, 1940–2. 3 July 1940 Jgykv – Hunnia Igazg-i ülésről, minutes 9. *MOL-Ó* – Hunnia, Z1123-r1-d2, 1940–2. 28 October 1940 Jgykv – Hunnia Igazg-i ülésről, minutes 3–4. *BFL* – Nb. docs re: László Balogh, Nb. 1699/1945. 'Hunnia Filmgyár Igazg-i üléséről', 3 September 1941.
44. See Lajta, 'A magyar film története', 176–96.
45. In June 1941, 27 companies distributed films in Hungary. During the war, as the number of films appearing on the Hungarian market declined, it is unlikely that the number of distribution licenses exceeded thirty. Specific annual statistics are scarce. Statistics from *Filmcompass 1941* (Budapest, 1 June 1941), 3–21.
46. *BFL* – MFSzSz XVII 647/1 – 287 B, Erdélyi István. Géza Kormos testimony, 28 June 1945, 'II. Jgykv', 25. Jgykv felvéve a Nb. által a Magyar Müvészek Szabadszervezet Filmosztályához [hereafter MMSzF-ához] kiküldött 289/b. szamú IB, 28 June 1945.

47. *BFL* – László Balogh, Nb. 1699/1945. 'Jgykv. az ONFB Budapesten 1942. évi március hó 3-ikán tartott üléséről'.
48. Sándor, *Őrségváltás után,* 107–9. Sándor quotes a letter found in *MOL-Ó* – Hunnia, Z1124-r6-d65. Ügyvezetőség iratai – Szinművészeti és Filmművészeti Kamara, Budapest – 1939–44.
49. Dezső Váczi, *Magyar Film's* managing editor until his troubled relationship with Kiss resulted in his dismissal in early 1942, asserted that Kiss dictated the Chamber's politics and had a hand in producing the most provocative, most anti-Jewish articles in *Magyar Film*. *BFL* – Dezső Váczi Nb. 1503/46. See the Dezső Váczi affidavit, September 1945. MÁBFPRO, G fcs, 14477/1945.
50. *BFL* – Ferenc Kiss, Nb. 4077/45, part 2, 26–7 November 1945, 19–20. Géza Staud testimony.
51. Bánky stated that Kiss' radical antisemitism was the reason for the defections from the High Film Committee. *BFL* – Victor Bánky [Bánki], Nb. 2540/1945. 'Bánki Gyula Viktor, Filmrendező', Nü 4959/1945, 8 August 1945, 7. Tőrey recalled events similarly in testimony before the Screening Committee of the People's Court on 18 September 1945, Nb.ügy XIV 991/1947-5. *BFL* – Nb. docs re: Magyar Hiradó, MFI Rt. XVII 789/6 – 399 A, Zoltán Tőrey [Taubinger] file. For others who attested to the excessive antisemitism of Kiss, see the testimony given at his postwar trial: *BFL* – Ferenc Kiss, Nb. 4077/45. Of note is the fact that this seems to partially contradict Bánky's version of the events at the time, which was that the Film Chamber was not a comprehensive enough organization empowered to fully deal with enforcing of the Jewish Laws. While it is likely that Bánky toned down his antisemitism in order to better his chances at trial, the fact that others confirm his later version of the story is of no small importance. Bánky's original account found in 'A Filmoperatőrök Társaságának jubiláris összejövetele', *MF* II/1 (6 January 1940), 3–4.
52. 'A Kamara átszervezés. Az előkészitő bizottság javaslata', *MF* II/6 (10 February 1940), 2–4.
53. *BFL* – MFSzSz XVII 647/1 – 287 B, Erdélyi István. Erdélyi testimony, 28 June 1945. 'II. Jgykv', 3, 18. Jgykv felvéve a Nb. által a MMSzF-ához kiküldött 289/b. szamú IB, 28 June 1945. While one might discount Erdélyi's testimony, tens of witnesses testified on his behalf and the court eventually found Erdélyi's wartime conduct exemplary.
54. The decision was made by people who had vested interests in both studios and likely in the presence of and with the active participation of János Bingert and Zoltán Tőrey, the respective directors of Hunnia and the MFI who regularly attended ONFB meetings.
55. *MOL* – MFI, K675-cs1-t3. VB-i jgyzk és iratok, 1928–44. 'A MFI VB-i ülésének napirendje és 41. számú jgykv-e. 12 October 1939, 4.
56. *MOL-Ó* – Hunnia, Z1123-r1-d2, 1940–2. Jgykv – 23 August 1940 Hunnia Igazg-i ülésről, minutes 7. Művelődés Film, owned by Count Géza Lipót

Zichy, completed the film in question, *The Belated Letter* [*Elkésett levél*], without crediting Béla Gaál. This meant either that Művelődés Film severed its connection with Gaál or merely had him work covertly. The EMK investigated Művelődés sometime during 1941, probably in the first half of the year. Kultsár's men concluded that Zichy's company continually violated the Jewish Laws, and proceedings against the company were halted only upon the death of Zichy. From discussion of Kultsár's investigations of all film production companies at the 3 March 1942 meeting of the ONFB. 'Jgykv. az ONFB Budapesten 1942. évi március hó 3-ikán tartott üléséről', 2–3. In *BFL* - László Balogh, Nb. 1699/1945.

57. *BFL* - László Balogh, Nb. 1699/1945. 'Jgykv. az ONFB Budapesten 1940. évi szeptember hó 19-én tartott üléséről', 7. Szlatinay testified in postwar hearings that he continued to compose music for Erdélyi-produced films. *BFL* -MFSzSz XVII 647/1 - 287 B, Erdélyi István. Sándor Szlatinay testimony, 6 July 1945, 'III. Jgykv', 10–11. Felvéve a Nb. által a MMSzF-ához kiküldött 289/b. szamú IB, 6 July 1945.
58. *BFL* - Károly Nóti testimony, 6 July 1945, 'III. Jgykv', 17. Felvéve a Nb. által a MMSzF-ához kiküldött 289/b. szamú IB, 6 July 1945. MFSzSz XVII 647/1 - 287 B, Erdélyi István.
59. Nóti wrote numerous films illegally, some for which he was actually credited, others either as an assistant or under an assumed name such as 'Rodriquez'. D. Ákos Hamza claimed that Nóti wrote *This Happened in Budapest* [*Ez történt Budapesten*], a film released in Hungary during the German occupation in June 1944. Other scripts credited to Nóti from the 1939–44 period are *Queen Elizabeth* [*Erzsébet Királyné*], 1940; *Military Hat and Jacket* [*Csákó és kalap*], 1941; and *White Train* [*Fehér vonal*], 1943. See 'Élő Filmtörténet: Hamza D. Ákos a Magyar Filmintézetben', 2 June 1987, 5; Varga (ed): *Játékfilmek*; Péter Gál Molnár, *A Páger-ügy* (Budapest, 1988), 152; Manchin, 'Fables of Modernity', 210. The blackmail of Bánky is discussed in chapter 7.
60. *BFL* - Victor Bánky [Bánki], Nb. 2540/1945. Bánky testimony in 'Bánki Gyula Viktor, Filmrendező', Nü 4959/1945, 8 August 1945, 79–80, 83. Bánky further claimed, and others supported this assertion, that he employed numerous other 'Jewish origin actors, costume designers, transcribers, lighting workers, transporters, traders, etc'. He testified that Békeffy, long a target of the antisemitic right for his cabaret satire, either had a hand in the writing or played an acting role in at least three of Bánky's 20 films. That Jewish ghost-writing was an open secret known to contemporaries by late 1940 is clear. See 'Nem szerepelhetnek többé a titkos zsidó filmírók', *Új Magyarság* (21 November 1940).
61. Period posters found in the OSzK *Kisnyomtató és plakat gyűtemény;* films found in the Hungarian Film Archive. In 1943, Békeffy wrote and was credited for the film *Taming of the Shrew* [*Makacs Kata*].

62. Szatmári helped write an additional three films, often cooperatively with Nóti. One of the few Jewish members of the Film Chamber, Szatmári noted that Bánky employed him and other Jewish writers in direct contravention of the Jewish laws. See Szatmári testimony at the Bánky hearings. *BFL* – Victor Bánky [Bánki], Nb. 2540/1945. 'Jgykv.' Nb.VI.2.540/1945/16, 27 June 1947, 202.
63. 'A magyarországi negyvenkét filmkölcsönző hatvanhét százaléka zsidó', *A Nép* (9 May 1940). The following week, *A Nép* proclaimed that 63 percent of Budapest's smaller theaters were still run by Jews. See Géza Alföldi, 'Késik a mozgószinházban a tisztogatás', *A Nép* (16 May 1940).
64. A sampling of the articles that appeared in a number of rightist journals in the second half of 1940 include: 'Meddig tart Castiglione úr diktatúrája', *Pesti Ujság* (3 June 1940); 'Gerő István a magyar filmszakma I. számú közellensége', *Nemzetőr* II/28 (5 July 1940), 5; 'Így köt szerződést Gerő István – a magyar filmszakma I.számú közellensége', *Nemzetör* II/29 (July 15 1940), 5; 'A Gerő likvidál', *A Nép* (18 July 1940); 'A zsidó Gerő ismét beszivárgott a magyar gazdasági életbe', *Pesti Ujság* (2 August 1940); 'A zsidókkal együtt a romlott szellemet is el kell távolítani a film- és moziszakmából', *Á Nép* (19 September 1940); 'Oh, azok a mozi ruhatárak!' *Magyarország* (8 November 1940); 'A zsidó filmvállalkozók ismét be akarnak kapcsolódni a magyar filmgyártásba', *Függetlenség* (9 November 1940); 'Filmgyártásunkból kiszoritják a titkos zsidó társszerzőket', *Új Magyarság* (13 November 1940); 'Kík a filmszakma titkos "technikusai"', *Új Magyarság* (15 November 1940).
65. 'Elég volt!' *MF* II/28 (13 July 1940), 1.
66. Yehuda Don, 'Economic Implications of the Anti-Jewish Legislation in Hungary', in David Cesarini (ed), *Genocide and Rescue* (Oxford, UK, 1997), 66–8.
67. *MOL-Ó* – Hunnia, Z1123-r1-d2, 1940–2. Jgykv – 2 October 1940 Hunnia Igazg-i ülésről, 12.
68. 57.750/1939 B.M. sz. rendelete signed by Keresztes-Fischer on 22 July 1939 raised the quota to 20 percent. 70.200/1940 B.M. sz. rendelete of 1 July 1940 upped it to 25 percent, and 192.000/1941 B.M. sz., issued 1 July 1941, mandated 33.3 percent of all feature presentations be Hungarian.
69. *TH* – Zita Szeleczky (Mrs. Gyula Haltenberger), file code V-102293. István Békeffy testimony, in 'Tanúvallomási Jgykv', 16 September 1946, Magyar Államrendőrség Budapesti Főkapitányságának Politikai és Rendészeti Osztálya [hereafter MÁBFPRO], 11202/1946. For a contemporary account of Szeleczky's views, see 'Látogatás Szeleczky Zitánál', *Magyarság* (13 July 1941). Szeleczky later fled Hungary for Argentina.
70. *TH* – Zita Szeleczky (Mrs. Gyula Haltenberger), file code V-102293. Pless, who had transferred all of his shares to his lawyer, vitéz Károly Rostaházy, in

October 1939, continued to run Harmónia until he was arrested in late 1941. *BFL* – Harmónia Filmipari és Filmforgalmi Vállalat Kft – File Cg35111.

71. György Guthy, 'Megteremtik a keresztény nemzeti filmszakmát!' *Mozi- és Filmvilág* III/3 (February 1939), 1.
72. 'A felvidéki mozirevízió és a trösztösítési törekvések', *MF* I/12 (6 May 1939), 1–2.
73. Sándor, *Őrségváltás után,* 96.
74. *MOL-Ó* – Hunnia, Z1123-r2-d4, 1940–4. 5 January 1940 Hunnia Igazg-i ülésről Jgykv, 15.
75. 'Így köt szerződést Gerő István...'' *Nemzetőr* II/29 (July 15 1940), 5. The article alleged Gerő collected hundreds of thousands, if not millions, of *pengő* annually through secret contracts with Christian movie theater license holders. The author charged that Gerő retained ownership of his theaters, leased them to Christian companies and charged 20–30 percent of the theaters' monthly ticket sales.
76. *MOL* – KMI, K429-cs59-t2. Vegyes MTI levelezés 1938–9. Anonymous, 'Helzet a magyar moziszakmában'. The text also alleges that Kálmán Tomcsányi in the Interior Ministry, János Bingert at Hunnia, and other officials were crucial protectors of Gerő.
77. *BFL* – László Balogh, Nb. 1699/1945. 'Feljegyzés magyar filmek világexport lehetőségeiről', 10 January 1941.
78. Ibid. If Gerő's plan did not succeed, Horthy Jr. intended to establish a distribution company in Rio de Janeiro to accelerate the global export of Hungarian film. This suggestion was also proactive. Should America join the war, Rio or Buenos Aires would serve as the neutral setting where film exchange could occur.
79. *BFL* – László Balogh, Nb. 1699/1945. Hóman paraphrased by Balogh, 'Jgykv. felvétetett az ONFB-nak Budapesten 1941.évi január hó 13-án tartott üléséről'.
80. *BFL* – László Balogh, Nb. 1699/1945. ONF, 1941, file # 20. Issue: Entrusting I. Gerő with film exports.
81. Ibid.
82. *BFL* – László Balogh, Nb. 1699/1945. Gerő to Paikert, Beograd, 23 February 1941.
83. *BFL* – László Balogh, Nb. 1699/1945. Gerő to Balogh, Beograd, 18 March 1941.
84. *BFL* – László Balogh, Nb. 1699/1945. Balogh to Kultsár, 17 March 1941.
85. 'Miért akart Gerő István kiutazni Délamerikába?' *Nemzetőr* III/10 (10 March 1941).
86. *BFL* – László Balogh, Nb. 1699/1945. Kultsár to Balogh, 28 April 1941.
87. Mester's partners, Barla László and szalontai Kiss Miklós, also bought large shares of Gerő's business.
88. Ferenc Keresztes-Fischer explained how this would happen in questioning before the Lower House of Parliament on 5 February 1941. See 'A belügyminiszter válasza a Rajniss-féle mozi-interpellációra', *MF* III/7 (15 February 1941), 2–3. I was unable to determine Gerő's ultimate fate. In 1943 he resurfaced in a bid

to arrange Hungarian film exports to Spain. I believe he fled Hungary, possibly before war's end.

89. Andor Lajta, 'Az év története', *Filmművészeti Évkönyv* (1942), 179.
90. Law XV: 1941, or the Third Jewish Law, forbade marriage between Jews and non-Jews, and racially redefined Jews, adding, according to Randolph Braham, nearly 100,000 Hungarians to the rosters of those with 'Jewish origins'. Braham, 'The Holocaust in Hungary. A Retrospective Analysis', in R. L. Braham and A. Pók (eds), *The Holocaust in Hungary: Fifty Years Later* (Boulder, CO, 1997), 293.
91. During this time, the Hungarian right stepped up attacks on the 'strawman system' and introduced plans for a third round of anti-Jewish legislation. Popular actors who publicly opposed discrimination against Jews, such as Pál Jávor, were heckled during film showings. Sándor, *Őrségváltás után,* 139.
92. Katzburg, *Hungary and the Jews*, 161.
93. Braham, *The Politics of Genocide,* Vol. 1, 178.
94. *Süß* premiered in late-January in Budapest and ran into April in some countryside theaters. An advertisement in *Magyar Film* proclaimed 'Never before seen success! Never before seen crowds!' and that in one Budapest premier theater alone, *Jud Süß,* within 25 days, had sold 75,600 tickets and produced HP 130,000 in income [equivalent to the cost of making a film in Hungary in 1940]. *MF* III/8 (23 February 1941), 9. German reports claimed excited Hungarian audiences called for the blood of Budapest Jews during the Süß execution scene. Susan Tegel, *Nazis and the Cinema* (London, 2007), 146. See also Sándor, *Őrségváltás után,* 146.
95. A sample of the articles calling for enforcement of anti-Jewish laws in early 1941 include: 'Az értelmiségi munkanélküliség kormánybiztosának figyelmébe!' *MF* III/6 (8 February 1941), 1; 'Keresztény mozifront előtt', *Magyar Film* III/17 (26 April 1941), 1; and a series in *A Nép* between February and March 1941 comparing *Jud Süß* with 'Jud Gerő' and others.
96. 'Rajniss Ferenc interpellációjában a zsidótörvény erélyesebb végrehajtását sürgette', *MF* III/6 (8 February 1941), 12.
97. 'A belügyminiszter válasza a Rajniss-féle mozi-interpellációra', *MF* III/7 (15 February 1941), 2. Keresztes-Fischer's response was also excerpted in a number of Budapest dailies.
98. 'Az OMME felhívása tagjaihoz a filmkölcsönzővállalatok érdekébén', *MF* III/33 (16 August 1941), 2.
99. *BFL* – László Balogh, Nb. 1699/1945. 'Jgykv. az ONFB f. [1941] évi augusztus hó 19-én tartott üléséről', 2.
100. *BFL* –MFSzSz XVII 647/1 – 287 B, Erdélyi István. Erdélyi testimony, 28 June 1945. 'II. Jgykv', 20–1. Jgykv felvéve a Nb. által a MMSzF-ához kiküldött 289/b. szamú IB, 28 June 1945.

101. 'Az OMME felhívása tagjaihoz a filmkölcsönzővállalatok érdekében', *MF* III/33 (16 August 1941), 2.
102. *BFL* – László Balogh, Nb. 1699/1945. Untitled OMME memorandum, 8319/1941 sz.
103. 'A zsidótörvény végrehajtása', *MF* III/33 (16 August 1941), 7.
104. 'Szakmai körúton', *MF* III/42 (20 October 1941), 1.
105. Universal, for example, was hit hard by the forced implementation of the Jewish laws in 1941. *BFL* – Universal Film Rt. – File Cg20315 – 3427. All American company branches were assigned government trustees in 1942. See *BFL* – Warner Brothers-First National Vitaphone Pictures – File Cg25822. M. kir. KKM, 14040/III sz., 9 February 1942.
106. *NARA* – RG59 [State], M-1206, Roll 15, 864.4016/Motion Pictures/82. Telegram from H. Pell to Sec State, No. 575, 14 October 1941.
107. 'Internálták hét szakmabelit', *MF* III/47 (24 November 1941), 3. By his own admission, Radó was working under an assumed name. *BFL* – Testimony of István Radó, 18 October 1946, 'Budapesti Népügyeszség 1945.Nü.10886/16 – Jgykv'. *BFL* – Dr. Nagy Sándor – Nb 3482/1946.
108. *BFL* – László Balogh, Nb. 1699/1945. 'Jgykv. az ONFB-nak a m.kir. VKM-ban 1941. évi november 18.-án tartott üléséről', 7.
109. 'Ismét internálták hét szakmabelit', *MF* III/51 (22 December 1941), 9.
110. 'Ultima ratio', *MF* III/48 (1 December 1941), 1.
111. *BFL* – László Balogh, Nb. 1699/1945. Kultsár to Wlassics Jr., 28 November 1941, 220.128/1941 sz., ONF Ikt. Sz. 408/1941.
112. *BFL* – László Balogh, Nb. 1699/1945. Jgykv az ONFB Budapesten 1941. évi december hó 10. napján tartott üléséről', 1. As EMK, Kultsár had the authority to review all ONFB decisions regarding Jewish issues.
113. *MOL-Ó* – Hunnia, Z1123-r1-d2, 1940–2. 17 December 1941 Jgykv – Hunnia Igazg-i ülésről, minutes 3.
114. Ibid., 3.
115. Ibid., 4.
116. Ibid., 5.
117. That the ONFB was uncomfortable with Kultsár's new assertions of authority is clear in its December 1941 meeting minutes. *BFL* – László Balogh, Nb. 1699/1945. Jgykv az ONFB Budapesten 1941. évi december hó 10. napján tartott üléséről', 3.
118. Ibid., 1.
119. *BFL* – László Balogh, Nb. 1699/1945. Jgykv az ONFB Budapesten 1942. évi április hó 20. napján tartott üléséről', 3.
120. Tibor Sándor claims that Kultsár established an independent film bank in January 1942, but my evidence, coming from ONFB meeting minutes, indicates that through April 1942, Kultsár was still negotiating with the Finance

Ministry for funding for the bank project. In fact, István Erdélyi and the OMME came out against the film bank plan, worrying that it would thwart production by making producers pay back all loans before beginning their next project. See István Erdélyi, 'A filmbank küszöbén', *MF* IV/4 (26 January 1942), 6–7. Even at the close of 1943, VKM Szinyei-Merse harped on the need to 'sound the alarms for Christian capital' and mobilize it for film production. He made no mention of a film bank. *BFL* - László Balogh, Nb. 1699/1945. 'Jgykv. felvétetett az ONFB 1943. december 10-iki üléséről', 1. For Sándor's assertion, see Sándor, *Őrségváltás után*, 163. See also chapter four of this text for discussion of earlier iterations of a non-Jewish funded film bank.

121. *OSzK - Keresztes-Fischer Ferenc Levelestár*. Keresztes-Fischer to F. Herczeg, 13 February 1942.
122. 'Smolka Jánosról, a magyar filmszakma III.sz közellenségéről, a stromanképző intézet igazgatójától', *Nemzetőr* II/33 (August 12, 1940), 5.
123. Italics mine. *BFL* - MFSzSz XVII 647/1 - 287 B, Erdélyi István. A. Brody and K. Nóti testimony, 6 July 1945, 'III. Jgykv', 17. Felvéve a Nb. által a MMSzF-ához kiküldött 289/b. szamú IB', 6 July 1945.
124. Lajta, 'A magyar film története', 49. In his contemporary account of the arrests, Lajta claimed most of the jailed film professionals were released after 3–4 months. Lajta, 'Az év története', *Filmművészeti Évkönyv* (1943), 212. News accounts indicate that Bródy, Ungár, Pless, and Várdai were released in January 1942.
125. Lajta, 'A magyar film története.', 81.
126. Erdélyi speech, quoted in 'Az OMME 31. évi rendes közgyűlése', *Magyar Film* III/51 (22 December 1941), 2–10, esp. 6.

6

The War[1]

[E]ver since Trianon…Hungary could and can have only one objective: to make one whole once more of our country ripped to bits and pieces and make it recover the weight and role in the Danubian Basin which befits it.[2]

István Bethlen

The Magyar nation in Southeast Europe is the balancing force whose useful and beneficent effects no new power constellation can do without…No one can deny that the Magyars have a calling…[and] it is an undeniable fact that…[through Hungary] Western Christian culture continues to radiate most [brightly] in Southeast Europe…within the boundaries of historical Hungary.[3]

Sándor Márai

It goes without saying that we cannot rule Europe economically if we do not also make ourselves supreme in the cultural field. Cultural hegemony, however, can only be achieved with the help of a large number of technical aids. And, in this respect, film is one of our major resources.[4]

Joseph Goebbels

Introduction

While domestic factors wreaked havoc on the creation of a durable Christian national system, war created new opportunities and dangers for those seeking to refashion the film world and its creations. In 1939, while Europe tottered, international instability set the stage for a more stable and productive Hungarian film industry. Through the early war years Hungary's industry grew stronger and more self-assured, yet paradoxically less autonomous as it became increasingly bound to Nazi Germany.

Nazi aggression, particularly the absorption of Austria and the destruction of Czechoslovakia and Poland, drastically altered European relations, film markets, and the continent's filmmaking calculus. French and British movies, not to mention the products of Hollywood, found it increasingly difficult to penetrate the European continent. The 1940 German victory over France temporarily halted all French movie production.[5] Several hundred motion pictures vanished, pictures integral to cinema programming throughout Europe. Desperate to satisfy their publics' desires, to provide a cinema of distraction in an age of fear, European distributors looked elsewhere, including to Hungary. War and instability thus worked to Hungary's advantage, permitting unmatched access to and acclaim in foreign markets.

Nazi-coerced Central European land transfers also benefited Hungary. In November 1938, the First Vienna Award gave Hungary a large slice of Slovakia and part of Ruthenia, the remainder of which Hungarian troops occupied in March 1939. The addition or 'return' of these lands to Trianon Hungary added between 58 and 65 cinemas and nearly HP 10,000 more in average ticket sales per Hungarian film.[6] The Second Vienna Award, which granted a large swath of Transylvania to Hungary in August 1940, brought 32 more cinemas into the Hungarian fold.[7] Finally, Hungarian participation in the invasion of Yugoslavia in April 1941 and the subsequent incorporation of the Bácska and its surroundings meant that Hungary's domestic market became some ten percent larger, augmented by 73 additional theaters.[8]

Besides conquest, natural growth meant that scores of theaters opened throughout Hungary in the early 1940s. The overall increase between 1938 and 1943 was staggering. In 1938, there were approximately 511 licensed

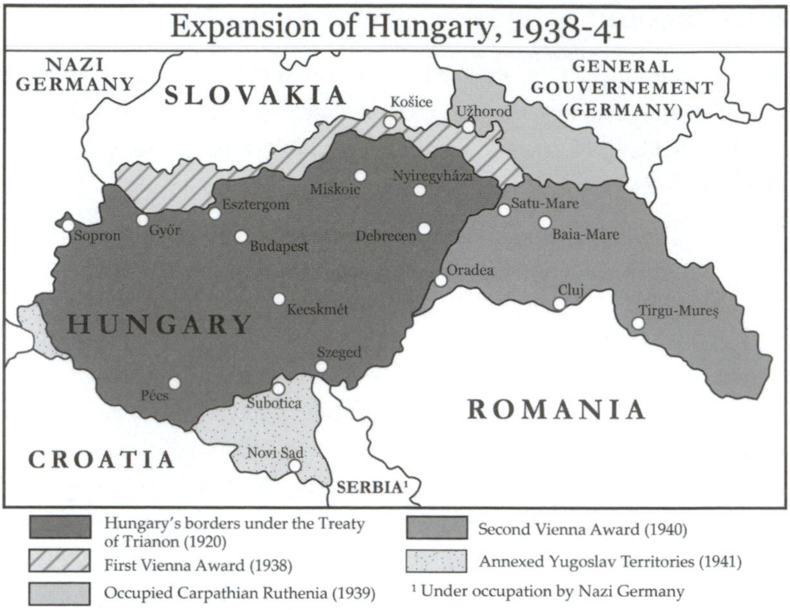

Figure 11: Hungarian Expansion, 1938–41. Copyright © 2015 Nathan E. McCormack and David S. Frey.

theaters operating in Hungary. In 1943, there were a minimum of 784, plus hundreds of additional centers or groups licensed to show narrow (16 mm) movies, including schools, associations, and clubs.[9] With this expansion came a rise in the number of screenings, a jump in ticket sales, and increased earnings.[10] Total revenue from tickets and concessions as much as tripled in three years, rising from HP 24 million in 1939 to 33.3 million in 1940 to between 61 and 90 million in 1942.[11]

The concurrent growth of Hungary's domestic market and export possibilities ignited the national film industry. Shortly after the disastrous production collapse of early 1939, Hungary astoundingly found itself on the podium of continental European feature producers, outpacing all but Germany and Italy. Hungary's elites interpreted this as a return to the natural equilibrium Europe had abandoned in 1918. Hungary's position was representative of its equivalent importance as a top tier culture-producing nation, one of the superior 'races' of Europe. 'In the framework of the Second World War,' bragged Hunnia chief János Bingert, 'Budapest occupies Paris' place...'[12] More than a mere vindication of the intrinsic value of the Hungarian people, Hungarian cinema's success was a validation of the nation's special position in Central Europe and proof of the illegitimacy of the Trianon judgment.[13] *Magyar Film* trumpeted this view in a flurry of articles. Celebrating the release of *Pista Dankó* [*Dankó Pista*], Hungary's 200th sound feature, the magazine's editors crowed:

> The two-hundredth Hungarian sound film not only means that alongside foreign film there are two hundred Hungarian creations to be enjoyed by our great audiences. It also means that our strengthened film art has found itself; that it fills such an important calling in Hungarian cultural life that few of the great culture-producing states can similarly boast. The cultural and economic importance of Hungarian film has today already spilled beyond our borders and...proclaims the words of St. Stephen's thought and the leading vocation of the Hungarian mentality in the Danubian basin.[14]

Intoxicated, the editors lauded the dissemination of Hungarian film as a continuation of 'Hungary's one-thousand year cultural mission in Central Europe and the Danubian basin,' and proof of Hungary's cultural supremacy.[15] The journal *Nation-builder* [*Országépítés*] proclaimed that by 1941: 'our film production...fills...an irreplaceable piece in the cupboard of international cultural necessities'.[16] The German-language Budapest daily *Pester Lloyd* triumphantly declared that the appeal of Hungarian film 'was of the highest national, cultural, and economic significance' to New Order Europe, illustrating that 'perhaps the day of the aesthetic revolution is not far ahead, in which the little nations find a way of making film which...finds a way into the hearts of the larger nations...'[17]

In a May 1940 letter to his friend Antal Ullein-Reviczky, the Foreign Ministry News Bureau Chief, Miklós Kozma gushed about the financial and cultural implications of new film orders from Greece, Bulgaria, Sweden, and the Baltic states.[18] *Hungary* [*Magyarország*] editor Pál Szvatkó, in a Christmas 1940 article, echoed many of Kozma's sentiments. Giddy over

Hungarian achievements and hopeful that Hungary's Southeast European successes would be replicated elsewhere, Szvatkó bragged:

> We have wondrous authority in the Balkans. We have prestige. Not primarily political, rather social [and] cultural and in the wake of these political authority is just now beginning to be created – and this is important. You would not believe how many excellent examples of Hungarian film there are in Belgrade, Sofia, and Istanbul. Our stars are of Hollywood rank: [Zita] Szeleczky, [Pál] Jávor are recognized the same way Greta Garbo and Clark Gable are.[19]

The achievements of Hungarian film reverberated in official corridors. Hungarian Garbos and Gables were the means by which their nation would re-establish its pre-1918 cultural, political, and economic authority in Central Europe and the Balkans.

Convinced of its new-found prestige and might by the warm reception its films received abroad, Hungary began to act with film star swagger. In late 1939, its film elites convened a film week in the resort town of Lillafüred, a chance to display new products, proclaim the quality of Hungarian goods, rally Christian nationalists, and create a parade of celebrities that would surpass all previous attempts to market Hungary's screen favorites. In 1940, Film Week organizers predicted their festival would soon become an international gala, the 'Venice Film Biennial of the small states,' destined to expose 'previously unseen opportunities' for Hungarian film.[20] Film elites were elated by the critical success of their films in international competitions in 1940 and 1941.[21] What these achievements illustrated, Hungary's leading film figures claimed, was that once the strait-jacket of Trianon was removed and Hungarian culture unshackled, truly liberated film audiences everywhere would give Hungarian goods their due.[22] The free market of culture would accurately reflect Hungary's true position in Europe.

How had Hungary ascended into the upper echelon of film producing states while purging much of its own talent and barely surviving its self-inflicted production crises? The short answer is exports. Shipments of film abroad propped up a fragile industry, providing, by some estimates, 30 percent more income than had been possible in 1939.[23] The lynchpin was a single country: Yugoslavia. Success in the Yugoslav market produced global interest in the products of Hungary's studios. Its confidence bolstered, the film elite continued to transform Hungary's national sound film realm. Hungary's Balkan foothold also had a critical influence on relations with Germany. While Hungary's political leaders dared to dream of a special place in New Order Europe, Nazi functionaries quickly identified Hungarian film as a threat on multiple levels. In addition to losing market share, the Nazis viewed Yugoslav consumption of Hungarian film as a preference for Jewish culture over Nazi ideology. Moreover, the Nazis recognized how Hungarian cinematic success might further impede the de-Jewification of the Hungarian film industry. Frustrated by Hungarian intransigence and independence, Reich authorities intervened in Hungarian film affairs with increasing vigor. Utilizing the limited tools they had, foremost the rhetorical concept of the Christian national film industry, Hungarian film elites pushed back. Conflict, cooperation, competition, and coercion thus characterized wartime Hungarian-German film politics, and further transformed Hungary's nation-building industry.

Yugoslavia, Springboard to Success

The nearly three years between the 1938 collapse of the Little Entente and the April 1941 demise of Yugoslavia were marked by a noticeable improvement in Hungarian-Yugoslav relations. This thaw was characterized by a surprising degree of cultural interaction and cooperation, and the unexpected triumph of Hungarian film was a major factor in this shift. The roots of the Hungarian-Yugoslav détente trace to the spring of 1936.[24] Italian and German initiatives in the Balkans persuaded Hungary to reach out to its neighbor to the south.[25] Hungary felt it had a unique imperialist mission: first, to act before either Italy or Germany gained hegemonic influence; second, to endeavor to separate Yugoslavia from its Little Entente allies; and third, to secure cultural autonomy for ethnic Hungarians within Yugoslavia's borders.[26] Hungary's political elite understood that culture, perhaps even film, could help mobilize the ethnic Hungarian minority in Yugoslavia and therefore might lead to greater political influence and new business opportunities. Film industry representatives believed access to the Yugoslav market, nearly equal in size to Hungary's, might permit their own industry's expansion, reduce unemployment, and establish Hungary as the gateway to the Balkans in New Order Europe.

Opportunity appeared in 1937, when the political climate changed and the Yugoslav government overhauled its censorship apparatus. Until that point, Yugoslav censors had banned all films containing the Hungarian language. In mid-1937, Censors approved the exhibition of *Skylark* [*Pacsirta*], a Hungarian-made German language film starring Marta Eggerth, despite the fact that Eggerth spoke some Hungarian in the picture.[27] This led Hungarian film authorities to believe that an improvement in cultural relations was imminent.[28] A number of proposals and projects reflected this new optimism. In mid-summer, a group of Hungarian filmmakers announced plans to construct a new sound studio in the southwestern city of Szeged. This facility, they felt, could establish Hungary as a cultural bridgehead to Southeastern Europe by producing pictures specifically for export to Yugoslavia and other Balkan states.[29] Later that autumn, Hungarian film luminaries, led by István Székely, proposed the founding of a Budapest-headquartered Hungarian-Yugoslav film society.[30] Simultaneously, representatives of the Yugoslav Film Center approached Hunnia and proposed that Hunnia help Yugoslavia create a national sound film industry. Rumors circulated in Budapest circles that Hunnia would make four Serbian-language films in 1938 and that Hungarian film companies would undertake projects in Yugoslavia. Hungary's official trade journal, *Filmkultúra,* ran two major articles on film life in Yugoslavia in October and November 1937, touting the potential of the Yugoslav market and the 'great cultural and moral significance' of the showing of Hungarian culture films in Belgrade.[31] Yugoslav authorities licensed exhibition of the first Hungarian-language feature film in March 1938 and in principle consented to import many more.[32]

Prime Minister Kálmán Darányi's March 1938 Győr pronouncement, which addressed rearmament, industrialization, and Hungary's 'Jewish question', understandably frightened Hungary's neighbors. Consequently, plans for Hungarian-Yugoslav cultural cooperation stalled. Yugoslavia rescinded import licenses and indefinitely suspended all Hungarian film purchases. Frustration just as suddenly morphed into anticipation in August 1938, when Hungary signed the Bled agreement, renouncing offensive use of force against the Little Entente states. This resuscitated possibilities for closer Yugoslav relations. By September 1938 select Hungarian circles, including those around Miklós Kozma, were again touting the vast

Figure 12: Marta Eggerth, the Hungarian-born, Hungarian-speaking star of *Skylark*.

political, cultural, and economic opportunities represented by closer association with Yugoslavia. Some even floated the idea of a Warsaw-Budapest-Belgrade-Rome alliance, a non-Nazi counter to Germany.[33] Although it opposed any alliance that did not include Berlin, the Hungarian Foreign Ministry did share the Kozma circle's enthusiasm for Yugoslavia. The Foreign Ministry encouraged the foundation of a Hungarian-Yugoslav League to mobilize 'people who could work in the interest of achieving closer Hungarian-Yugoslav ties'.[34] Vienna's *Neue Freie Presse* correctly identified this as part of a new cultural-political offensive, one designed to deepen ties with the countries of Europe, to correct misunderstandings about the Hungarian nation, and to secure Hungary's proper place in Europe's New Order.[35]

This offensive was imperative because German aggression had shattered the few foreign lands previously accessible to Hungarian film. The *Anschluß* deprived Hungary of its best European customer: Austria. Germany's unwillingness to open its market, which now included the Czech lands, frustrated Hungary's film establishment. Continued adversarial relations with its other neighbors limited the potential for regional sales potential. Hungary's 1939 production crisis, aggravated by these geopolitical conditions, meant the palpable improvement in Hungarian-Yugoslav relations and the promise of a new market could not come at a better time.[36] The second half of 1939 saw the signing of bilateral trade agreements, the opening of an airlink between Belgrade and Budapest, and Yugoslavia's assent to the distribution of Hungarian cultural products. Hungarian theater troupes performed in Belgrade and Zagreb, and in December, the first entirely Hungarian-language film played in Yugoslav theaters, followed by a handful more before the year's end. What began as a trickle in 1939 became a veritable torrent in 1940, one that totally altered, at least temporarily, the landscape of popular culture in Yugoslavia and permanently impacted the evolution of the Hungarian film industry.

On the Ascent: Hungary as European Film Power

By December 1939, Hungarian-Yugoslav cultural interactions were burgeoning.[37] *Magyar Film* gloated that Hungarian sound film equipment had recorded the signing of the Cvetković-Maček Compromise in August 1939.[38] Since Yugoslavia lacked the capacity to make its own sound film, the filming had extensive repercussions. Pleased with the product, the head of Jugoslavija Film, Mika Đorđević, determined to 'bring Hungarian experts and sound equipment to Yugoslavia, to lay the foundations of Yugoslav national film' instead of turning to American or German technology and expertise.[39] In December, Yugoslav and Hungarian delegations led by the Hungarian actor of Yugoslav origin, Szvetiszláv Petrovics, began preliminary discussions about the manufacture of dual language features.[40] Three months later Miklós Kozma traveled to Belgrade to discuss the prospects for Hungarian film, radio, and news services. He commented on a new Yugoslav appreciation for Hungarian culture, cheerfully reporting that 'the Yugoslav public, without taking nationality into account, indeed graciously receives the creations of the Hungarian film industry…' Hungary's movie exports, he announced, were a 'great moral and financial success'.[41] Soon after Kozma's visit, Yugoslav film professionals reciprocated, arriving in Budapest for special courses offered by the Hungarian Film Office.

All these developments signified high-level interest in Yugoslavia and an unprecedented amount of cultural cooperation.

By far the most substantial driver behind improved relations was Hungarian sound film. Hungary exported four features in 1939, 143 in 1940, and another 23 in the first months of 1941, meaning that in the 16 months between December 1939 and March 1941, Yugoslav cinemas exhibited nearly every Hungarian sound feature ever made. Dubbed or subtitled, Hungary's bourgeois musicals, Budapest comedies, and historical romances screened throughout Yugoslavia. Their sales yielded between 2.5 and 5.2 million Yugoslav dinar in profit in 1940 alone, slightly less than Hungary's total profits for all film exports for the entire decade of the 1930s.[42] Hungarian authorities described the results as 'absolutely striking'. Hungarian films set box-office records and averaged higher per film earnings than their German and American competition.[43] The success of Hungarian motion pictures in Yugoslavia prompted Andor Lajta to report 'with glee that the quality of Hungarian films has risen greatly, and [Hungarian film] is now recognized for its good reputation in the farthest foreign states. In the past year, Hungarian film has become a serious *export item*'.[44]

Feature film success led to greater cultural exchange, with exports of newsreels and radio broadcasts experiencing considerable growth in mid-1940, concomitant with a minor explosion in the number of Yugoslav Hungarian-language journals.[45] The flood of Hungarian film also stimulated the expansion of Yugoslav film distribution networks, spawning the birth of numerous Yugoslav companies that competed to win contracts to import Hungarian film. Croats and Serbs formed companies specifically to distribute Hungarian products, competing with Hunnia and groups of ethnic and expatriate Hungarians.[46] The Hungarian ambassador to Yugoslavia reported that these companies could be found throughout Yugoslavia and even in urban areas lacking large ethnic Hungarian populations, signifying the astonishingly wide appeal of Hungarian works.[47] Yugoslav popular culture, primarily its motion picture production and distribution but including radio and print culture, became rapidly and increasingly dependent on Hungary. This active intervention into how Yugoslavs could imagine their cultural spaces was of great significance, particularly at a time when Hungarian leaders repeatedly spoke of the connection between assertions of cultural superiority and eventual territorial hegemony.[48]

For the first time, Hungary had captured a prominent share of a market that was not Hungarian speaking. Full houses in Belgrade engendered confidence and convinced distributors in other states that they too should import Hungarian features.[49] Predictions that Hungarian products would become the natural replacements for disappearing French, British, and American pictures allowed Hungary's film establishment to believe its Yugoslav bonanza could be replicated worldwide. The Foreign Ministry reinforced this fantasy, actively seeking contacts in Mexico, Chile, Argentina, Uruguay, Brazil, Sweden, Spain, and Turkey in July 1940 '… so that the distribution and progress of not only Hungarian film, but of Hungarian culture be served…'[50] These efforts were somewhat redundant, as suitors pursued Hungary:

> After Yugoslavia – Sweden, Denmark, Norway, Finland, Holland, Germany, Switzerland, Italy, Spain, Bulgaria, Turkey, Egypt, Syria, and Persia as well have already put in for Hungarian films – … it is up to Hungary's filmmakers to fill an international role similar to that which it held in the days of silent film, to achieve a similar position from Scandinavia to the Middle East.[51]

Between 1939 and 1940, Hungary experienced a meteoric rise in exports, cinema prominence, and cultural power.[52] Sales to Bulgaria shot up from three in 1939 to around 50 in 1940. Greece's Hungarian imports rose from zero in 1939 to approximately 20 in 1940. Eleven Hungarian features premiered on Italian screens in 1940.[53] Interest spread well beyond Southeastern Europe, as Scandinavian, northern European, and Middle Eastern countries purchased substantially higher numbers of Hungarian film or contracted for future imports.[54] Despite all impediments, even exports to the United States shot up during the second half of 1940.[55] These accomplishments convinced top government officials that film was a legitimate cultural commodity and powerful political tool. They imagined they possessed a juggernaut which would help Hungary reestablish the 'natural' order Europe had shunted aside after the Great War.

The Paradoxes of Success

The Hungarian film sensation in Yugoslavia produced expected and unexpected aftershocks. Success righted a wobbly industry. Hungary's animated film establishment began to behave as a European film power. To rake in profits, maintain its place on the podium of culture producers, and reclaim the glorious position Hungary had occupied in the early days of the moving picture, leading figures became convinced that they should ramp up domestic production. Hungary's first priority, argued *Magyar Film,* should be to secure continuous production of features in order to solidify its newly attained shares of export markets.[56] Yet, producing as much as possible as quickly as possible created multiple paradoxes. It complicated the process of de-Jewifying the business. It led to the creation of lower quality, less palatable films. Perhaps most importantly, it aggravated Hungary's relations with Germany and to a lesser extent Italy, just as ties with both countries were deepening. Tragically, Hungary's motion picture upswing contributed directly to the industry's downfall.

In a period characterized by unparalleled domestic need and unmatched export opportunities, a growing number of constituencies in Hungary's film world found themselves overtly acknowledging the internally inconsistent nature of the concept of a thriving Christian national cinema. Wartime conditions created perverse enticements to retain Jewish employees with experience and connections in production and distribution. Veteran film chiefs thus proved unwilling to rush the Christianization of companies involved in production and export because they believed it would jeopardize the foundation of the movie business' success and its future viability. The antisemitic wing of the profession vociferously disagreed. B-grade professionals clustered in the Film Chamber saw the opening of new markets as a rare opportunity to prepare a new generation of Christian Hungarian culture makers and to purge Hungarian culture, once and for all, of its 'pervasive' Jewish influence. The time to train new Christian professionals, they argued, had never been more ideal. Unable to resolve this internecine dispute, Hungary's film elites defaulted to a middle track, allowing the removal of all but the most vital Jews involved in production and distribution, exploiting the expertise of those who remained, and rapidly cultivating new Christian craftsmen. Without Jewish film agents at home and

Figure 13: Bulgarian language poster for *Crossroads*. *Crossroads* was a Hungarian production starring Klári Tolnay, pictured above. Source: MaNDA, Budapest. Note: MaNDA is no longer able to locate this poster.

abroad, Hungary would not have profited from its exports to the degree it did in the early 1940s. Conversely, with the global dearth of competitive feature films, Hungary also had the opportunity to produce at a record pace, giving hundreds of new Christian film professionals a chance to learn the trade.

A second challenge deriving from the Yugoslav windfall concerned the structural organization of the Hungarian film industry. In elite Yugoslav circles and some sections of the Yugoslav popular press, opposition to Hungarian imports surfaced in late 1940, resulting in the construction of a new Yugoslav licensing regime biased against Hungary's features.[57] Hungarian diplomats blamed Budapest film distributors who, placing profit above propaganda value, had sent too many older, lesser quality films to Yugoslavia, thereby sullying the reputation of their nation.[58] This occurrence strengthened the corporatist movement within the Hungarian film establishment. Corporatists of all stripes believed a central agency empowered to control film export could avoid the error of oversupply, and better balance propaganda and profit goals.[59] While radical corporatists grouped around István Kultsár and Ferenc Kiss called for Nazi-style nationalization of the industry, moderates among the ONFB and Hunnia felt monopolistic decision-making should be restricted to matters of export. Between mid-1940 and early 1941, Hunnia's Board occasionally discussed establishing film centers in Zagreb and Belgrade to guide Hungarian imports through the labyrinth of Yugoslav regulations and prevent the import of low-quality films.[60] When this project was finally undertaken it involved the omnipresent István Gerő, whose efforts collapsed under the weight of Hungary's April 1941 attack on Yugoslavia. Gerő's participation is itself demonstrative of the wide chasm between the circle of radical nationalizers, for whom Gerő's mere scent evoked disgust, and the more conservative, limited corporatists, who always kept an eye on the bottom line.

It was not until January 1943 that the corporatists resolved the issue of centralized control over exports, and it was without a doubt a victory for the moderates. Rather than nationalization, Hungary's film powers once again chose to link the country's state authorities and its private business interests without actually fusing them. Neither the Film Chamber nor the National Film Committee was party to the arrangement. Instead, an agreement between the OMME, Hunnia, and the MFI created the Hungarian

Film Export Cooperative [*Magyar Film Kiviteli Szövetkezet*] 'to distribute Hungarian films abroad and to make sure that Hungarian films are distributed by the best foreign film companies'.[61] From 1943–4, the new cooperative served as an export clearinghouse, collecting data about foreign markets, contracting with foreign distributors, and negotiating entrée into markets.

Part of the reason Hungary's corporatists did not act sooner to consolidate control over exports was that Hungarian films were doing incredibly well across Europe, all borne of success in Yugoslavia. Private film companies, such as Béla Csepréghy's Filmimpex, sold Hungarian films to Sweden, Bulgaria, Croatia, Italy, and Greece in 1941.[62] The Mária Hausz Film Distribution and Production Company alone exported 20 Hungarian films to Italy in the first half of 1941.[63] When Italy announced its intention to show only Italian, German, and Hungarian films during the 1941–2 film year, Hungarian officials' wildest 'fantasies' had come true.[64] The August 1941 visit of an Italian delegation headed by Serafino Mittiga, the director of Italy's Film Export Consortium [*Consorzio Esportazione Filmi Italiani*], produced banner headlines. Papers proclaimed a deepening of Hungarian-Italian film links and proudly highlighted Italy's promise to show 100 Hungarian features over the next three years.[65]

Even the German occupation of Serbia in April 1941, though it deprived Hungary of a large part of the former Yugoslav market, did not quash the continued demand Balkan audiences expressed for Hungarian movies. Officials in both Croatia and Bulgaria reported that in 1941 the Hungarian film *Flames* [*Lángok*] was a huge hit. In Croatia, it earned more than any previous film, save another Hungarian feature, *Dangerous Spring* [*Halálos tavasz*] and an unnamed American film. Hungarian film achieved similar results elsewhere in Europe. The Norwegian delegate to the International Film Chamber [IFK] informed his Hungarian counterparts that the two biggest blockbusters in Norway in 1941 were *Pista Dankó* and *Vision by the Lakeside* [*Tóparti látomás*]. Finnish officials requested expedited contracts for Hungarian films and the Belgian IFK delegate reported that he had already begun discussions with German authorities to allow Belgium to screen at least 15 Hungarian films.[66] In Italy and Bulgaria, average Hungarian films could count on income of HP 25,000 in rental/licensing fees in late 1941.[67] Export revenue, which totaled an unheard of

HP one million in 1940, doubled to two million in 1941.[68] Germany purchased several Hungarian features and even allowed two to show in the Czech Protectorate.[69] All signs pointed to faster growth in 1942, and there appeared to be little need for state control of exports to gain entrance into foreign markets. Hungary's private distributors were doing just fine, and even the corporatists recognized this.

The awesome production opportunity afforded by the combination of a 50 percent increase in the number of domestic theaters and the explosion of export venues quickly created a shortage of studio space. Hunnia was booked solid in 1940 and the MFI compensated as best it could, making twelve features, rather than the seven to which it had been limited by earlier 'gentlemen's agreements'.[70] In September 1940, MFI executives mulled building a studio in Kolozsvár [Cluj], the recently reincorporated hub of Hungarian Transylvania.[71] Hunnia's Directorate discussed expansion of its own productive capacity, encouraged by interest among private domestic filmmakers.[72] German and Italian signals that they wished to make films in Hungary inspired even greater optimism. Barely a year after production had temporarily skidded to a halt, Hungary was overwhelmed by demand and desperately seeking additional studio space.

The most logical location was the bankrupt Star studio on Pasareti Street in Budapest. Like the Corvin/Hunnia studio, the Star had been one of Hungary's most storied private studios during the silent era.[73] Unused since the late 1920s, it had been repossessed and in limbo until the army commandeered it in 1940 and converted it into a truck depot and repair facility.[74] The government rejected Hunnia's late 1940 request that the Film Production Fund purchase the facility and turn it over to Hunnia.[75] However, Hunnia's quest got a big boost from the German government in early February 1941.[76] The German-Hungarian Culture Committee [*Deutsch-Ungarische Kulturausschuss*], a bilateral committee made up mainly of government officials, determined that studio space must be made available immediately so that Germany could begin making features in Budapest, in cooperation with Hunnia.[77] In early 1941, most Hungarian officials, even if anti-Nazi, assumed German pockets to be deep and German control of continental Europe secure. Hungarian film figures hoped this contract would commence a new era of cooperation and goodwill in German-Hungarian film relations. Authorities at Hunnia, the

FIF, and the ONFB now conceived of the Star purchase as a 'can't miss' deal. German investment, including promises to make up to six films and provide Hunnia with HP 200–250,000 of profit, would fund the expansion of Hungary's productive capacity. It would reprise the Osso arrangements of nearly a decade earlier, benefitting every segment of Hungarian film.[78] Beguiled, the Hunnia Directorate abandoned its request that the FIF underwrite the sale. Instead, Hunnia agreed to take on HP 600,000 worth of debt to pay the estimated HP 825,000 price for purchasing and rehabilitating the Star facility.[79] German officials intervened to speed up the transfer in April 1941, prompting the Prime Minister's Council to approve the deal.[80] In mid-September 1941, the Star was reborn.

What inspired Hunnia to get serious about renovating the Star was more than German promises; it was domestic competition, specifically rumors of the imminent MFI construction of a Kolozsvár studio.[81] Once Hunnia finalized the Star purchase, the MFI and its partners decided to shelve their plans.[82] Nevertheless, the prospect of a Kolozsvár atelier remained alluring to the most patriotic film professionals, who viewed the city as the seedbed of ancient Hungarian culture.[83] Talk of founding this studio, perhaps in cooperation with Italy, reached its zenith in late 1942 and persisted through the war.[84] Although a Transylvanian studio never materialized, the addition of the Star pavilion and the emphasis on making films for the Balkans was clear evidence of the appeal of Hungarian film in Southeastern Europe. This startled not only the Hungarian film industry, but the German and Italian film establishments as well. These states now devoted more attention to Hungary and its cinematic products. Unfortunately for Hungary, the type of attention it received was not always that which it desired.

Sovereignty and Hegemony: Hungarian-German *Filmpolitik*

The experience in Yugoslavia did more than boost the rapid ascent of Hungarian film. It also set in motion the industry's dramatic decline. While standing at Europe's filmmaking pinnacle with Italy and Hungary, Nazi film bureaucrats found they preferred to be alone, or at least to have their Italian and Hungarian counterparts on a tighter tether. Beginning in 1940,

Germany took steps to guarantee itself continental film hegemony, in part in response to Hungarian cinema's gains in Yugoslavia. Efforts to constrain the Hungarian film industry were various, ranging from negotiation to interference to outright coercion.

The abrupt alteration of the European film trade between 1938 and 1940 alone, German observers understood, could not explain Hungarian film's success, especially in Yugoslavia. At first, Hungarian accomplishments evinced grudging German admiration. An article in the 25 April 1940 edition of the official Nazi film journal *Film-Kurier* publicly complimented Hungary, noting that 'a good Hungarian film earns as much as a good German film'.[85] But behind the scenes, German authorities quickly became jealous of the appeal of Hungarian pictures. As early as the spring of 1940, German analysts expressed discomfort with Hungary's sustained good fortune, protesting that it endangered Germany's share of the Yugoslav market. The *Nachrichten für Aussenhandel* blamed Hungarian films for the drop in German exhibition income, implying that the rapid rise in Hungarian exports allowed Yugoslav theater owners to boycott German film.[86] The *Deutsche Akademie* reported to the Foreign Ministry that Hungarian movies were screening non-stop and averaging full houses of ethnically mixed spectators. The report's author made the astounding claim that one day's income from a good Hungarian film was equal to one month of German receipts. This crushing reality compelled theater owners to order Hungarian rather than German products, hampering the dissemination of Nazi ideology.[87] The second half of 1940 proved this assessment correct. Hungary displaced Germany in the Yugoslav market, ranking first in overall revenue and second, behind only Hollywood, in the number of films screened.[88] Hungary's best sold nearly as many tickets as pictures from the United States, and regularly outperformed German competition.[89]

In early June 1940, an agent working for Germany's Tesla-Film in Belgrade confirmed the *Deutsche Akademie's* analysis, emphasizing that Hungarian films out-earned German films not only in the Vojvodina, where the Hungarian population was high, but also in Belgrade, Zagreb, and other areas with little ethnic Hungarian presence.[90] In certain regions, the agent fretted, even ethnic Germans were willing to pay more to see Hungarian films than German ones. The writer attributed this, in part, to the persistence of a pre-Great War legacy of Hungarian cultural superiority shared by Yugoslavia's 'better circles'. Among elderly Yugoslavs, especially among Jews, wrote the anxious Tesla representative, preference for Hungarian film was clear, even when Hungary was no longer exporting its highest quality works.[91] Choices by Yugoslav audiences were proving the validity of Hungarian culture and, by implication, the inferiority of Nazi goods.

By the summer, the German Foreign Ministry joined the chorus of negativity, framing its critique in ideological and imperial terms. The head of the Cultural Section of the Foreign Ministry informed the German Embassy in Budapest that the spread of Hungarian film in Yugoslavia caused definite damage to German film, film companies, and more generally, German *Kulturpolitik*. The entire Yugoslav public, particularly its ethnic German component, was under threat from 'Jewish-distributed Hungarian film' because Hungarian 'film-Jews' were the 'chief carriers' of anti-German propaganda in Southeastern Europe.[92] The letter urged the German ambassador to convince Hungarian authorities that competition in the Yugoslav film market be 'carried out in a loyal way in accordance with German-Hungarian political relations'.[93] Put plainly, Germany demanded Hungary limit its exports to Yugoslavia. In a more complete sense, Germany desired greater control over Hungarian film, which it now saw as a legitimate threat to its rapidly evolving concept of New Order Europe.[94]

To counter the threat to its continental dominion, the German government took action on several levels, progressively cranking up the pressure in the early 1940s. Nazi authorities maneuvered to restrict sales of Hungarian films in Yugoslavia, negotiating fixed import fees two to three times lower than those levied on Hungarian films.[95] Who, precisely, convinced Yugoslav authorities that 'there were too many Hungarian language films playing in Yugoslavia' was unclear to some Hungarian officials, but not to Nándor Jenes, a vice-president in the Hungarian Film Office and a state secretary in the Manufacturing Ministry.[96] Jenes concluded that, 'Every [exported] Hungarian film takes the place of a German film,' and that this was the reason that Nazi apparatchiks felt the need to suppress Hungarian film in the Balkans.[97] In 1940 and 1941, nearly 50 percent of Germany's foreign film income came from Southeastern Europe, and Nazi authorities were not about to surrender the honey pot to Hungary.[98] The

Nazis had to have foreign audiences in order to successfully amortize their films,[99] and only they could be the 'usufructuaries of the economic process'.[100] Consumer-driven free trade in culture was simply not compatible with Nazi New Order designs.

This, however, was the most innocuous of Germany's attempts to stem the tide of Hungarian movie exports and command the development of Hungarian cinema. Despite the July 1940 extension of the German-Hungarian Film Agreement, film links between the two countries were no better in 1940 than they had been in previous years. Repeated promises to open the German market to Hungarian film had come to naught. German authorities were aggravated by Hungary's creeping de-Jewification, the state's unwillingness to seize control of the movie business, its refusal to allow German film into Transylvania, and German film's failure to win a larger share of the Hungarian market as other imports declined. Determined to reshape German-Hungarian film relations to better conform to Nazi priorities, Germany raised these issues in organized and *ad hoc* negotiations and diplomatic communications between mid-1940 and late 1941.[101]

The main obstacle to better relations was, unsurprisingly, the Jewish 'problem'. In nearly every discussion of cultural transfer, German officials conveniently conflated the Hungarian preference for anything Western and non-German, in other words anything that might compete with German cultural products, with Jewish tastes. Therefore, the Nazi campaign to accelerate the demolition of Hungarian Jewry went hand-in-hand with the German effort to dominate the continental film world. Continued Jewish participation in Hungarian film making, dissemination, and exhibition became the excuse for direct German intervention into the Hungarian motion picture business' sovereign affairs, especially after mid-1940. In bilateral film discussions, *Gauhauptstellenleiter* Otto Melcher blamed the continued domination of American, English, and French film in Hungary on the 'yet unbroken influence of the Jews on the Hungarian public's tastes'.[102] Dr. Karl Fries, a high ranking staff member of Goebbels' *Reichsministerium für Volksaufklärung und Propaganda* [RMVP], charged that continued Jewish manipulation of premier theater scheduling in Hungary constrained German features. Frequently, he claimed, only one premier theater, Ufa's Urania, headlined German movies before they were shunted to the countryside, depriving Germany of profit and propaganda opportunities.[103] German papers delivered diatribes echoing Fries' critique, savaging Hungarian film production and distribution for its continued employment of Jews.[104]

Neither did Hungarian film distribution policy in Transylvania please the Nazis. When the Hungarian army occupied northern Transylvania in August 1940 and declared it a military zone, it prevented the showing of anything but Hungarian films in local theaters. Years of being stymied by forked-tongued Nazi authorities had taught Hungary well, as its authorities promised that when the time was right, the films of other nations would be exhibited. This uppity behavior annoyed *Reichsminister* Goebbels, for one. His representative instructed the German Foreign Ministry and the German Ambassador in Hungary to rectify the situation immediately. 'This region,' wrote Fries, '...represents a particularly good market, and therefore the measures of the Hungarian military authority can no longer be endured'.[105]

Leading up to the 1941 Cultural Agreement talks, German frustration with the situation in Transylvania and with the distribution of films in the rest of Hungary boiled over. Since 1936, Germany had encouraged Hungary to raise its domestic exhibition quotas to help make the industry more 'national' and nationalistic. At the same time, to protect its interests, Germany pressured Hungary to guarantee an ever-increasing minimum number of German film imports. With the onset of war, these requests mutated into demands. Until 1939, Hungarian authorities had consistently avoided making specific concessions, believing that Germany had not honored its part of the bargain. However, escalating German arm-twisting, an emergent pro-German constituency in the government, and a shrinking number of alternatives pushed Hungarian officials toward the path of least resistance. They ostensibly submitted in late 1939, promising to slash the quantity of American, British, and French movies Hungary purchased from a total of 134 to half that amount in 1940, and to import more German and Italian films. The *Berliner Börsen Zeitung* applauded the decisions, claiming they took 'into account the wishes of the wide masses of Hungarian movie-goers who have, in the last years, been forced by Jewish film distributors to watch four times as many shallow American and French films as German'.[106] In July 1940 discussions,

German *Reichsfilmkammer* officials secured a Hungarian pledge to import 85 German films for the 1940–1 season and to reduce the number of American pictures on the Hungarian market.[107]

Hungarian authorities may have been obfuscating, or they may have been sincere in their commitment. Hungarian film importers, however, had other ideas, and only 42 German films premiered in Hungary during the 1940–1 season.[108] This irked German officials, who by December 1940 realized they would not achieve their export goals. They voiced dissatisfaction in diplomatic exchanges and meetings of the German-Hungarian Culture Committee during the first half of 1941.[109] At these gatherings, convened to rework the bilateral Cultural Agreement, Hungary's representatives did not sit silently. Their mushrooming domestic market and export success gave them the self-assurance to reply to their German opposites as somewhat equal partners. 'When the Hungarian government determines to…support the progressive expansion and valuable use of German film, it believes it obligatory to request of appropriate German authorities similar support in the interests of Hungarian film.'[110] While Hungary eventually promised Germany the right to send 85 films to Hungary, they complained, 'not a single Hungarian film, neither this year nor last year, has been shown in German territory, even though Germany's film leaders… have been to Budapest many times, and have taken notice of those movies which are acceptable from an Aryan standpoint.'[111] It was high time, crowed Hungary's embolden authorities, that the Third Reich, if it were serious about helping Hungary make its industry more national, honor the perennially unfulfilled pledge to permit at least five Hungarian-made films into Germany.[112]

The mixed Culture Committee discussed these divisive topics alongside less prickly ones, including securing studio time for German companies in Budapest.[113] For months, Hungarian negotiators tried to placate Nazi bureaucrats while sidestepping concrete assurances. Ultimately the talks failed, derailed by German insistence that its films be guaranteed exhibition space and/or a specific percentage share of the Hungarian market, requirements that even the pro-German Minister of Religion and Education found 'unrealistic'.[114] Thus, as Hungarian negotiators fought for 'national' interests, i.e. to retain domestic sovereignty, the most contentious film issues remained unresolved, particularly those tied to reduction of the quantity of American movies shown in Hungary and the purge of Jews from the Hungarian profession. The Nazis signaled their frustration by diverting attention to 'atrocity film'. The original 1936 Film Agreement, when ratified by both sides in 1938, included a clause stating that neither country would distribute or screen any film that offended the 'feelings' [*Gefühl, érzelem*] of the other or its inhabitants.[115] Once World War II began, Germany exploited this clause, attempting, in essence, to persuade Hungary that nearly every American film was a *Hetzfilm,* or inflammatory propaganda offensive to the German *Gefühl.* German authorities complained to their Hungarian counterparts, for example, that between 1–8 May 1941, 15 American films appeared in Budapest theaters and 'no less than 11 came from producers who were known to be writers of particularly hateful atrocity films against the German Reich.... This situation is hardly in accordance with the political union of the German Reich and Hungary and its continuation is intolerable'.[116]

This hard line approach also proved unconvincing. The Hungarian film establishment refused to capitulate and dump 'so-called [American] atrocity films'.[117] Appealing to arguments of sovereignty, Hungary's bureaucrats issued two rejoinders in the course of the 1941 negotiations. First, they insisted their censors were intelligent enough to determine which films were atrocity films. Second, they defined American film as a constituent element in the construction of their Christian national industry.[118] Premiers of newly imported American films had been dropping, down from 83 in the 1939–40 season to 67 in 1940–1. This was due mainly to war-related transport problems, currency restrictions, and Hungary's own de-Jewification program, which disrupted links with American distributors. Still, Hollywood outpaced Babelsberg 67 to 42 on the Hungarian market. Slicing this number or completely banning American film exhibition, argued Hungary's negotiators, would grievously injure Hungary's domestic Christianization effort.[119] Turning the Nazis' own antisemitic logic against them, Hungarian delegates explained that further cuts would deprive two important segments of Hungary's fledgling Christian industry of their livelihoods. Many distributors owned the rights to American films, and banning showings would mean certain bankruptcy. Christian-owned and managed small rural and late-run theaters relied on Hollywood features to fill programs.[120] Proscribing American film might have the counter-productive impact of reducing the future profitability of both Hungarian and German film by decreasing the number of theaters in which the films could be screened. Antál Ullein-Reviczky, then a high-ranking aide to the Prime Minister and a top Foreign Ministry official, concluded that 'it does not appear [that Hungary] could fulfill German desires in full measure' without causing its own film industry significant 'financial damage' and thus harming its own national, and Christian, interests.[121]

Unable to craft a new protocol, the two sides defaulted in July 1941, merely agreeing to extend the original 1936 Film Agreement, which Germany promptly violated.[122] In addition, its officials found alternative, less comprehensive methods of exerting influence over film in Hungary. They pursued unilateral actions: buying Hungarian theaters and licenses, renting Hungarian theaters for special free showings of German features, founding German film companies in Hungary, and dominating the Hungarian narrow film and *Kulturfilm* markets.

As early as 1938, German authorities had considered purchasing better access to the Hungarian film market. In October 1939, the Foreign Ministry and the Propaganda Ministry again raised the prospect of rental or purchase of at least a small theater in Budapest where German films and newsreels could be shown.[123] A year later, Nazi officials became convinced that they must acquire at minimum one large Budapest first-run cinema, or at least place it contractually in German hands.[124] Ufa's Urania theater, which exhibited 20 German films a year, was already at its maximum capacity. The remainder of Budapest's premier theaters, claimed Nazi bureaucrats, were still controlled by Jews who strangled German premier opportunities.[125] Multiple Reich emissaries advocated for the purchase of a second premier theater so that German pictures could be adequately exhibited.[126]

Nazi officials, acting through the German cinema conglomerate Tobis and its Hungarian proxy, Antal Schuchmann, purchased the Corvin cinema in September 1941.[127] This placed the two largest first-run theaters in Budapest in Nazi hands. It also permitted the Reich to consolidate other Hungarian companies in which it had financial interests.

Figure 14: The Corvin theater in 1941. Courtesy of Roel Vande Winkel.

Distributor Tibor Walter and his eponymous Tibor Walter Company became a partner in the Corvin deal and the legal purchaser of the movie theater. Because he was a Hungarian citizen, Walter was a crucial link, assuring Tobis unfettered operation in Hungary.[128] When Tobis fused with Ufa in 1942, the unified Nazi film industry possessed the means to directly control its Hungarian assets, and, if it so desired, to make films in Hungary.

Even before German officials had completed the Corvin deal, they had initiated a variety of filmmaking, exhibition, and distribution ventures in Hungary designed to better spread the Nazi *Weltanschauung*. So concerned were German authorities that their movie products and ideology penetrate Hungary, especially into Hungary's ethnic-German enclaves, that they arranged numerous free exhibitions of new German motion pictures, *Kulturfilme*, and newsreels, beginning in earnest in 1941.[129] This same year saw the proliferation of German-funded Hungarian film companies. In July, Erich Lübbert, the chairman of a Berlin-based transport company, garnered support for the creation of a film production company in Hungary from the German Foreign Ministry, the Propaganda Ministry, and the German Embassy. The company, originally envisioned as an incubator for film projects in the 'interests of the propaganda of both peoples,' transformed into a for-profit partnership. The Hungarian-German Society in Budapest, the German-Hungarian Society in Berlin, Lübbert, and the German Embassy in Budapest were equal partners. Using language that must have won the hearts of many Hungarian officials, Lübbert claimed that 'Hungarians were a very artistically gifted *Volk* and, in the framework of the Europe of the future, especially predestined for this sort of work'.[130] Pointing out that filmmaking costs were cheaper in Budapest than in the Reich, he won RM 100,000 in capital from the *Reichswirtschaftministerium* to set up the company that helped facilitate some of the German-funded pictures made in the new Star studio in late 1941. Ultimately, however, the enterprise proved neither long lasting nor influential.

German funds also helped establish Léna Film, an offshoot of the Mária Nagy Company, in 1941. Nagy specialized in the distribution of German film and newsreels in Hungary, and Léna's specific purpose was to enable the production of German films in Budapest. It proved more durable than Lübbert's endeavor, though most of the films it financed were Hungarian projects. Initially, German control over Léna was so tight that its director, Felix Szentirmay, resided in Berlin. In the summer of 1941, despite Lena Film's somewhat minimal role in Hungarian film production – it had yet to make one film – German authorities persuaded Hungarian officials to name Szentirmay as part of Hungary's delegation to the International Film Chamber [*Internationale Filmkammer* – hereafter IFK]. This ensured that the Hungarian delegation would contain at least one pro-German voice.

More influential were German ventures in Hungary concerned with the creation and circulation of narrow gauge (16 mm) films. This type of film included a range of products, from features and cultural films to advertisements. Both Hungarian and German film elites believed narrow film to have important persuasive power. Cultural film, educational film, news reports, tourism films, and advertisements in particular offered what authorities on both sides believed to be 'real' glimpses of their cultures.[131] They were vehicles for glorifying national or racial achievement, for visualizing and promoting national characters, ways of life, and ideologies.[132] Nazi bureaucrats also believed narrow film was especially suited for use abroad. Small enough to be transported by one person and exhibited on inexpensive projectors, narrow film could be screened by virtually any school, political, or social organization, especially in areas of German settlement outside the Reich. Normal features could be copied into narrow formats and distributed throughout Europe, avoiding licensing fees and circumventing Jewish-owned cinemas. Narrow film exhibition was also a means of generating buzz about normal German features. This was the logic behind the establishment of Descheg [*Deutsche Schmalfilm Exportgesellschaft*].[133] In October 1941, Descheg opened branches all over Europe, including one in Hungary called Kefifor [*Keskényfilm Forgalmi Kft.*]. Kefifor received the largest initial outlay of cash of any European office.[134] It also received monopoly status as the only distributor of German narrow film in Hungary, disseminating German culture film, newsreels, feature films, and projectors.[135] By 1942, Kefifor had secured agreements with Hungária Film Rt., the distributor of all Hunnia and MFI film, guaranteeing that more often than not, when Hungarian narrow film showed, so did the German variant.[136] By the middle of the war, as a result of Descheg's and Kefifor's efforts, there were hundreds of German films and thousands of narrow projectors in Hungary.[137]

By 1942, in total, German motion picture assets in Hungary were extensive. They included the Urania and Corvin premier theaters; several smaller theaters; two distribution companies, the Budapest Film Company and the Walter Tibor Film Company, both of which also held licenses to make movies in Hungary; the Kefifor narrow film company; and, of course, the Budapest branch of Tobis/Ufa. These fixed holdings were one element of the German effort to inject more of its film products into Hungary. A second element was the use of mobile sound film trucks [*Tonfilmwagen*], fully equipped cinemas-on-wheels capable of travel to remote areas for propaganda purposes or to serve troops. Hundreds of these vehicles held daily showings inside Germany and tens more took to the roads outside the Reich. The *Auslandsstelle des Lichtbilddienstes der Filmabteilung* [*Auli*], the adjunct of Goebbels' propaganda ministry charged with distribution and exhibition outside of the Reich, had planned to send nearly 40 sound film trucks to Hungary. In November 1940, when *Auli* sent its second *Tonfilmwagen* to Hungary, the Hungarian Foreign Ministry registered reservations.[138] When Germany tried to ship three more trucks in September 1941, Hungary thwarted the enterprise, denying permission for the trucks to cross its borders. As a result, only two mobile theaters served German interests in Hungary.

Despite the limit on the number dispatched to Hungary, the *Tonfilmwagen* had a substantial impact. In the first three months of

1943, they held 118 showings which attracted nearly 70,000 residents of Hungary.[139] German authorities hoped that this method eventually would provide exposure to much larger audiences, as it had in Italy where, in the same three months, film trucks reached almost one million viewers.[140] Taken together, sound film trucks, narrow film, and the purchase of theaters and film companies all were invasive attempts to manipulate the Hungarian film market and alter the content Hungarian audiences saw. Hungarian authorities rightly understood these efforts as part of a wider German attempt to control the matter and distribution of culture in Hungary and even as part of a larger Nazi plan to whittle away at Hungary's sovereignty. Perhaps the most outrageous example of this was the Nazi campaign to force Hungary to expunge American film from its market between 1941 and 1942.

Removing Hollywood from Hungary

Germany and the United States had battled for control of the world's sound movie markets since the conflict over sound formats in the late 1920s. After the Nazi seizure of power, Hollywood gained the upper hand. Unwilling to accept Hollywood's victory in Europe, Germany determined to use the war to alter the balance of power. Naturally, the conflagration devastated European film markets, destroying some and completely reorienting others. In Hungary, the war produced a relatively sharp and steady decline in imports from Great Britain, France, and the United States. This should have worked to Germany's advantage, but with the exception of a big increase from 1938–9 to 1939–40, the main beneficiary of this drop off was not the Reich, as Tables 1–4 show. In fact, after the 1939–40 film year, German film exports to Hungary never reached the 50 film per annum minimum determined by the 1936 bilateral Film Agreement, making the 1941 German demand that Hungary import 85 films seem utterly delusional.

German authorities explained this predicament by blaming continued Jewish conspiracies among theater managers and distributors to prevent screenings and deny access to larger Hungarian theaters. Hungarian statistics seem to tell another story that Nazi authorities were loath to accept: Hungarian audiences still preferred the products of Hollywood.[141] The numbers indicate that even while the United States lost market share in

Table 1: Feature Films Premiering in Hungary, 1938–43[142]

	1938–9	1939–40	1940–1	1941–2	1942–3
American	107	80 (78)	67 (68)	26	0
French	40	34 (36)	21	11 (13)	13
British	4	7 (8)	3	0	0
German	35	48	44 (42)	46	48
Austrian	1	1	0	0	0
Italian	1	1	8	17	29
Hungarian	33	27	38	36	51
Other	11	0	0	10	16
Total	223	198	181 (180)	146 (151)	157

Table 2: 1939–40 Film Exhibition Statistics[143]

	Number of Films	Total Number of Days Shown	Number of Days Average Film Shown	Percentage of Total Films	Percentage of Total Days Shown
American	78	1547	19.8	39.4	41.7
French	36	543	15.1	18.2	14.7
British	8	213	26.6	4.0	5.7
German	48	698	14.5	24.3	18.7
Italian	1	19	19.0	0.5	0.5
Hungarian	27	696	25.8	13.6	18.7
Total	198	3716	18.8	100.0	100.0

Table 3: 1940–1 Film Exhibition Statistics

	Number of Films	Total Number of Days Shown	Number of Days Average Film Shown	Percentage of Total Films	Percentage of Total Days Shown
American	68	1742	25.3	37.9	46.0
French	21	265	12.6	11.7	7.1
British	3	41	13.7	1.7	1.1
German	42	533	12.7	23.4	14.2
Italian	8	79	9.9	4.4	2.1
Hungarian	38	1110	29.2	21.1	29.5
Total	180	3752	20.8	100.0	100.0

Table 4: 1941–2 Film Exhibition Statistics

	Number of Films	Total Number of Days Shown	Number of Days Average Film Shown	Percentage of Total Films	Percentage of Total Days Shown
American	26	1001	38.5	17.2	24.8
French	13	200	15.4	8.6	5.0
Other	10	148	14.8	6.6	3.6
German	46	549	11.9	30.5	13.6
Italian	20	300	15.0	13.3	7.4
Hungarian	36	1839	51.1	23.8	45.6
Total	151	4037	26.7	100.0	100.0

Hungary, its features outdrew German competition by a wide margin, played longer, and generated greater daily income. Even as late as 1942, the theater runs of Hollywood pictures were longer than all but their Hungarian competition. That year, Hollywood products averaged a 38.5 day tenure, over three times as long as the typical German film, which lasted less than 12 days. Although there were twice as many German than American features on the Hungarian market, new German products accounted for less than 14 percent of theater time, whereas Hollywood pictures accounted for 25 percent.

In terms of absolute numbers, a transformation did occur in the summer of 1941. For the first time since 1932, the Hungarian market featured more German than American movies. This change was concurrent with Hungary's entry into the global conflict and with Germany's plan to film in Hunnia's new Star studio. That it occurred as the 1941 Hungarian law precluding Jewish participation in business partnerships came into effect and as the Commissioner for Unemployed Intellectuals István Kultsár intensified his extended effort to cleanse the Hungarian film profession of its Jewish interests was no coincidence. Yet this was also a time during which Hungarian-German film relations came under greater strain. Nazi authorities actively blocked the export of Hungarian features and culture films to German occupied lands, dismissing Hungarian protests by contemptuously claiming, for example, 'the time was just not right' for the exhibition of Hungarian culture film in Belgium.[144] Germany heavily curtailed or banned Hungarian movies in the parts of Yugoslavia the *Wehrmacht* occupied after April 1941.[145] And in the supposed interests of 'European film,' a concept that will be described in the following chapter, Nazi bureaucrats claimed a reduction in the numbers of American films exhibited, even if drastic, was not enough. In the summer of 1941, Nazi officials began insisting that Hungary cease screening any film originating in the United States.

In addition to the German-Hungarian Culture Committee, diplomatic exchange, direct investment, and incursions in the Hungarian film industry, Nazi authorities sought to exert leverage over Hungary through the International Film Chamber. Originally created in 1935, the IFK was inconsequential in film affairs until its reorganization and expansion in mid-1941. German and Italian authorities led the 17 member organization, whose purpose was, theoretically, to support the technical and artistic development of European filmmaking, to mediate disputes between members, and to promote the distribution of its members' movies.[146] By 1941, its actual function was to enable Nazi authorities to corporatize and dominate continental European film production, distribution, and exhibition.

Initially, Hungarian representatives had high hopes for the revived IFK. At their July 1941 gathering, IFK leaders recognized Hungary's newly earned film prominence and selected Hungary's László Balogh, the ONFB's second-in-command, as one of the organization's four vice-presidents. Top Hungarian officials, such as Balogh's boss Gyula Wlassics Jr., expressed guarded optimism that membership in the IFK would permit negotiations to secure the 'unmolested playing' of Hungarian movies in German theaters.[147] *Magyar Film* viewed the organization as the avenue for European-wide distribution of Hungarian film, a gateway to those parts of the continent Hungarian works had never reached nor been allowed, such as France, the Czech Protectorate, Romania, and German-occupied lands.[148]

Hungarian goodwill toward the IFK quickly dissipated. In the July 1941 meeting of the Chamber in Berlin, Nazi authorities forced all member countries to affirm Article 23 of the IFK's founding charter, which required that decisions made by its Arbitration Committee be binding upon all member countries. Hungarian officials recognized how Germany, as the dominant power in the IFK, would exploit this technicality. Kálmán Tomcsányi of the Interior Ministry marshaled representatives from six government ministries and three film organizations to discuss the implications of Article 23 on 20 November 1941. The Justice Ministry representative worried that acceptance of the IFK dictates might jeopardize Hungarian autonomy. If trumped by decisions of an international organization, Hungarian law might be contravened and national interests endangered.[149] Without a supplement to the paragraph stating that decisions made by the Arbitration Committee were 'only suggestive in character,' the IFK rules could represent a 'danger to state sovereignty'. Those gathered unanimously agreed that Hungary should interpret IFK decisions as recommendations, not requirements. Second, they decided to disregard entirely any IFK resolutions which violated Hungarian law.[150]

At the Munich gathering of the Chamber only four days later, Hungarian officials raised these and other objections to the IFK's basic rules. Their positions resonated with other countries, and they succeeded in affecting revision of the paragraph concerning the Arbitration Committee and its role in resolving disputes. The change took into account anxieties about national law, but did not render Arbitration Committee rulings suggestive as the Hungarians wished. Instead, the new language read: 'all disputes between members of the IFK must be decided through the Arbitration Committee in so far as [the decisions] do not contravene the laws of individual lands as they exist at the time'. This resolution was consistent with the overall tenor of the meeting, during which Hungary found most of its apprehensions about IFK conventions disregarded.[151] Hungarian representatives, for instance, lost the argument over atrocity films. The Hungarians favored the position that the designation of an atrocity film and subsequent action taken to ban it should be a matter negotiated by two sovereign states. The IFK committee, heavily influenced by its Nazi members, disagreed, asserting that all member states were obliged to prevent exhibition of atrocity films once the IFK designated them as such.

Other IFK decisions further accentuated the dangers Nazi supremacy posed to the Hungarian film establishment. In September 1941, the IFK met during the Venice Film Biennial. At this meeting, IFK policymakers, guided by a German outline devised before the conference, determined that only 25 of the movies produced by Hungary in 1941 would be deemed 'exportable,' a decision which severely disappointed the Hungarian delegation and curtailed Hungary's export horizons.[152] After the November decisions, it became increasingly obvious to Hungarian authorities that the IFK was little more than a front to enable Germany to tighten its grip on the European movie market. István Kutassy, an Assistant Secretary in the Ministry of Trade and a Hungarian representative on several IFK committees, eventually made these sentiments public. 'In the course of every [IFK] discussion...a strong German predominance was perceptible,' he wrote. So stifling and blatant was German pressure that he and other delegates instantly saw through the 'general stress on European solidarity,' which merely 'served the interests of German film production and film placement...' and the German desire to initiate the immediate removal of American film from European cinema houses.[153]

Hungary had long resisted German attempts to coerce it to cut American imports and, within the IFK, Hungarian representatives adopted the same defensive position. According to Pál Morvay, who helped formulate the Hungarian tactics for IFK meetings, industry representatives and Interior Ministry bureaucrats instructed Hungarian delegates to the import-export section to 'evade' all proposals for eliminating American film, despite the fact that Hungary and the United States already were opponents in history's costliest military conflict. The Hungarian stance attracted support from Italy and a group of neutral states. When Germany announced at the March 1942 distribution-import-export section meeting that the agenda of the general IFK meeting scheduled for April would include a ban on all American movies, consolidated and 'unexpected' opposition arose.[154] Like audiences in Hungary, Italian viewers had a voracious appetite for Hollywood fare.[155] Film officials of both countries, later joined by Spain, Sweden, Denmark, and Romania, felt the sudden withdrawal of these products from their respective national markets would be catastrophic for cinemas and distributors. When Kutassy spoke to the IFK general assembly in April, he created quite a stir. His government's position, he explained, was that only it had the right to forbid American films and the Hungarian delegation to the IFK lacked the authority to affect such decisions. This display of cheek 'enraged' Nazi film officials. Other IFK delegates, parting company with the Germans, 'roared with joy' as 'no one else was interested [in banning American-made film]'.[156]

Disenchantment with the IFK spread rapidly. At the March meeting, the import-export section established how many new features each member country required for the upcoming season, and from where those movies would come. Hungary, still the third leading producer of motion pictures on the continent, was awarded a much smaller share of the foreign markets than it felt it deserved. The IFK granted Hungary permission to send a mere four films to Italy and Spain, and 20 to Croatia. Hungary received no specific allotment for any other IFK member country.[157] This German-orchestrated put-down stung the Hungarian delegation and produced a backlash in Hungary's leading film circles, which manifested itself in continued Hungarian obduracy regarding Hollywood films.

In April, Germany ignored the objections of Hungary and its partisans and strong-armed the adoption of a rule legitimizing sanctions against any IFK members who failed to exclude US features from their markets by 31 December 1942. These sanctions included denying offending countries the raw materials necessary to produce film. The decision makers casually brushed aside the recommendation of an IFK expert committee on imports which had recommended US films be permitted to show until mid-1943.[158] Aggreived, Hungary's envoys formally objected and attempted to delay implementation of the decision. When this failed, Iván Kőszeghy, a ministerial advisor in the Trade Ministry, recommended that Hungary drag its feet in direct negotiations with Germany. He suggested developing a strategy to postpone the final date for elimination of American film, and/or to secure more time for exhibition of Hollywood pictures already in the possession of Hungarian distributors.[159]

At this point, German officials lost patience. Due to a stipulation of the Treaty of Trianon, Hungary could not produce its own raw stock.[160] By 1941, Hungary relied on Germany for at least three-fifths of its raw film, with Italy providing the remainder.[161] This reliance left Hungary in a vulnerable position, one which quickly proved untenable due to wartime shortages.[162] As of early 1942, Hungarian filmmaking was still proceeding at breakneck speed and European supplies of raw film could not keep pace. When Nazi officials threatened, in May 1942, to cut off Hungary's raw film supply and shut down its filmmaking capacity if Hungary did not comply with Nazi and IFK dictates, the situation became dire. The German representatives demanded Hungary agree that:

(1) No new American film [will] be imported or distributed, effective immediately.

(2) American films already in the possession of Hungary's distributors [will] be shown only until 31 December 1942.
(3) Reprises of American films [will] immediately cease being shown in premier theaters.[163]

The Hungarian subcommittee responded that it was fundamentally prepared to accept the German ultimatum, but only if its stipulations did not violate Hungarian law. Again, the excuse of sovereign legal rights – the right of a state to shape its own national film industry regulations – was offered as a method of postponing the imposition of Nazi mandates.

In June and July, Hungary and Germany held a series of bilateral meetings, during which they negotiated a schedule for delivery of German raw film. Hungary again beat the tired horse of the Reich's promise to import five Hungarian films, and German officials again dissembled. Hungarian representatives then broached the issue of obtaining a license for a premier theater in Berlin to promote the exhibition of Hungarian film in Germany. In reciprocation for the Corvin theater in Budapest, this must be seen as a reasonable request, suggested the Hungarian negotiators. Instead of outright rebuffing the proposal, the German contingent merely promised to forward the idea to the proper authorities.[164]

Increasingly convinced that they were unlikely to win any concessions from Germany whatsoever, Hungarian film officials sought alternatives to the German supply of film stock, while at the same time taking the necessary steps to appear to fall in line with the German dictates. By the summer the Hungarian Film Office was begging its raw film supplier, Italy's Ferrania, to up the amount of stock it sent the studio. In October 1942, Nándor Jenes, an employee of the MFI and an official in the Manufacturing Ministry, travelled to Milan to negotiate the acquisition of one million meters of Ferrania's raw film. Jenes hoped that the deal would go through, despite expected German opposition.[165] So while Hungarian-German film relations festered, Hungary redoubled its efforts to improve relations with Italy.

Hungary was not about to pin its film industry's future on Italy alone, however. The Ministry of Manufacturing also searched for additional solutions to the raw film crisis, and it began negotiations with a European subsidiary of Kodak regarding the construction of a raw film factory in

Hungary during the summer of 1942.[166] Although I have found no evidence that anything came of the initial inquiries, that government officials were willing to entertain thoughts of an American-owned company building on their lands against the wishes of Germany while Hungary was officially at war with the United States indicates how desperate Hungarian film figures believed the situation to be, and how important filmmaking was for Hungarian popular culture.

Even among the most pro-German segments of the Hungarian film profession, there was resentment over the Nazis' heavy-handed behavior. By August 1942 a spate of articles had appeared in the official film profession journal, *Magyar Film,* calling for the sparing of raw film and its recycling. In late September 1942 the Hunnia Directorate warned of a 'catastrophic lack of raw film'. The Directorate said it would only consent to dispatching one of its best cameramen, István Eiben, to complete work on the Hungarian-German co-production *Karneval in Rom* if Germany were to first ship the promised raw film.[167] Only days later, *Magyar Film's* 5 October 1942 lead article urged the Hungarian film profession to start making stock for itself, a call echoed in other journals.[168] As of October 1942, Hungarian production was on the verge of a complete stoppage, a result of Germany's failure to deliver most of Hungary's raw film allotment.[169] Filmmaking projects were delayed, and directors hoarded supplies, limiting the number of takes. It became harder to edit. Contemporary observers predicted a decline in future production and noticed a commensurate drop in the quality of pictures made near the close of 1942.[170] In October and November, there was a flurry of movie-related diplomatic activity, last ditch attempts to resolve differences before the IFK's Supreme Council had its tri-annual gathering, by chance scheduled for Budapest at the end of November. At that meeting, Hungary's new Minister of Religion and Education, Jenő Szinyei Merse, declared that the provision of raw film had been 'the great' and most vexing question faced by the Hungarian industry in 1942. He also noted, in an oblique jab at Germany, that the raw film shortage threatened the well-being of foreign film in Hungary, as Hungary risked not having enough material to copy the films and assure their distribution.[171]

As of early December, while the Budapest IFK meeting was in progress, Germany still had not shipped 721,000 of the 930,000 meters of film stock it had pledged to provide Hungary by 31 October 1942. This tactic forced

Hungary to fold. At the December meeting, its representatives agreed their state would end showings of American films by 31 December. As the IFK gathering closed, Hungarian Trade and Interior Ministry officials began rescinding exhibition licenses and forbidding all American film imports.[172] As of the 31 December 1942 deadline set by Germany, Hungarian authorities had, for all intents and purposes, banished most American film from Hungarian screens. The Nazi campaign appeared to win a full and final victory on 16 January 1943, when a trade ministry official announced that licenses for all American feature films had been cancelled.[173] In addition, Hungary would no longer decide how many films it would produce. Germany would do so, via the IFK. In 1943, the Chamber gave Hungary a quota of 45 features and awarded it enough raw stock to make these films.[174] The Faustian bargain made, Hungary's national film industry avoided complete collapse but only by surrendering some of its sovereignty.

Although the Trade Ministry's submission may have been in accordance with Nazi wishes, the Keresztes-Fischer-led Interior Ministry continued to behave independently. It refused to rescind all short film permits, because to do so would deprive United States film companies with offices in Budapest of all ability to function.[175] In other words, these companies, with nothing to trade, would have to dissolve, inflating the ranks of the jobless. At an inter-ministerial conference concerning US movies, the participants from several ministries decided to protect the American companies' presence in Hungary and to communicate this to the Germans.[176] Whether or not this particular decision provoked a German reaction is not borne out by the extant archival records. However, soon after, Germany canceled a contract for Wien Film to make two features at Hunnia's new studio, using the excuse of exchange rate questions and a new German prohibition on making films abroad.[177]

Did Hungary become the object of Nazi ire simply because its film industry officials were willing to oppose a few of its decrees and keep Hollywood in Hungary? Some film figures felt there was more to the story. By 1942, Hungary had established itself as a major film power, particularly in Southeastern Europe. Hungary ranked second as a provider of films to Croatia, cornering nearly 25 percent of its neighbor's market by sending 55 features in 1942.[178] Even into 1943, Hungarian films still 'densely' populated Croatian theaters, and Hungarian starlet Zita Szeleczky's face adorned many a Croatian newspaper.[179] Hungary was also the number two supplier of film to Bulgaria, shipping 42 features in 1942, nearly doubling the amount sold by third place Italy.[180] A 1942 Hungarian-Bulgarian co-production titled *Opportunity* [*Alkalom/Izpitanie*] and made at the MFI studio was a smash in Bulgaria, deepening the cultural connection between the two states.[181] Hungarian features continued to make their way to Greece and a handful even began to appear on Turkish screens. Significant numbers of Hungarian movies also opened in Finland, and distributors in Denmark, Norway, and Sweden continued to demand Magyar movies.[182] Two Hungarian films, *People on the Alps* [*Emberek a havason*] and *To the Fourth Generation* [*Negyediziglen*] won top awards at the 1942 Venice Film Biennial. As late as February 1944 Hungarian radio boasted that 'in every country in Europe they are showing Hungarian films and everywhere they…are very popular. The Hungarian film industry is the most outstanding…in all of Europe.[183] The MFI's Nándor Jenes believed it was this combined success that most irritated Germany. This was the reason that Nazi film officials felt the need to suppress Hungarian film in the Balkans, to cut off access to export markets, and to limit the growth of Hungarian film by 'decreasing shipments of raw film'. These were the 'most sure-fire indirect tools' they could use to achieve their goal: 'to subdue Hungarian production'.[184]

Removing Hungarian Film from Europe

To shackle Hungary and advance the process of constructing an autarkic European film system based on National Socialist principles, German authorities both limited Hungary's production possibilities and actively undermined Hungary's ability to export.[185] In addition to using the IFK rather than the market to determine the number of Hungarian films to be distributed throughout Europe, Reich lackeys controlled the actual distribution methods as well by means of the Transit Film Company.[186] The Germans did not establish Transit with the goal of containing Hungarian production. Rather, Transit was one part of the Goebbels-led effort to unify all German film production, distribution, and exhibition. Specifically, the Nazis used Transit to coordinate all European film trade, tying together the disparate regional distribution companies they had previously established.

By May 1942, Nazi authorities had made Transit the sole firm legally allowed to import movies for the German market.[187] Transit's domain included currency exchange, censorship, and quota questions for the Reich and all German-occupied lands.

Even before Transit was fully operational in its expanded capacity, Hungary's film authorities realized the threat it posed to the health of the whole industry. László Balogh sounded the alarm in a March 1942 meeting of the National Film Committee, explaining that: 'Hungarian film going beyond [Germany] to other countries…would have to use the Transit path. Thus, to sell to Bulgaria, to Croatia, etc., the individual countries' central film services would grant licenses only to Transit'. Balogh further warned that the Nazis would use Transit to enforce the unfair export quotas imposed at the International Film Chamber meetings, and that 'the expansion of Transit might bind up [*gúzsbakötné*] the entire manufacture of Hungarian film, so that… [Hungary would be permitted] no greater motivation than being backwater, creating products suitable for satisfying the demands of the Hungarian countryside. This would definitely lead to the regression of Hungarian film'.[188]

Given this danger, Hungary's ONFB resolved to establish an outpost for Hungarian film in Berlin, empowering a representative who would deal directly with Nazi authorities rather than through Transit. ONFB members still imagined shipping 30 films to Germany each year, and raking in annual profits of RM 2–3 million.[189] István Kutassy, who became the ONFB emissary, had no more luck wringing concessions from Germany than any previous Hungarian envoy. According to Kutassy, Transit could scuttle all Hungarian film trade, and the idea of creating a state-supported Hungarian Film Center in Berlin was a non-starter. Such arrangements had not aided other states. 'Germany had obstructed every one of our attempts…[to sell film on the German market]', and thus Hungary's only viable options were to cooperate with Transit and to work within the IFK to get its viewpoints heard, possibly in concert with Italy. Otherwise, it may as well withdraw from the IFK and 'renounce exports, in principle, to the German market'.[190]

In 1943, Hungarian fears proved correct. Transit became the sole pathway into Germany and all the countries German forces occupied. Although it was within its rights to export to neutral or independent lands allied with the Axis, Hungary was quite limited in its options. Hungarian exporters only successfully circumvented Transit in Italy and in lands directly bordering Hungary, such as Croatia. Trade with Bulgaria, Switzerland, Turkey, and the other Scandinavian countries did continue, but wartime conditions and German interference constricted these markets. The Nazis, for example, forbade Vichy France from importing Hungarian film.[191] István Erdélyi claimed that if Reich censors in Berlin did not clear films for showing in Germany, Germany would refuse to grant export licenses through Transit.[192] This policy made it nearly impossible to provision Hungarian minorities in occupied lands with Hungarian film. It also meant that in nearly all areas of German occupation, Hungarian film was excluded. To wit, in 1943 German authorities not only forbid the screening of Hungarian films already in Serbia, but in the first three months of the year, Transit rejected 20 of 21 Hungarian requests for Serbian export licenses for new Hungarian works.[193]

Behind the long Transit shadow, however, there were glimmers of the schizophrenic hope that characterized German-Hungarian film politics. In March 1943, the head of Transit visited Budapest and viewed nearly 30 Hungarian films. He selected seven to exhibit in Germany and the occupied lands, and told Géza Paikert, an Advisor to the Minister of Religion and Education, that he wished to raise the quota of Hungarian films shown in Germany. This induced the usual paroxysm of fantasy on Paikert's part. He predicted that Germany and Hungary would finally iron out the details of their bilateral agreement, the 'future would naturally be better' and Hungary should expect up to RM 400,000 in annual film sales to the Reich.[194]

Of course Paikert, like the long line of optimists before him who had wistfully dreamt of Hungarian exhibitions in Reich theaters, was wrong. Measurable exports to Germany never materialized. Film relations with Germany continued to be marred by mistrust, jealousy, and the fight over markets. Virtually every film exchange and continued Nazi meddling in the affairs of Hungarian film caused strain. The German Embassy kept tabs on many of the new Hungarian production companies.[195] Ufa monitored all Hungarian filmmaking activity, keeping lists of Hungary's leading actors, actresses, and directors.[196] Nazi authorities in fact fired Hungary's leading male, Pál Jávor, from a role he had contracted to play in a German version of *The Danube Sailor* [*A dunai hajós*] once they discovered his

Jewish wife and his open opposition to the Jewish laws.[197] Germany also forbade the import of any films starring Jávor, constituting nearly one of every six Hungarian wartime works, including some of the most successful films of the era. This was an obvious slap-in-the-face to the Hungarian film industry.

Rebuffed by Germany at every turn, Hungarian film professionals and bureaucrats naturally found themselves drawn toward Italy, Europe's second most powerful film producer. The attraction was mutual. Relations between Hungary and Italy grew warmer for a variety of reasons, but primary among them was their shared opposition to German continental movie hegemony.[198] At the Rome meeting of the IFK in April 1942, delegate Pál Morvay reported that Hungary and Italy were the most vocal supporters of independent action, free of German dictates. Morvay detailed long talks with Alessandro Pavolini, the top Italian IFK representative. Since the primary purpose of the IFK was to promote the German film industry and the sale of its products, complained Morvay, 'the positions of Italy and Hungary in particular suffered shipwreck'.[199] He and Pavolini agreed that they must stand together, whether or not they were joined by other IFK member states, in opposition to German imperialist ambitions.

Soon after this discussion, bilateral Italian-Hungarian film relations palpably improved. *Magyar Film* published a number of articles on the budding 'Hungarian-Italian film friendship'.[200] In June 1942, Italian film companies expressed interest in making co-productions at the Hunnia studios and, within a month, screenwriters were already scribing screenplays.[201] At the November 1942 meeting of the IFK, Eitel Monaco, the official in charge of cinema in the Italian Ministry for Popular Culture, announced that Hungary and Italy would make between three and six co-productions annually.[202] Italy opened a branch of the Esperia Company in Budapest to facilitate these joint film projects and to advance bilateral film exchange. Hungary reciprocated, directing the Hungarian Film Export Cooperative to augment Hungarian film sales in Italy.[203] Hungary also turned to Italy to bolster its dwindling supplies of raw film as the conflict with Germany came to a head. Several Hungarian actors, including its biggest stars Pál Jávor and Zita Szeleczky, and two of its top directors, Ákos Ráthonyi and Géza Radványi, traveled to Italy to act in and direct Italian movies. Scores of Hungarian actors studied at the Film

School in Rome until the program abruptly ended when Germany occupied Italy in 1943.[204]

It would be inaccurate to declare Hungarian-Italian film relations a success story with a happy-ending. Hungarians complained that films sent to Italy received short runs in lesser theaters.[205] Tensions arose over the comparative success of Hungarian films in the Balkans particularly while Italy, which also considered the former Yugoslav film market a 'vital space' and like Hungary had revisionist claims on Yugoslav lands, was unable to gain a significant share of the market.[206] Co-productions were censored by both sides. Hungary, for example, outright forbade the showing of the *100,000 Dollar Girl* [*Száz-ezer dolláros leány/La ragazza dei 100.000 dollari*], an Esperia product filmed in Budapest, because of its 'sharply anti-Hungarian' characters.[207] Friction also developed when Hungarians groused that Italy's alleged pro-Romanian bias surfaced in its films.

It is reasonable to conclude, however, that despite these disagreements Hungarian-Italian film ties, virtually non-existent until the late 1930s, flourished during the wartime period, especially in contrast to Hungarian-German relations. Even after Italy's revolution and Nazi imposition of a puppet government, the popularity of Hungarian film in Italy continued to rise. In the first six months of 1944, 21 different Hungarian films premiered in Fiume alone, according to the Hungarian Consul, and many of the films appeared in other Italian towns and cities.[208] Italian producers continued to scour Hungary is search of talent, and film production heads promised even closer ties.[209]

Friendly Hungarian-Italian movie relations irritated Nazi film officials, who rightly viewed the two countries as united against Nazi interests through 1943. The German occupation of Italy put a stop to most Italian-Hungarian opposition. But according to German officials, even without its Italian partner in crime, Hungary continued to misbehave. Put simply, its offense was independence. For example, its public and its distributors had the temerity to express a continued preference for American film. Although Hungary did not officially import a single American film in 1943 or 1944, Hollywood's products retained their allure. According to US sources, surreptitious showings were the norm by 1943, sometimes even in regular movie theaters.[210] A report from the US Office of War Information, probably written in early 1944, stated that:

> American motion pictures are still widely tolerated in Hungary. A recent survey of all motion pictures shown over a period of six weeks revealed that 40 percent of them were of Hungarian origin, while another 40 percent was [*sic*] made in the U.S., Germany represented by only 13 percent, 7 percent coming from France.
>
> Despite obvious difficulties, American movie representatives are still permitted to maintain offices in Budapest.... the firms now release old movies, old hits, and it may be stated that such revivals are widely applauded by the general public.[211]

Hungarian and German sources confirmed this report, indicative of Hungary's a complete abrogation of the deal concluded with Germany at the end of 1942.[212] Joseph Goebbels' assistant Karl Fries, just days before the German occupation of Hungary in March 1944, disclosed that a few Disney films had recently opened in Budapest. Hungary, he noted with dismay, was demonstrating 'an increasingly strong tendency to allow American films to be shown'.[213] In the final analysis, it appears that even war could not quell Hungary's love for Mickey Mouse and Mickey Rooney. Only when it put tanks on the streets of Budapest was Germany able to finally drive a wedge between Hungary and Hollywood and fully subordinate Hungary's film community.

Conclusion

Hungary's membership in the pantheon of European filmmaking was short. By 1944, the flame of success flickered as the course of the war extinguished export markets. In the West, Germany had long denied Hungary access to most states, occupied or not, and it cut off official Hungarian movie sales to Italy in 1943. In the East, first Bulgaria, then Croatia was lost to the Red Army and the Partisans. By the time German forces deposed the Horthy government and Hungary's domestic industry fell into the hands of the Arrow Cross in November 1944, feature film production had all but ceased. Film export was but a fleeting memory.

Wartime conditions created space for Germany and Hungary to cooperate, compete, and come into conflict over cinema. This complex international and transnational dynamic meant that developments outside of Hungary heavily influenced the evolution of Hungary's film industry and thus the construction of, and character of, Hungary's 'national culture'. In early April 1941, just as Nazi and Hungarian forces were poised to invade Yugoslavia, Hungarian filmmakers trumpeted the arrival of a 'golden age' of Hungarian film.[214] This cinematic heyday, due primarily to the triumph of Hungarian exports to Yugoslavia and beyond, was a mirage, an illusion conjured and then destroyed by Nazi Germany. Whether its films were purchased because they were genuinely liked by audiences, viewed as a proxy for anti-German sentiment, or necessary simply to fill programs, Hungary was a major beneficiary of the unprecedented paucity of film in continental Europe resulting from the German-initiated war.[215] This demand, in turn, had the ironic impact of extending the professional lives of many of Hungary's 'film Jews', whose talents Hungary required to insure sufficient production. For all of this, Hungary had the Third Reich to thank.

An unintentional benefactor, Germany was doubtless an intentional competitor. Neither altruistic nor benevolent, it was the primary obstacle blocking Hungarian film from making greater inroads into foreign markets. Nazi authorities came to believe that Hungary's success on the silver screen undermined their imperialist mission in the Balkans and in other European venues. German bureaucrats perceived the threat to the Reich's economic, ideological, cultural, and political dominance to be substantial enough to merit a multi-faceted effort to restrain not only Hungarian exports to Yugoslavia, but the Hungarian film industry as a whole. The continued presence of Jews in Hungarian film production and distribution rankled Nazi authorities, and they used this issue as an excuse to directly intercede in Hungarian affairs.

From persuading Yugoslav authorities to impose higher tariffs on Hungarian film imports to purchasing theaters and film companies in Hungary to using the International Film Chamber as a mechanism to manipulate Hungary, the Germans were quite creative in their meddling. This sort of action became so pathological that Nazi officials went so far as to protest the manner in which the Hungarian film *Landslide* [*Földindulás*] portrayed an ethnic German landowner, asking that the film be forbidden everywhere, including *inside Hungary*.[216] The Nazi penchant for micro-managing is cliché, so the size of the dossier of incursions into Hungarian film comes as no surprise. Neither is it shocking that the Nazis eventually concluded that Hungary had done nothing to help Germany create its New Order, had failed to contribute to the 'war to defend European popular culture,' and as a country was biased against German film.[217]

From 1942 through the cessation of Hungarian filmmaking after October 1944, German actions largely determined how many films Hungary could make and where those films could be exported. German authorities also influenced the content of Hungarian newsreels and the exhibition of narrow films.[218] Yet it would be unfair and incorrect to conclude that Hungary was a complete pawn of the Nazis or that its film establishment had no control over its own industry's evolution. Hungary produced nearly twice as many features between 1940 and 1944 as it had in the previous decade. Persuaded of the cultural, political, and economic importance of the movie medium, Hungary's filmmakers and politicians were confident in their cinema and in the importance of Hungarian culture within Europe.

Thus fortified, Hungarian officials endeavored to stake out their own ground, walking a fine line between appeasement of and resistance to German demands. Perturbed by Nazi treatment of their country as the ugly step-sister, film authorities became willing to challenge German motion picture supremacy. An unsigned letter, probably written by Zoltán Tőrey, the Managing Director of the MFI, was emblematic of latent anti-German sentiment. 'German capitalism has stepped into the place of international Jewish capitalism' and this economic subjugation represented a 'great problem for the future of Hungarian economic life'. German economic hegemony will serve to promote its domination of all facets of European life, culture in particular. Given this new framework, Hungary must think hard about how it must appear to play the role of the 'good boy' [*jó fiu* (*sic*)] while protecting its national interests.[219] And therein lay the dilemma. As they forged a defiant, semi-independent path by exercising the limited agency they possessed, Hungarian film elites again took contradictory stances regarding perceived national interests, further complicating their own attempts to nationalize and Chrisitanize their industry.

Notes

1. Parts of this chapter originally appeared in 'A Smashing Success? The Paradox of Hungarian Cultural Imperialism in Nazi New Order Europe, 1939–1942', *Journal of Contemporary History*, 51/3 (2016), 577–605. DOI: 10.1177/0022009415622804
2. Bethlen, quoted in Bethlen & I. Bolza, *Hungarian Politics during World War Two: Treatise and Indictment* (Munich, 1985), 16.
3. Márai's 1942 quote in I. Romsics, 'From Christian Shield to EU Member', *The Hungarian Quarterly*, 48/188 (Winter, 2007), 22.
4. Goebbels quoted in M. S. Phillips, 'The German Film Industry and the New Order', in Peter Stachura (ed), *The Shaping of the Nazi State* (London, 1978), 263.
5. For Hungary, the repercussions were significant. In the 1939–40 film year, Hungarian theaters exhibited 198 features, including 36 from France and eight from Britain. The 44 total films represented 24 percent of all movie premiers. In the 1940–1 film year, of the 180 films Hungarian theaters screeened, 21 came from pre-defeat France and a mere three from Britain, reducing the French/British share of the Hungarian market to 13 percent. In the 1941–2 film year, only 13 French and no British films were imported. Frigyes Pogány, 'Utóhang a nemzetközi filmkongresszushoz', *MF* IV/51 (21 December 1942), 2. German and Hungarian film authorities agreed that Hungary would only import French films made before May 1940. See *AAA* - GsB, Fach 27, Kult 12, Nr 4a, 'Ungar. Filmprop in Yugoslawien'. Erdmannsdorff to AA, 28 October 1940, Nr.3218.
6. Henrik Castiglione and István Erdélyi estimated the addition of the *Felvidék* and Ruthenia would add 68–85 theaters and a five to 14 percent increase in per movie ticket sales. Castiglione, 'A magyar mozipark új térfoglalása. A visszacsatolt Felvidék bekapcsolódása a magyar filmélet vérkeringésébe', *MF* I/3 (4 March 1939), 7–10; Castiglione, 'Ruszinszkó és a magyar mozipark', *MF* I/19 (24 June 1939), 8; Erdélyi's remarks in 'Az OMME közgyűlése', *MF* I/20 (1 July 1939), 7.
7. 'Hány mozi működik jelenleg Erdélyben?' *MF* II/39 (28 September 1940), 4.
8. 'A délvidéki mozgószinházak jegyzéke', *MF* III/19 (10 May 1941). Andor Lajta, later downgraded this estimate to 60 theaters. See Lajta, 'Az év története', *Filmművészeti Évkönyv* (1942), 178.
9. Statistics from Langer, 'Fejezetek', 207 and Géza Ágotai, '400 Schmalfilmtheater in fünf Jahren', *Interfilm: Blätter der internationalen Filmkammer* Heft 6, (August 1943), 76. Ágotai asserts that Hungary had over 800 normal theaters and over 400 narrow film theaters in operation in 1943. By the end of 1943, there were over 500 narrow film theaters. Andor Lajta, 'Az év története', *Filmművészeti Évkönyv* (1944), 254.

10. Lajta, 'A magyar film története', 72. The 53 percent increase domestic cinemas closely mirrors the 58 percent increase in population Hungary experienced between 1938 and 1943.
11. The math is fuzzy. The 1939 ticket numbers are from Géza Matolay, 'Sürgősen és egyszerűen kell megoldalni a magyar filmgyártás problémája', *Függetlenség* (16 July 1939), 5. The 1940–2 figures are from several sources, primarily Langer, 'Fejezetek', 210, who, without a citation, reports that theater revenue rose to HP 61 million in 1942. An Interior Ministry memo confirms Lánger's 1940 estimate of HP 33 million. See *MOL-Ó* - Hunnia, Z1124-r1-d21. Német kivánságok...1941–2. Keresztes-Fischer to Hóman, BM 163.442/1941 V.sz., 20 February 1941, 63. László Balogh told his fellow ONFB members in a December 1943 meeting that Hungary's approximately 800 operating normal film theaters took in nearly HP 90 million in 1942. *BFL*– László Balogh, Nb. 1699/1945. 'Jgykv. felvétetett az ONFB 1943. évi december hó 10-iki üléséről', 4. Both Langer's and Balogh's estimates could be exaggerated, as new films shown in Hungary in 1943 only yielded around HP 32.5 million in income according to the Reich Nazi paper *Neue Ordnung* Nr. 146 (21 May 1944). *BADH* - R4902/5972. DIA, Zeitungsausschnitssammlung. That reprise showings and perhaps concessions could account for an additional HP 60 million of income seems unlikely, but possible, as the HP 90 million number was repeated numerous times 1943–4. See Gy. Wlassics, Jr., 'A megszállottság bűvöletétől kisérve készül a magyar film, aminek lekicsinyelése igazságtalan volna', *MF* V/48 (1 December 1943), 1–2.
12. J. Bingert, 'Kvalitás és export', *MF* II/51 (21 December 1940), 3.
13. 'Kis nemzetek a filmversenyében', *MF* III/20 (17 May 1941), 3.
14. '200 magyar hangosfilm', *MF* III/2 (11 January 1941), 1.
15. 'Szent István ünnepére', *MF* III/33 (16 August 1941), 1.
16. 'A harmadik helyet foglaljuk el Európában', *Országépítés* (undated, 1941), 14. From *MOL-Ó* - Hunnia, Z1124-r1-d5. A filmgyártás helyzete...1941.
17. 'Zukunftsfragen des ungarischen Films', *Pester Lloyd* 26 (1 February 1941).
18. *MOL* - KM, K66-cs474, 1940, III-6/c, 404. Kozma to Ullein-Reviczky, 19 May 1940.
19. Pál Szvatkó, quoted in 'A hazai film ünnepe', *MF* III/3 (18 January 1941), 3. *Magyarország*, the government's paper, was one of Hungary's leading dailies.
20. Miklos Kádár, 'A magyar film export-lehetőségei', *MF* II/14 (6 April 1940), 2. To create hype and establish the festival as the third-ranking European annual event behind Venice and Berlin, Hungarian officials chose to move the 1941 gathering to Margaret Island in Budapest. The plan ended in disaster.
21. *Alter Ego* [*Alteregó*] and *Europe Doesn't Answer* [*Europa nem valaszól*] won widespread acclaim at the 1941 Biennial and a third film, *Flames* [*Langók*], played well in Italy. Some Hungarian diplomats criticized their country's entries in the competition as 'not suited for today's ideologies espoused

by the Italian and German publics'. Antal Páll, Hungary's cultural attaché in Venice, remarked that 'All the other films from other countries, without exception, offered views of their histories and their national and volkish lives. We Hungarians, however,...came to Venice with a film with a theme more-or-less about the coming apart of a marriage and a boat leaving America'. He urged better future selections with more emphasis on propaganda. *MOL* – KM, K99 [Rome Embassy]-cs103, 1941. Páll to ME Bárdossy, Venice, 19 September 1941.

22. 'Magyar siker Velencében', *MF* III/38 (20 September 1941), 1.
23. 'Túlzott igények', *MF* III/12 (22 March 1941), 1. See also 'Die ungarische Filmproduktion im Spiegel des Jahresberichts der Budapester Handels- und Gewerbekammer', *Pester Lloyd*, 197 (30 August 1941).
24. Hungarian-Yugoslav relations reached their nadir in 1934, when Yugoslavia accused Hungary of having a role in the assassination of King Alexander I, and in retribution expelled thousands of ethnic Hungarians. On the Yugoslav crisis and its after-effects, see Mária Ormos, *Merénylet Marseille-ben* (Budapest, 1984).
25. Pál Pritz, *Magyarország külpolitikája Gömbös Gyula miniszterelnöksége idején 1932–36*, (Budapest, 1982), 224–5.
26. *MOL-Ó* – Microfilm # 10939. AA Politische Abt., Akt. betr.: Politische Beziehungen zw. Jugoslawien und Ungarn von 12 Juni 1936 bis 13 Juni 1941. GsB (von Mackensen) to AA, A. Nr. 114 P 31, 7 May 1937.
27. *MOL* – KM, K66-cs336, 1937, III-6/c, 383, 387. MKK Belgrád (Rothkugel) to AA, Nr.7349/1937, 25 September 1937.
28. *MOL-Ó* – Hunnia, Z1123-r1-d1, 1938. 14 March 1938 Hunnia Igazg. jgyek.
29. 'Nyilatkoznak a 'Tisza Filmgyár' alapítói', *A Film. A Magyar Mozi és Filmszakma Lapja* IV (July-August 1937), 1. The Tisza film studio was not constructed until late 1943 and never produced a feature.
30. 'Magyar-jugoszláv filmtárgyalások', *Mozgó Élet* I/4 (30 September 1937), 4.
31. 'Magyar Kulturfilm Bemutató Belgrádban', *FK* X/10 (1 October 1937), 6.
32. *MOL-Ó* – Hunnia, Z1123-r1-d1, 1938. 14 March 1938 Hunnia Igazg. jgyek, 18–22.
33. Kozma quoted in Pál Pritz, *Magyar diplomácia a két háború között* (Budapest, 1995), 221; Ormos, *Egy magyar médiavezér*, 592.
34. *MOL* – KM, K66-cs418, 1939, III-6, 49. KM memo, ad 60.794/1938, 18 January 1939.
35. 'Blick nach dem Südosten. Kulturpolitische Ziele Ungarns', *Neue Freie Presse* (5 December 1939).
36. The primary reasons for this change were the replacement of Milan Stojadinović with Dragiša Cvetković as Yugoslav premier, the dismemberment of Czechoslovakia and the subsequent collapse of the Little Entente. See Foreign Ministry, *Külpolitikai adatok az 1939.évről* (Budapest, 1941), 92–3.
37. 'Tőrey Zoltán dr., a M.F.I. ügyv. igazgatója, a Filmművészeti Kamara alelnöke eladása', *MF* I/17 (10 June 1939), 5–6.

38. The Cvetković-Maček Compromise or *sporazum* was in essence a Yugoslav *Ausgleich*, granting Croatia significant political, economic, and cultural autonomy. See 'A Pulvári-felé hordozható hangosfilmfelvevőgép és első szereplése a szerb-horvát kiegyeszés alkalmából Zágrábban', *MF* II/7 (17 February 1940), 4.
39. Ibid. Although unable to construct a Yugoslav sound feature film industry prior to the South Slav state's annihilation, Hungarian experts played an outsized role in the production of Yugoslav and later Croatian sound newsreels well beyond 1940.
40. *BFL*– László Balogh, Nb. 1699/1945. 'Jgykv. az ONF-nak a m.kir. Miniszterelnökségen 1939, évi december 7-én tartott üléséről', 573.
41. 'Kozma Miklós a jugoszláv-magyar filmkapcsolatokról', *MF* II/11 (16 March 1940), 2.
42. Estimates are scant and varied. For the first half of 1940, Hungarian films brought in 2.5 million dinar, outselling all other films shown in Yugoslavia. German films made 2.0 million dinar and American films 1.3 million. 'Jugoslawiens Filmeinfuhr', *Auslandsstimmen für die deutsche Wirtschaft*, Nr. 611, (21 August 1940). The most authoritative full year estimate is István Langer's ('Fejezetek', 253). He determined that Hungarian films earned 4.8 million dinar (approximately HP 700,000) in 1940, likely drawing from 'A harmadik helyet foglaljuk el Európában', *Országépítés* (n.d., likely 1941), 13, which puts the exact figure at 4,876,430 dinar. German estimates round the figure to 5.0 million dinar. 'Ungarn', *Südostecho Wien*, 9 (28 February 1941).
43. *MOL* – KM, K66-cs474, 1940, III-6/c. Kozma to Ullien-Reviczky, 19 May 1940.
44. A. Lajta, 'Az év története', *Filmművészeti Évkönyv* (1941), 178. Italics in original.
45. *BADH* – R4902, DIA, Zeitungsausschnitssammlung, Nr. 9680, Ung-Jugo Kulturbeziehungen, 1939–41.
46. Some ethnic Hungarians won Hungarian government subsidies for film distribution. One example is Mucsi-Lendvay, a Yugoslav distributor opened by two Hungarian engineers with no film experience. The KM and the Minorities Department in the Prime Minister's office promised Mucsi-Lendvay extensive support. *MOL-Ó* – Hunnia, Z1123-t1-d2, 1940–2, [1940]. 7 June 1940 Hunnia Igazg. jgyek.
47. *MOL* – KM, K66-cs474, 1940, III-6/c. György Bessenyey to KÜM, MKK Belgrád, 6353/1940 sz, 1 May 1940.
48. In 1940, the parliamentarian János Makkai wrote: 'Now every great European nation is clear on the fact that Hungary has to be considered in the coming organization of Europe'. Former Prime Minister István Bethlen, still a force in Hungarian politics, expressed similar sentiments, stating 'There can be no doubt that at the end of the current war, the victor will redraw the map of Europe to create a New European Order. For that reason, Hungary must do all that it can so that when the time comes, the victor honors the interests of the Hungarian nation'. Quoted in Case, *Between States*, 62–3.

49. *MOL* – KM, K66-cs474, 1940, III-6/c. Kozma to Ullein-Reviczky, 19 May 1940.
50. Various papers in *MOL* – KM, K66-cs475, 1940, III-6/c. Quote from Bingert to KM, 26 June 1940.
51. 'Skandináviától Kis-Ázsiáig', *MF* II/41 (12 October 1940), 1–2.
52. The statistics are inconsistent. The article, 'A harmadik helyet foglaljuk el Európában', op. cit., 13, indicated Hungary exported 30 films to the United States, 21 to Italy, 20 to Sweden, and 18 to Greece in 1940. A.W. Just, in 'Der ungarische Film' in the *Krakauer Zeitung*, 243 (16 October 1941), marveled at the 'surprising development' of Hungarian film, noting that in less than a year, Hungary's film industry become a net exporter of film, sending a total of 239 films abroad in 1940. Just's numbers differed somewhat from *Országépítés*. He claims 48 Hungarian films went to Bulgaria, 21 to the US, eight to Greece, seven to Sweden, six to Italy, and a handful of others to Norway, the Czech Protectorate, and Finland. The official journal of the IFK puts Hungary's 1940 feature film exports much lower, totaling only 25. *Interfilm. Blätter der internationalen Filmkammer* Heft 3 (November 1942), 31.
53. *MOL* –BM, K150, V.Kútfő, t15, cs3590 – Mozgoképüzemi ügyek. Filmgyártás. OMB ügyei. Hóman to Keresztes-Fischer, 19 February 1941, 160.611/1941 sz. – VII.2. ügyosztály.
54. Lajta, 'Az év története', op. cit., 178. 'Skandináviától Kis-Ázsiáig', *MF*, op. cit.
55. See series of 'Kimutatás…[a] bemutatott magyar filmekről' in *MOL* – KM, K66-cs475, 1940, III-6/c; K66-cs517, 1941, III-6/c; and K66-cs653, 1944, III-6/c. Correspondence from December 1940 indicates that in the 1939–40 film year, exports to the US dropped from around 25 in the previous year to only five, disrupted by the outbreak of war. In 1940, exports to the US rebounded, totaling 21–30 features (sources vary).
56. Miklós Kádár, 'A magyar film export-lehetőségei', *MF* II/14 (6 April 1940), 2.
57. *MOL* – KM, K66-cs475, 1940, III-6/c. Rihmer (Belgrade) to MKK Belgrade, 4 October 1940, 551/K.
58. *MOL* – KM, K66-cs475, 1940, III-6/c. K. Tomcsányi to OMB President, 8 June 1940, BM 66.167/1940 – V. Tomcsányi reached this conclusion based on a report by the Hungarian Ambassador in Belgrade, György Bessenyey, MKK Belgrád, 6353/1940 sz, 1 May 1940. See also *MOL* – KMI, K429-cs59-t3. Vegyes MTI levelezés 1940. Kozma to Erzébet Balogh, 6 May 1940.
59. This group's main proponents included Bingert and László Balogh. Supporting concentration of power when it came to exports, they steadfastly opposed nationalization of the entire business, for obvious reasons. Hunnia-controlled exports would bring profits and not impinge at all on the power bases Balogh and Bingert occupied. Full government control over filmmaking would likely mean the reorganization of the entire industry under the aegis of the Film Chamber. This would mean that the ONFB would be responsible to a higher, and perhaps more radical, authority and that Hunnia's Bingert would find his

relatively free hands tied. And there were other, less selfish reasons to oppose government centralization. Some were stated by Balogh, Bingert, and others when István Kultsár raised the matter at the 17 December 1941 Hunnia Directorate meeting. Christians might lose jobs, standardized production might result in artificial and less appealing products, the number of movies might decline, etc. See 'Magyar filmek értékesítése külföldön', *MF* II/19 (11 May 1940), 4 and *MOL-Ó* – Hunnia, Z1123-r1-d2, 1940–2. 17 December 1941 Jgyek. – Hunnia Igazg. jgyek ülésről, 3–5.

60. *MOL-Ó* – Hunnia, Z1123-t2-d4, 1940–4. 2 April 1940 Hunnia Igazg. jgyek, 5–6.
61. §32 – A szövetkezet feladata, *A Magyar Film Kiviteli Szövetkezet Alapsabályai,* Bp., 20 January 1943.
62. *MOL-Ó* – Pesti Magyar Kereskedelmi Bank Rt. Okmánytár, Z41, dosszié sz. 10590. Filmbehozatali és Kiviteli Kft.
63. Problems exporting *Vision by the Lakeside* [*Topárti látomás*] and *Silenced Bells* [*Elnémult harangok*] discussed in *MOL* – KM, K66-cs517, 1941, III-6/c. Wlassics, Jr., to L. Bárdossy, 11 July 1941, VKM 161.854/1941.sz – VII.2.ügyosztály.
64. 'Balogh László dr. előadása az OMME-ben', *MF* III/13 (29 March 1941), 2–3.
65. The 27 August 1941 issues of Hungary's leading dailies and some its lesser, more rightist papers, including *Esti Ujság, Magyar Nemzet, Magyarság, Nemzeti Ujság, Pester Lloyd, Pesti Hirlap,* and *Új Magyarság,* all contained articles about the success of the Mittiga visit and the prospects for export to Italy. *Nemzeti Ujság* predicted that 100 Hungarian films would soon premier in Mussolini's fascist state. '3 év alatt 100 magyar filmet mutattak be Olaszországban', *Nemzeti Ujság* (27 August 1941).
66. *BFL*– László Balogh, Nb. 1699/1945. Unsigned report (probably Balogh), to Hóman, 2 December 1941, concerning the 24–7 November 1941 IFK meeting in Munich. The Belgians apparently took a liking to Hungarian musicals.
67. Numbers offered by István Erdélyi in a speech before the OMME. 'Az OMME 31. évi rendes közgyülése', *MF* III/51 (22 December 1941), 3–4.
68. 'Die Budapester Tagung der Internationalen Filmkammer', *Nachrichten für Außenhandel* 286 (7 December 1942).
69. Estimates of the number of films Germany purchased in 1941 range from one to nine. The low estimate comes from *MOL-Ó* – Hunnia, Z1124-r1-d23. Nemzetközi Filmkamara, 1941–3. Deutsches Institut für Wirtschaftsforschung, 'Zur öffentlichen Vorführung zugelassene lange Spielfilme in wichtigen Ländern europas 1940/41', (Sept 1942). This document also attests to Hungarian film exports to the Protectorate. The number of nine, undoubtedly too high, comes from 'Die Budapester Tagung der Internationalen Filmkammer', op. cit.
70. Langer, 'Fejezetek', 262–3. Langer reports that MFI income from filmmaking increased from HP 1.9 million in 1939 to HP 5.16 million in 1942, almost

entirely due to its expansion of feature production. By 1941, the MFI was making 16 films annually. *MOL* - KM, K66-cs474, 1940, III-6/c. Kozma to Ullein-Reviczky, 19 May 1940.

71. *MOL* - MFI, K675-cs1-t3. Végrehajtóbizottsági jegyzökönyvek és iratok, 1928–44. A MFI Végrehajtóbizottsági ülésének napirendje és 39. számú jgykv-e, 24 September 1940.
72. *MOL-Ó* - Hunnia, Z1123-r1-d2, 1940–2 [1940]. 21 November 1940 Jgykv. - Hunnia Igazg-i ülésről, minutes 9.
73. The MFI Board of Directors had considered purchasing the Star studio as early as 1935. See *MOL* - MFI, K675-cs1-t3. Végrehajtóbizottsági jegyzökönyvek és iratok, 1928–44. A MFI Végrehajtóbizottsági ülésének napirendje és jgykv., 27 June 1935.
74. *MOL-Ó* - Magyar Általános Hitelbank Rt, Z58, Ipari osztály, cs261, t1224, Corvin-Star-Kamara - Ügyvezető igazgató, 1921–40. 'Jgykv.' Felvétetett 1940. március hó 14-én d.e. 11 órákor a Star Filmgyár telepén.
75. Sándor, *Őrségváltás után,* 148.
76. *MOL-Ó* - Hunnia, Z1123-r1-d2, 1940–2 [1940]. 21 November 1940 Jgykv. - Hunnia Igazg-í ülésről, minutes 9.
77. *AAA* - GsB, Fach 27, Kult 12, Nr 4a, 'Filmwesen Ungarn.' Folder 'Star Atelier', Aufz. I, 16 April 1941, reporting on 10–13 February 1941 meeting of Deutsch-Ungarische Kulturausschuss. The two sides signed a 'Niederschrift über die Kulturverhandlungen vom 10.-13. Februar 1941' according to the German Ambassador to Hungary, Otto von Erdmannsdorff.
78. *MOL-Ó* - Hunnia, Z1123-r1-d2. Igazg-i jgyek., 1940–2 [1941]. 18 March 1941 Jgykv - Hunnia Igazg-í ülésről, minutes 3–9, esp. 5–6.
79. The FIF provided Hunnia HP 150,000 cash and 600,000 of credit to buy and refurbish the Star.
80. *AAA* - GsB, Fach 27, Kult 12, Nr 4a, 'Filmwesen Ungarn.' Folder 'Star Atelier', telegraph from the RFK to GsB (Köhler), received 15 April 1941 at 15:59, 993 Berlin/15 2614 155/147 15 1309. *MOL* - K-27 Minisztertanácsi jgyvk xeroxmásolatai. Az 1941. április 28-i minisztertanácsi ülés anyaga. Quoted in Sándor, *Őrségváltás után,* 149.
81. *MOL-Ó* - Hunnia, Z1123-t1-d2, 1940–2 [1940]. 21 November 1940 Hunnia Igazg. jgyek., 9. See also Lajta, 'A magyar film története', 53.
82. The MFI established a Kolozsvár office mainly to distribute newsreels and features in Transylvania.
83. The founders of Transylvánia Kulturfilm wrote that the company's purpose was to 'inform and prove to the countries of the world that the culture of Transylvania was always Hungarian…from ancient times forward…' *MOL* - KM, K66-cs609, 1943, III-6/c. Transsylvánia Kulturfilm to KM Department of Culture, Kolozsvár, 20 Oct. 1943.

84. Langer, 'Fejezetek', 248. Various American, German, and Hungarian sources confirm that until the Nazi occupation in March 1944, Hungary's film community clung to the dream of a Kolozsvár studio. *NARA* –RG 262, Records of the Foreign Broadcast Intelligence Service [FBIS], Transcripts of Monitored Foreign Broadcasts, 1941–6, Budapest 1941–6, Box 153, File June 1942. Budapest short wave 25 June 1942, Record 08471-2. *BADH* – R63, Fach 352, Kb. aus SOE, 14.2.44–26.3.44. 'Pressestelle. Vertrauliche Mitteilung'. 127. Kb. SOE, 21.2–27.2.44. 'Bingert János a Hunnia Filmgyár vezérigazgatója nyilatkozik 1943 új filmtermeléséről, a szines filmről és a kolozsvári filmgyár megépitésének tervéről', *Mozi Ujság* III/1 (6–12 January 1943), 5.
85. *Film-Kurier* article quoted in 'Magyar filmek Jugoszláviában', *MF* II/18 (4 May 1940), 2.
86. 'Jugoslawien – Die Filmeinfuhr seit 1935', *Nachrichten für Aussenhandel* (16 July 1940). The article also blamed Czech films.
87. *AAA* – GsB, Fach 27, Kult 12, Nr 4a, 'Ungar. Filmprop. in Yugoslawien'. Ahrens to Gs. Belgrad, 18 April 1940, Kult K 4521.
88. 'Jugoslawiens Filmeinfuhr', *Auslandsstimmen für die deutsche Wirtschaft* 611 (21 August 1940).
89. 'A magyar film térhódítása Jugoszláviában', *MF* III/1 (4 January 1941), 2.
90. Hungarian films were extremely popular in the Vojvodina, and Hungarians such as Pál Jávor were among the most popular movie stars in all of Yugoslavia. Dejan Kosanović, *A Short History of Cinema in the Vojvodina* (Belgrade, 2012), 44–5.
91. On average, according to the Tesla report, Hungarian feature films earned 70–100,000 dinar (approximately HP 10,000–14,000), while German films earned only 40–80,000 dinar. *AAA* – GsB, Fach 27, Kult 12, Nr 4a, 'Ungar. Filmprop. in Yugoslawien'. Unsigned Tesla-Film-A.G. report, Belgrade, forwarded from Gs. Belgrad to AA Berlin, 3 June 1940.
92. *AAA* – GsB, Fach 27, Kult 12, Nr 4a, 'Ungar. Filmprop. in Yugoslawien'. Kolb to GsB, 30 June 1940, Kult K 6753/40.
93. Ibid.
94. For more on New Order Europe, see D. Frey, 'A Smashing Success? The Paradox of Hungarian Cultural Imperialism in Nazi New Order Europe, 1939–1942', *Journal of Contemporary History* 51/3 (2016), 577–605.
95. Nazi authorities may have acted in concert with US film agents and diplomats. On this subject, see *MOL* – KM, K66-cs475, 1940, III-6/c. Hunnia to KM, 30 August 1940; and David S. Frey, 'Competitor or Compatriot? Hungarian Film in the Shadow of the Swastika, 1933–44', in R. Vande Winkel & D. Welch (eds), *Cinema and the Swastika. The International Expansion of Third Reich Cinema* (London, 2007), 159–71.
96. *MOL* – KM, K66-cs475, 1940, III-6/c. M.kir.Külkereskedelmi Hivatal Belgrád to MKK Belgrád, 4 October 1940, 551/K.

97. *MOL* – MFI, K675-cs3-t16, 1942–3. 'Jelentés a Jenes Nándor olaszországi utjáról', 3 November 1942.
98. *BA* – RWM R3101 (Alt R7), Nr. 3067. M. Winkler to Goebbels, 8 March 1943. By June 1941, all American income from the former Yugoslavia had ceased. See also Herrick Library – MPAA General Correspondence Files, Reel 6 [1941]. Acting Foreign Manager MPPDA to Chairman, Committee for Reciprocity Information Tariff Commission, 10 June 1941, 2. According to A.O. Ritschl, although it was a minor part of Germany's overall trade picture, film was one of the few components of foreign trade that consistently brought in needed hard currency between 1938 and 1941. See Ritschl, 'Nazi Economic Imperialism and the Exploitation of the Small: Evidence from Germany's Secret Foreign Exchange Balances, 1938–40', *The Economic History Review*, New Series, 54/2 (May 2001), 328.
99. Klaus Kreimeier, *Die Ufa Story. Geschichte eines Filmkonzerns* (Munich, 1992), 394–5.
100. Stefan T. Possony, 'National Socialistic Economics: The Contradictions of the New Order', *The Journal of Politics* 4/2 (May 1942), 151.
101. There are striking parallels between Germany's treatment of Hungary as a wartime competitor and the United States' methods of restraining the Mexican national film industry. On Mexico, see the works of Seth Fein.
102. *AAA* – Kult. Pol – Generalia, Sig. R 61430, Kult V.1, Deutsche Kulturverträge, Ungarn – 1939. Melcher to GsB (Overbeck), 19 February 1940, Kulturamt V/2/40/32 Kult 2 N = 2. Vorschläge für Kulturabkommen.
103. *AAA* – GsB, Fach 27, Kult 12, Nr 4b, 'Filmwesen ung. dt. Beziehungen, 1939–41'. RMVP (Fries) to AA – Kult. K., F 5842/5.11.40/506–56, 1, 18 November 1940.
104. E.g. 'Ungarische Filmfragen', *Berliner Börsen-Zeitung* 55 (2 February 1941); 'Die Lage in der ungarischen Filmindustrie', *Die Bewegung* (27 May 1941). 'Ausländische Filme in Ungarn', *Südostecho Wien* 22 (30 May 1941), 9. The *Berliner Börsen-Zeitung* and *Südostecho Wien* articles in particular appear to have raised the hackles of Hungarian authorities, as several ministries discussed these articles and their implications.
105. *AAA* – GsB, Fach 27, Kult 12, Nr 4b, 'Filmwesen ung. dt. Beziehungen, 1939–41'. RMVP (Fries) to AA – Kult. K., R 5842/23.9.40/948-40,4, 18 Okt. 1940.
106. 'Ungarn drosselt Filmeinfuhr', *Berliner Börsen Zeitung* 476 (10 Okt. 1939).
107. *MOL* – KM, K66-cs475, 1940, III-6/c: Filmügyek. 'Aufzeichnung' n.d.
108. Hungarian authorities rightly faulted the German-formulated clearing agreement that governed all German-Hungarian trade. Because currency exchange was difficult, and Germany insisted that its films be paid for in cash not kind, it took months, claimed the pro-Nazi Hungarian ambassador to Germany Döme Sztójay, for even 'a few hundred Marks' to become available.

This hindered Hungarian purchases of German cultural products, especially movies, books, and art. To alleviate the problem, PM Teleki established a special HP one million fund specifically designated for purchases of culture in late November 1940. *MOL* – ME – Tájékoztatási Osztály K30-cs4, év nélkül, B/45.b.t: A németországi magyar propaganda szervezésével kapcsolatos iratok, 495–6. Sztójay to D. Szent-Iványi, 7 November 1940, MKK Berlin, 850/ biz.-1940.

109. *BFL* – Dr. Nagy Sándor – Nb. 3482/1946. 200.002/1945 sz. 'Kivonat: Hunnia Filmgyár rt. ügyből'. The German-Hungarian Culture Committee was a working group containing a range of officials. It was established in accordance with the bilateral cultural agreement signed by the two countries in October 1936 in order to facilitate artistic and intellectual exchange and resolve disputes over cultural matters.
110. *MOL* – KM, K66-cs517, 1941, III-6/c: Filmügyek, 33. Drafts and copies of Verbal Note delivered to the GsB by the KM, 27 June 1941.
111. *MOL* – KM, K66, cs517, 1941, III-6/c: Filmügyek, 82. 'Feljegyzés. A magyar filmeknek Németországban való előadása tárgyában. Undated, unsigned, attached to meeting minutes of German-Hungarian cultural subcommittee, 10 February 1941.
112. In the 1941 meetings, Hungary asked that Germany import six to eight Hungarian films annually.
113. For the March/April 1941 meeting of the German-Hungarian Culture Committee, see *MOL-Ó* – Hunnia, Z1124-r1-d21. Német kivánságok német filmek átvételével és amerikai filmek behozatalának csökkentésével, 1941–2. For other German-Hungarian diplomatic correspondence and cultural committee meetings, see *MOL* – KM, K66-cs517, 1941, III-6/c: Filmügyek.
114. Germany demanded either licenses to three to four Hungarian premier theaters or a quota requiring 15 to 25 percent of Hungarian screen time be devoted to Nazi products. It also demanded a 50 percent share of the Hungarian narrow film and *Kulturfilm* markets, and VKM documentation suggests that they asked that 50 percent of the premier market be reserved for Axis-made film. For the demands and Hungary's reaction to and rejection of them, see *MOL* – KM, K66-cs517, 1941, III-6/c: Filmügyek. GsB A Nr. 200 Verbalnote, delivered to the KÜM, 26 May 1941; *MOL-Ó* – Hunnia, Z1124-r1-d21. Német kivánságok német filmek átvételével és amerikai filmek behozatalának csökkentésével, 1941–2. *MOL* – BM, K150, V.Kútfő, t15, cs3591, 1941–2 – Mozgoképüzemi ügyek. Filmgyártás. OMB ügyei; BM Keresztes-Fischer to the KM and VKM, BÜM 188285/1941 sz., 10 June 1941. The VKM's response is found in *BFL* – László Balogh, Nb. 1699/1945. Response to German Verbal Note 200/1941, VKM 29.241/Eln.C., 16 June 1941.

115. *MOL-Ó* – Hunnia, Z-1124-r1-d20, Art. 4, 'Abkommen zwischen dem Deutschen Reich und dem Königreich Ungarn zur Regelung von Fragen über das Filmwesen' effective through 30 July 1940.
116. *MOL* – KM, K66-cs517, 1941, III-6/c: Filmügyek. Copy of GsB A Nr. 171 Verbalnote, delivered to the KM, 8 May 1941.
117. *MOL* – KM, K66-cs517, 1941, III-6/c: Filmügyek. Verbalnote 728/3/1941, KM to GsB, 13 May 1941.
118. *MOL* – KM, K66-cs517, 1941, III-6/c: Filmügyek. Drafts/copies of Verbalnote KM to GsB, 27 June 1941.
119. Ibid.
120. *MOL-Ó* – Hunnia, Z-1124-r1-d21, 1–19 – Német kivánságok német filmek átvételével és amerikai filmek behozatalának csökkentésével. Untitled notes and drafts from March/April 1941 meeting of mixed German-Hungarian Culture Committee. Other notes indicate that there were 266 total American features on the Hungarian market in 1940. This number was to be reduced to 160 in 1941 (46 new movies, 114 older films with unexpired licenses). See also *MOL-Ó* – Hunnia, Z1124-r1-d21, 81–2.
121. *MOL* – BM, K150, V.Kútfő, t15, cs3590, 1941 – Mozgoképüzemi ügyek. Filmgyártás. OMB ügyei. Ullein-Reviczky to Keresztes-Fischer, KÜM 789/res/3 – 1941, 9 June 1941. See also *MOL-Ó* – Hunnia, Z1124-r1-d21. Ullein-Reviczky recommendation re: German Verbalnote 200/1941, 9 June 1941.
122. In October 1941, German authorities announced they would import only 'two to four specific Hungarian films' yearly, a severe insult to Hungary's elites and a violation of the 1936 promise of a minimum of five. 'Wachsende Filmfreudigkeit', *Deutsche Allgemeine Zeitung* 489 (12 Okt. 1941).
123. *AAA* – GsB, Fach 27, Kult 12, Nr 4b, 'Filmwesen ung. dt. Beziehungen, 1939–41'. 'Abchrift', AA to RMVP, AA Nr. Kult.Gen.C.3432, 4 Okt. 1939.
124. German and Hungarian sources indicate that Germany wanted to purchase or control up to three Budapest premier theaters. Kálmán Tomcsányi, an assistant to BM Keresztes-Fischer whose portfolio included cinema issues, claims to have thwarted German efforts to buy more than one theater. See his testimony in *BFL*– Kálmán Tomcsányi, Nb. 839/1946. 'Tanúvallomási Jgykv', MÁBFPRO, 21389 sz./1945, 26 January 1946, 3. On the Europe-wide campaign to purchase large theaters, see R. Vande Winkel & D. Welch, 'Europe's New Hollywood? The German Film Industry under Nazi Rule, 1933–45', in Vande Winkel & Welch (eds), *Cinema and the Swastika,* 19.
125. *AAA* – GsB, Fach 27, Kult 12, Nr 4b, 'Filmwesen ung. dt. Beziehungen, 1939–41'. RMVP (V[F]ries) to AA – Kult. K., F 5842/5.11.40/506–56, 1, 18 Nov. 1940.
126. *AAA* – GsB, Fach 27, Kult 12, Nr 4b, 'Filmwesen ung. dt. Beziehungen, 1939–41'. Erdmannsdorff to AA, 3608, 28 Nov. 1940.

127. *BA*– Ufa R109 I, Nr. 505. 'Corvin-Kinotheater AG, Bp. Bericht über Bilanz und Ergebnis per 31. Mai 1942. Gegeben von der Revisions-Abteilung der UFA, 1–4. Ufa took over the Corvin Theater when it fused with Tobis in June 1942. According to this report, the purpose of the purchase was 'to strengthen German film interests [in Hungary]'. Budapest Film was founded in May 1937. Its original HP 150,000 of basis capital belonged to Ufa-Berlin after the consolidation of German film in June 1942, but for legal purposes, its shares were held in Budapest by Erich Matyasfalvy-Glatz (HP 135,000) and Antal Schuchmann (HP 15,000). An ethnic German born in Hungary, Schuchmann took German citizenship and the name Anton sometime prior to 1941. *BA* - Ufa R109 I, Nr. 504. Budapest Film AG, Bp. 1941–3: Bericht über das Geschäftsjahr 1941/42.
128. *BFL*– MFSzSz XVII 647/5 - 287 B, Walter Tibor. József Antal letter to the MFSzSz, 14 August 1945.
129. *MOL* –BM, K159-cs1-tA, 1941: Mozgóképek ellenőrési ügyei. 'Német kovetség különböző cimü filmjei'.
130. *AAA* – GsB, Fach 27, Kult 12, Nr 4b, 'Filmwesen ung. dt. Beziehungen, 1939–41'. Erich Lübbert to RWM, 18 July 1941.
131. For an early contemporary account of how the Nazis believed these types of films to have distinct propaganda value, see Hans Traub, *Der Film als politisches Machtmittel* (München, 1933), 26–32. For historical consideration of how the Nazis considered documentary films, culture films and newsreels as ideally suited for propaganda, including in neutral and occupied lands, see Hilmar Hoffmann, *The Triumph of Propaganda. Film and National Socialism 1933–45*, trans. J.A. Broadwin and V.R. Berghahn, (Providence, 1996), 132–213; Felix Moeller, *Der Filmminister. Goebbels und der Film im Dritten Reich* (Berlin, 1998), 347–402; Roy Armes, 'Cinema of Paradox: French Film-making during the Occupation', in G. Hirschfeld & P. Marsh (eds), *Collaboration in France: Politics and Culture during the Nazi Occupation, 1940–44* (Oxford, UK, 1989), 126–9; and R. Vande Winkel & D. Welch, 'Europe's New Hollywood?' in Vande Winkel & Welch (eds), *Cinema and the Swastika*, 14–15. While newsreels and *Kulturfilme* possessed distinctive propaganda qualities, they also allowed audiences and even exhibitors distinct opportunities to express political opposition. See Jiří Doležal, *Česká kultura za Protektorátu. Školství, Písemnictví, Kinematografie* (Prague, 1996), 189.
132. Géza Ágotai, the second in command on the OMB, wrote of narrow culture film as an influential '*Vorkämpfer der ungarischen Kultur*'. G. Ágotai, '400 Schmalfilmtheater in fünf Jahren' *Interfilm. Blätter der internationalen Filmkammer* Heft 6 (August 1943), 76.
133. *BA* - RMVP, R55/493, Fiche 1. See Johannes Eckardt, 'Deutsche Schmalfilm - Exportgesellschaft', 30 April 1941; unsigned, undated

'Gedanken zur Schmalfilmexportgesellschaft'; and 'Vermerk', RMVP Abt H.Ref.: Dr. Gemündt, Berlin, den __ Juni 1941. Eckardt later became the Descheg's Director.

134. *BA* - RMVP, R55/493, Fiche 2. Deutscher Schmalfilm-Vertrieb GmbH 'Geschäftsübersicht ab Gründung bis einsschl. 31 Dez. 1941'.
135. Culture film in this sense included films for educational, instructional, scientific, and touristic purposes.
136. 1941 Hungarian legislation mandated that two-thirds of all narrow film programming be of Hungarian origin. Germany held a virtual monopoly on the remaining one-third. Although the German-Hungarian Culture Committee agreed to readjust the ratio of narrow films to 1:1 at a June 1942 meeting, the Hungarian Interior Minister refused to accept this, maintaining the 2:1 ratio. *MOL* – KM, K66-cs609, 1943, III-6/c: Film, szinházügyek. 'Jgykv.', Hungarian-German culture committee meeting, 5 June 1942. See also *MOL* - BM, K150, V.Kútfő, t15, cs3594, 1942–3 - Mozgoképüzemi ügyek. Filmgyártás. OMB ügyei. F. Inotay notes, BM memo BÜM 116.833/1943, 17 July 1943. Regarding Hungária Film, see *MOL* - MFI, K675-cs12-t29, 1943–4, Hungária Film Rt. *Előterjesztés* a Hungária Film Rt. és a Kefifor között létesitendő megállapodás tárgyában, 9 Dec. 1943.
137. 'Der Schmalfilm vor groβen Aufgaben', *Deutsche Stimmen* [*Pressburg*] (10 June 1944).
138. *AAA* – GsB, Fach 27, Kult 12, Nr 4b, 'Filmwesen ung. dt. Beziehungen, 1939–41'. AA (Kolb) to GsB, Kult K 11547/40, Berlin 30 November 1940.
139. *BA* – Reichspropagandaamt Ausland [RPAA] NS 18/12. 'Tätigkeitsbericht des RPAA für die Monate Januar, Februar, und März 1943'. RPAA, Pro 2005, 30 April 1943, 23. The majority of the audiences were ethnic German Hungarians.
140. Ibid., 11.
141. T. Pryor, 'By Way of Report', *New York Times* (20 April 1940), x4. Pryor wrote that Hungarian audiences, and even its censors, 'still have marked preference for the West....German films are tolerated, but not patronized to any extent'.
142. Data from Lajta, *Filmművészeti Évkönyv* (1941–4); Langer, 'Fejezetek', 212; and Pogány, 'Utóhang a nemzetközi filmkongresszushoz', 1–3. Numbers in parentheses are Pogány's. I cannot explain the discrepancies. A 1944 article by A.J. Richel in the Dutch paper *De Schouw* quoted by Ingo Schweck adds to the confusion, claiming that Hungary surpassed German feature film production in 1942, outproducing it 63:58. Ingo Schweck, '*[...] weil wir lieber im Kino sitzen als in Sack und Asche*' (Muenster, 2002), 271.
143. Pogány, 'Utóhang a nemzetközi filmkongresszushoz', Tables 2–4 all come from this source.
144. *MOL* – KM, K66–cs474, 1940, III–6/c. 'Verbalnote', AA Nr. Kult K F 1987/41, 22 April 1941, response to ambassadorial advisor in the Hungarian embassy, MKK Berlin, 3267/1941, 2 April 1941.

145. Immediately after Germany invaded, subdued, and partitioned Yugoslavia in April 1941, Nazi authorities cut off the import of Hungarian film to all but the Hungarian-annexed Bánát and Vojvodina, even restricting Hungarian exports to the supposedly independent state of Croatia. They did so by anointing Tesla Film the only company permitted to import films and by attaching extraordinarily high duties to Hungarian films. Film distributors in Serbia proper asked the Hungarian ambassador in Belgrade to protest this measure in June 1941, indicating that despite Hungarian participation in the invasion of Yugoslavia, Yugoslav film distributors still believed demand for Hungarian films existed. *MOL* - KM, K66-cs517, 1941, III-6/c. MKK Belgrade (Bolla) to KM L. Bárdossy, 6195/1941.sz, Belgrade 6 July 1941. *MOL* - KM, K66-cs517, 1941, III-6/c.
146. In 1941, the members of the IFK were: Germany, Italy, Hungary, Belgium, Bulgaria, Croatia, the Czech-Moravian Protectorate, Denmark, Finland, Holland, Norway, Romania, Slovakia, Spain, Sweden, Switzerland, and Turkey. Bogusław Drewniak, *Der deutsche Film 1938–1945* (Düsseldorf, 1987), 799–802; B.G. Martin, '"European Cinema for Europe!" The International Film Chamber, 1935–42', in Vande Winkel and Welch (eds), *Cinema and the Swastika*, 25–9.
147. *BFL*– László Balogh, Nb. 1699/1945. Wlassics Jr., attachment to VKM 1893 Eln. sz., 25 June 1941.
148. 'A harmadik évad küszöbén', *MF* III/32 (9 August 1941), 1.
149. *MOL* - BM, K150, V.Kútfő, t15, cs3590, 1941 - Mozgoképüzemi ügyek. Filmgyártás. OMB ügyei. Antal to VKM Hóman, Igazságügyminiszter 73.512 - 1941.I.M.VII. sz., 19 November 1941.
150. *MOL* - BM, K150, V.Kútfő, t15, cs3590, 1941 - Mozgoképüzemi ügyek. Filmgyártás. OMB ügyei. 'Jgykv. készült 1941.évi november hó 20-án, a belügyminiszteriumban, a Nemzetközi Filmkamara szabályzata tárgyában tartott értekezletről'. Representatives of the Interior Ministry, the Justice Ministry, the Ministry for Manufacturing, the Trade and Transport Ministry, the Foreign Ministry, the Ministry of Religion and Education, the OMME, the MMOE and Hunnia attended this conference.
151. *BFL*– László Balogh, Nb. 1699/1945. Meeting account found in Balogh to Hóman, 2 December 1941.
152. István Kauser column on the IFK in *MF* III/39 (27 September 1941), 9.
153. *BFL*– László Balogh, Nb. 1699/1945. I. Kutassy, 'Jelentés', 14 April 1942. Also Martin, '"European Cinema for Europe"', in Vande Winkel and Welch (eds), *Cinema and the Swastika*, 31.
154. *BFL*– László Balogh, Nb. 1699/1945. I. Kutassy, 'Jelentés', 14 April 1942.
155. In 1938, when the Italian government banned imports from the United States, American films constituted about two-thirds of Italian programming. Even after the ban, American movies continued to be distributed and attract

large audiences. Pierre Sorlin, *Italian National Cinema 1896–1996* (London, 1996), 71; Ricci, *Cinema & Fascism*, 70–1.

156. *BFL*- MFSzSz XVII 647/1 - 287 B, Erdélyi István. P. Morvay testimony 6 July 1945, 'III. Jgykv., 21–2. Felvéve a NB által a MMSzF-ához kiküldött 289/b. szamú IB, 6 July 1945.
157. 'IFK. Tagungen der Internationalen Filmkammer', published by the IFK, 1942. Minutes of Berlin (2–3 March 1942); Rome (8–10 April 1942); and Florence (10–11 May 1942) meetings.
158. *MOL* - BM, K150, V.Kútfő, t15, cs3593, 1942 - Mozgoképüzemi ügyek. Filmgyártás. OMB ügyei. BM (Perczel) 'Jelentés a Nemzetközi Filmkamara folyó évi április hó 7–10.-ig Rómában tartott 1942.ápr.római ülészszakáról', 18 April 1942. See esp. 3. Perczel noted that the meeting was the most argumentative of any IFK gathering, confirming Morvay's later testimony.
159. *MOL* - VKM, K507-cs91-t12, 1937–41 - Filmügyek. MKK (Kőszeghy) to VKM, 69.947/III sz. - 1942, 26 May 1942.
160. The chemicals required to make raw film included some used in the production of war material, which the Treaty of Trianon forbid. When Hungarians attempted a film stock manufacturing venture in the early 1920s, they failed because of the high price of importing these chemicals. The stock could not compete with less expensive imports from Germany, Italy, and even the United States. From that point on, Hungary had not made its own raw film. Germany, however, evaded the restrictions on the same chemicals. 'Miért szűnt meg a 22 évvel ezelőtt alapított magyar nyersfilmgyár?' *MF* IV/43 (26 October 1942), 5.
161. In 1942, three companies sold raw film to Hungary – Agfa (German), Ferrania (Italian), and Kodak (United States). The Kodak shipments were minimal. See Lajta, *Filmművészeti Évkönyv* (1942), 207–8.
162. 'A nyersanyaghiány és a magyar filmgyártás', *MF* IV/10 (10 March 1942), 3.
163. *MOL* – KM, K66-cs609, 1943, III-6/c: Film, szinházügyek. 'Jgykv.', Hungarian-German cultural committee, 5 June 1942. These strictures were added to previous German limits on use of its raw film for copies of American, British, and certain French films. *MOL-Ó* - Hunnia, Z1123-r1-d23. Nemzetközi Filmkamara, 1941–3. Agfaphoto to Hunnia, 7 March 1942 and IFK Gen. Sect. K. Melzer to Hunnia, Berlin, 30 March 1942
164. *MOL* – KM, K66-cs609, 1943, III-6/c: Film, szinházügyek. 'Jgykv.', Hungarian-German cultural committee, 5 June 1942.
165. *MOL* - MFI, K675-cs3-t16, 1942–3. 'Jelentés a Jenes Nándor olaszországi utjáról', 3 November 1942. As the details of this plan were worked out in late 1942 and early 1943, German officials caught the scent. According to Margit Simonyi, an MFI director who traveled to Berlin in order to sell Hungarian culture films in May 1943, high officials in the *Reichsfilmkammer* were very

interested in the details of the arrangement. In the same file, see 'Jelentés – Simonyi Margit rendező németországi utjáról', 22 May 1943.

166. The Manufacturing Ministry asked the MFI if it would be interested in joining the film factory venture, to which the MFI responded with a non-committal yes. *MOL*– MFI, K675-cs2-t5, 1942–4. Végrehajtóbizottsági jgykv-einek az MFI-re és filmmel kapcsolatos vállalatokra vonatkozó kivonatai és az azokhoz kapcsolódó iratok, 163. 1942.VIII.28 – MFI végrehajtóbizottsága a következő határozatot hozta.

167. *MOL-Ó* – Hunnia, Z1123-r1-d2, 1940–2. 23 September 1942 Hunnia Igagz. jgyek, 3ff. Discussion continued at 14 October and 11 November Directorate meetings. See 14 October 1942 Hunnia Igagz. jgyek, 6; 11 November 1942 Hunnia Igagz. jgyek, 4. Hunnia also requested Germany increase its annual film stock quota by 1.2 million meters.

168. 'Meg kell teremteni a magyar nyersfilmgyártást', *MF* IV/40 (5 October 1942), 1. Other film journals, including *Filmművészeti Évkönyv* and the newly established *Film Hiradó* and *Filmszinházak Közlönye* also contained articles discussing stock shortages and the need for Hungary to make its own supply.

169. The Germans determined, through mixed Culture Committee 'negotiations', that Hungary needed 3.6 million meters of raw film for 1943, shunting aside Hungary's own estimate of nine million. As a matter of perspective, the MFI alone used 3,429,449 meters of raw film in 1942.

170. The MFI predicted Hungary would not replicate its 1942 success. *MOL* – MFI, K675-cs2-t6, 1935–45, Igazg-i ülési jgyzk., 994. Complaints about quality and editing were aired in Geza Matolay's multi-article series 'A magyar filmgyártás mult évi gondjai és jövő évi kívánságai', in the January-March 1943 issues of *Magyar Film*. These complaints were reiterated in official circles and magazines throughout 1943, as films begun in late 1942 were completed and released. For example, see 'Filmankét. Erdélyi István dr. a magyar film sorsdöntő kérdéseiről', *Mozi Ujság* III/34 (25–31 August 1943), 4–5.

171. Szinyei Merse's speech in *Magyarország Évkönyve – 1942.* IX.évfolyam (Budapest, 1943), 104.

172. *MOL* – KM, K66-cs517, 1941, III-6/c. KKM Varga to KM Bárdossy, 164.098/III.sz. – 1942, 11 Dec. 1942.

173. *MOL* – BM, K150, V.Kútfő, t15, cs3594, 1942–3 – Mozgoképüzemi ügyek. Filmgyártás. OMB ügyei. BM 85350/1943, Bp. 22 January 1943. The KKM's Iván Kőszeghy explained, in a series of letters to Keresztes-Fischer, that his decision was a choice of lesser evils and that had Hungary not complied with the American film ban, Germany would have caused the total collapse of Hungarian production. *MOL* – BM, K150, V.Kútfő, t15, cs3594, 1942–3 – Mozgoképüzemi ügyek. Filmgyártás. OMB ügyei.

Letters 135.470/III. sz. - 1942, 13 October 1942 and 2.188/III. sz. - 1943, Bp. 7 January 1943.

174. 'Negyvenöt uj magyar film nyersanyagát biztositotta a nemzetközi filmkamara', *Film Hiradó* (3 September 1943), 1. The article makes it clear that IFK decisions are German decisions, not those made by consensus.
175. In February 1942, some two months after declaring war on the United States, the Hungarian government assigned trustees to all American film companies operating in Hungary. Surprisingly, they allowed Fox, Paramount, and Warner Brothers to continue to function with minimal interference. 'Megkezdték működésüket a gondnokok az amerikai filmképviseleteknél', *MF* IV/7 (18 February 1942), 9; and 'Az amerikai filmvállalatok likvidálása', *MF* IV/22 (2 June 1942), 7.
176. *MOL* - BM, K150, V.Kútfő, t15, cs3594, 1942–3 - Mozgoképüzemi ügyek. Filmgyártás. OMB ügyei. BM 85350/1943, 22 January 1943.
177. *MOL-Ó* - Hunnia, Z1123-r1-d3, 1943–5, 1943. 24 February 1943 Jgykv. - Hunnia Igazg-i ülésről, minutes 2–3.
178. *MOL* - MFI, K675-cs3-t16, 1942–3: A vállalat dolgozóinak jelentései külföldi tanulmányutakról. 'Jelentés Kalló Vilmos zágrábi útjáról', 1 June 1943.
179. Ibid.
180. *MOL* - KM, K66-cs609, 1943, III-6/c, 1031. Professor Géza Fehér to KM, August, 1943, forwarded to the National Security and Propaganda Minister, VKM, ONFB and Film Export Cooperative on 18 August 1943, 2. Hungary sent 40 films to Bulgaria in 1941, according to statistics in *Interfilm. Blätter der internationalen Filmkammer* Heft 3 (November 1942), 31.
181. *MOL*- MFI, K675-cs1-t2, 1928–44. Közgyülési jelentések, jgyzk., és iratok. *A Magyar Film Iroda Rt. igazgatóságának és felügyelőbizottságának Közgyűlési Jelentése az 1942-i tizenkilencedik üzletévről* (Budapest, 26 March 1943).
182. *MOL* - KM, K66-cs609, 1943, III-6/c, 629–31. For Finland: Frigyes Lasestky, 'OMME körlevél: Helyzetjelentés a magyar filmek európai piacairól', 820/1943 sz., 79 ssz., 18 September 1943. For Sweden, Denmark, and Norway: 'Die Budapester Tagung der Internationalen Filmkammer', *Nachrichten für Außenhandel* 286 (7 December 1942). This article predicted Hungarian exports would number at least 11 to each country.
183. *NARA* -RG 262, Records of FBIS, Transcripts of Monitored Foreign Broadcasts, 1941–6, Budapest 1941–6, Box 154, File February 1944. Latin American transcript 15 February 1944, 19:00.
184. *MOL* - MFI, K675-cs3-t16, 1942–3. 'Jelentés a Jenes Nándor olaszországi utjáról' op. cit., 10.
185. Martin, '"European Cinema for Europe!"', 33–7.
186. The role of Transit Film, for unexplained reasons, remains virtually unexplored by scholars.

187. *BA* - RMVP, R56 VI/31 - Transit Film GmbH. RFK Fachgruppe Filmaussenhandel Mitgliederrundschreiben, FFA Schw/Sn 830, 13 May 1942.
188. *BFL*- László Balogh, Nb. 1699/1945. 'Jgykv. felvétetett az ONFB-nak Budapesten 1942.évi március hó 27-én tartott ülésről'.
189. Ibid.
190. *BFL*- László Balogh, Nb. 1699/1945. I. Kutassy, 'Jelentés', 14 April 1942.
191. *MOL* - KM, K66-cs609, 1943, III-6/c. MKK Vichy to the KM, Vichy, 26 May 1943, Vid. 1989. From September 1942, Germany forbade Hungarian films in free Vichy territory - only Italian, German, and French films could be screened. The Hungarian Consul General in Paris reported to Budapest in May 1943 that due to the German-French Film Agreement, no Hungarian film would be allowed into the occupied part of France either. In the same file, see M.kir.Főkonzulátus to the KM, 14 May 1943, Var. 2937/1943.
192. *BFL*- Exculpation of István Erdélyi, in MFSzSz XVII 647/1 - 287 B. Erdélyi quoted in 'II. Jgykv', 12. Felvéve a NB által a Magyar Müvészek Szabadszervezete Filmoztályához kiküldött 289/b. sz IB, 28 June 1945.
193. *MOL* - KM, K66-cs609, 1943, III-6/c. György Szávich, 'Feljegyzés a megszállt Szerbia területére exportált magyar filmek helyezte tárgyában', 20 April 1943.
194. *MOL* - KM, K66-cs609, 1943, III-6/c. Paikert to KM M. Kállay, VKM 12.877/ 1943.sz. - Eln.C, 9 April 1943. Visit of Dr. Wiers.
195. *AAA* - GsB, Fach 27, Kult 12, Nr 4a, 'Filmwesen Ungarn'. German agents spied on top wartime production companies, such as Hamza, Harmonia, Imágo, Panorama, and others.
196. *BA* - Ufa R109 I, Nr. 5465. Liste der führenden Regisseure, Schauspielerinnen und Schauspieler Ungarns', Auslandsabteilung Hö, 24 June 1941.
197. Jávor describes this in his book *Egy szinész elmondja* (Budapest, [1946] 1989), which also details his odyssey fleeing the Gestapo after the German occupation in Hungary. See also, 'Jávor Pál', *Színészkönytár,* http://www.szineszkonyvtar.hu/contents/f-j/javorelet.htm; and Tibor Bános, *Jávor Pál, szemtől szemben* (Budapest, 1979), 223.
198. *BFL*- László Balogh, Nb. 1699/1945. I. Kutassy, 'Jelentés', 14 April 1942, 7–8.
199. *MOL* -BM, K150, V.Kútfő, t15, cs3593 - Mozgoképüzemi ügyek. Filmgyártás. OMB ügyei. Pál Morvay, 'Jelentés a Nemzetközi Filmkamara 1942.ápr.római ülésről', 15 April 1942.
200. E.g. 'Magyar-olasz filmbarátság', *MF* IV/6 (9 February 1942), 3; 'Magyar filmek, szinészek és rendezők sikere Olaszországban', *MF* IV/16 (21 April 1942), 6; István Erdélyi's two part 'Az olasz filmélet', *MF* IV/19 (13 May 1942), 3–5 and *MF* IV/20 (20 May 1942), 3; and 'Az olasz filmélet vezetői Budapesten', *MF* IV/24 (16 June 1942), 2–3.

201. *MOL-Ó* – Hunnia, Z1123-t1-d2, 1940–2, [1942]. 12 June 1942 Hunnia Igagz. jgyek, 9; and 8 July 1942 Hunnia Igagz. jgyek, 5.
202. *MOL* – KM, K66-cs609, 1943, III-6/c. Pro Memoria az olasz-magyar verziós filmek gyártása tárgyában, Iparügyi Miniszterium XII.szakosztály, undated [likely early 1943].
203. A. Lajta, 'Az év története', *Filmművészeti Évkönyv* (1943), 209; and untitled paragraph on the Magyar Film Kiviteli Szövetkezet, *Filmművészeti Évkönyv* (1944), 195.
204. Nemeskürty, *A magyar film 1939–1944*, 66–7.
205. Langer, 'Fejezetek', 249.
206. de Grazia, 'European cinema and the idea of Europe, 1925–95', 24. John Connelly and Aristotle Kallis propose that Southeastern Europe did not figure in to Hitler's *Lebensraum* plans and until 1943, the Nazis presumed it to be within the Italian sphere of influence. Connelly, 'Nazis and Slavs: From Racial Theory to Racist Practice', *Central European History* 32/1 (1999), 9; Kallis, 'A War within the War: Italy, Film, Propaganda and the Quest for Cultural Hegemony in Europe (1933–43)', in Vande Winkel & Welch (eds), *Cinema and the Swastika*, 180.
207. *MOL* – KM, K99 [Italian Consulate]-cs110, 1943, 231. NVPM I. Antal to KM [J. Ghyczy], 8 February 1943, 308–1943.N.P.M.-I.
208. *MOL* – KM, K66-cs609, 1944, III-6/c: Filmügyek. M.kir.K. Fiume to KM, Fiume, 1 July 1944, 467/1944 sz.
209. 'Olasz Filmvezér Budapesten magyar tehetségeket fedezett fel', *Film Hiradó* II/17 (27 April – 4 May 1944).
210. *NARA* – RG208 [OWI Overseas Branch], Bureau of Overseas Intelligence Central Files 1941–5. File: 'E: Balkans (Hungary) 6'. 'Response to Special Request', 27 December 1943, OWI BOI to Mr. Robert Riskin, Interest in American Motion Pictures in Hungary.
211. *NARA* – RG208 [OWI Overseas Branch], Bureau of Overseas Intelligence Central Files 1941–5. File: 'E: Balkans (Hungary) 6'. 'Listening Conditions in Hungary', n.d.
212. Imre Hecht, interview with author, 18 June 2007. Hecht, a Hungarian Jewish former MGM distributor, confirmed the exhibition of many American films in Hungary, 1943–4.
213. *BA* – RMVP, R55 (Alt R50.01), Nr. 20891- Deutsch-ungarische Verhältnis 3.44. V[F]ries to Naumann, Leiter A, Berlin, 7 March 1944. The import and screening of Disney films must have been done surreptitiously, as Hungarian exhibition statistics and censorship records give no indication that these films were on the market. Newer American films may have been smuggled through Turkey or Switzerland.
214. István Erdélyi, 'Magyar filmgyártók, vigyázzunk!' *MF* III/15 (12 April 1941), 3. An editor's note indicates the article was written before the invasion.

215. Nemeskürty, *A magyar film 1939–1944,* 65–6.
216. *MOL* - KM, K66-cs474, 1940, III-6/c. KM [Perczel] to OMB, 17 July 1940, KM 34.198/3 - 1940.
217. *BA* - RMVP, R55 (Alt R50.01), Nr. 20891- Deutsch-ungarische Verhältnis 3.44. Fries to Naumann, Leiter A, Berlin, 7 March 1944, esp. 3–5.
218. As of 1939, most foreign news came to Hungary via Ufa, although through 1941, MGM and 20th Century Fox newsreels were shown in Hungarian theaters. After 1942, only Luce and Ufa were regularly screened alongside Hungarian MFI reports. Much of the news of foreign countries presented in MFI newsreels was cut from footage provided by Ufa. See *MOL* - KMI, K429-cs59-t2. Vegyes MTI levelezés, 1938–9. 'A MFI Hiradó és Rövidfilm Kapcsolatai', November 1939. See also the weekly newsreel listings in *Magyar Film.*
219. *MOL* – ME Nemzetiségi és Kisebbségi osztály iratai, K28-cs265-t1, 1930–44, A. d. E. Flachbert collection. Undated (1942 or 1943), unsigned letter (likely Z. Tőrey) to the PM's Nationality and Minority Department. Historians have also noted the imperial nature of German capitalism and the 'virtual interlocking' of the German and Hungarian economies that Tőrey opposed. See Braham, *The Politics of Genocide,* vol. 1, 56.

7

The National Spirit Doesn't Stick to Celluloid: Hungary's Failed National Film Experiments

> We know that our contributions to the world are valuable only while we enrich the European culture with typically Hungarian creations....Being Hungarian is our only claim for our European destiny.[1]
>
> István Antal

Introduction

In a year-end assessment, Film Chamber President Ferenc Kiss pronounced 1939 a watershed year for Hungarian film. Hungarian audiences had affirmed 'the right of Hungarian film to exist and to succeed, if [film] took into account Hungarian racial character, customs, ways of life, and all other means of separating [the Hungarian] from other nationalities'. According to the actor, 1939 was the year the film profession recognized its products must grow from 'the roots of the popular spirit'.[2] Kiss' words echoed those of Imre Huzly, whose inaugural Lillafüred film festival address had welcomed the arrival of 'the new populist film'.[3] If we accept Kiss and Huzly's aspirational analyses, Hungarian filmmaking had begun a major transformation. Scholars, benefiting from distance, generally concur. According to Miklós Lackó, the topics addressed in Hungarian film changed 'radically' by wartime. No longer were Hungarian audiences offered fantastic bourgeois visions of the future or comedies in which 'urban middle-class attitudes confronted gentry provincialism'. Instead, continued Lackó, spectators were force-fed films which 'described the pseudo-democracy of fascism, the problems of the man-in-the-street, and the [sham resolution of] conflict between landlords and peasants'.[4] Anna Manchin sees a darker side, where slapstick replaced satirical comedy, melodramas expressed disillusionment rather than optimism, and faith in the individual was displaced by communal primacy.[5] Manchin, Lackó, Györgyi Vajdovich, and others agree that Hungarian filmmakers had broken with the past, finally accepting the premises that individuals have inner dispositions, or subjectivities, shaped by a national/ethnic accident of birth.[6] Wartime filmmakers thus designed works to promote a different national unity, one synchronized with the 'fascist-leaning' oligarchy of the day. They had started, it appeared, to craft a new, non-Jewish, nationalist aesthetic.

If only it were so clear. If Hungary were ever to develop a coherent national and nationalist identity transferable to film, the wartime period should have been the time to do so. Hungary had built the scaffold for a national film industry. The corporatist institutions were in place, specifically the Film Chamber and the National Film Committee, and all filmmaking processes were increasingly centralized. The legislative structures for narrowing the concept of nation, the Jewish laws, were more strictly enforced. Financial ruin had been averted and the industry was beginning to prosper. Domestic and export market growth provided unprecedented opportunities to make motion pictures. There appeared to be a consensus that a new national style should and would emerge organically from Hungary's newly reconstituted society.[7] Despite all this, Hungary's film elites struggled mightily when it came to the specifics of forging new national products. Of course, Hungary made numerous films celebrating nationalism and patriotism, including paeans about heroic soldiers and flyers, acclamations of revisionist conquest, or odes to *Heimat*. However, the political squabbles, contrary political and economic imperatives, talent shortages, cultural inertia, and international pressures that stymied the establishment's attempts to unite behind any lucid concept of a Christian national film system exerted the same contradictory pressures on efforts to radically transform film products. An occasional exemplar would emerge, touted as the true cinematic manifestation of the Hungarian essence. But in the aggregate, and while the aesthetics were tweaked, Hungary's pursuit of the golden fleece of an integrative national content and style during the wartime period was largely unsuccessful. Overall film industry output, in terms of the styles and genres, remained surprisingly constant from the mid-1930s through 1944. The elites themselves, especially those on the right, were unconvinced, and in some cases frightened, by their own efforts. Through 1944, they did not believe they had successfully forged an authentic, integrative, reproducible Christian national film form or style. At best, they succeeded at replacing what they termed *giccs*, meaning the trashy, superficial films of the 1930s, with nostalgic *kitsch*, simple message films crammed with stereotypical characters and scenes of *film noir*-like despair.

The persistence of heterogeneity, or at least films which undermined elite pursuit of a cohesive and coherent national narrative and a clearly demarcated national culture, should not surprise scholars, as it mirrors the conclusions of most of the recent generation of national film studies. However, by locating the reasons for this diversity and explaining why prewar forms and styles associated with Jews had resilience, this study breaks new ground. It considers the economy of the motion picture medium and finds contradictions in industry practices that eclipse the micro-politics of production. By studying genres and archetypes, rather than a miniscule selection of a country's production, this chapter also explains why Hungarian film elites, given their greatest opportunity to nationalize and Christianize, had such difficulty visualizing the national spirit in celluloid.

Wartime Christian National Film Rhetoric and Practice

When Hungarian film began integrating populist ideas and retreating from the glamorization of middle-class modernization in the latter 1930s, it appeared that the industry had found its direction and subject. The metamorphosis might have been simple and on some levels, it was. Hungary's new Christian vanguard embraced the art of filmmaking from a national standpoint with stunning naïveté. If filmmakers were merely to draw from sources that were 'intrinsically' Hungarian, such as history and literature, then a distinctive national film style would arise. Fantasy producers, it seems, could easily lose touch with reality. Others were only slightly more pragmatic, urging Hungary's film elite to add populist perspectives to their works. Vociferous advocates of this position included József Daróczy and Antal Güttler, the chiefs of Hajdu Film and Palatinus Film, two of Hungary's leading Christian-owned production companies, and the critic Ádám Szücs.[8] Agreeing that Hungary deserved more populist, less puerile films, Szücs wrote 'The exhibition of the spirit of the people does not begin and end with behavior, dances, and customs....The distinctive characteristics of the Hungarian mentality [portrayed] in a procession of kicking, whooping *csárdás*-es; whip-cracking, loose-panted young men; ribbon-covered, singing young women; etc.' was not representative of Hungary. These types of images, created by 'urban-scented filmmakers', did not access the 'spirit of the people'. Rather, Szücs and his companions concluded that Hungary's new national style, reflecting the 'unique conditions of Hungarian life, morality, and art', would be 'mediated by the blood of those who prepare [films]' and would 'spring unadulterated from the national mentality'.[9]

These visceral declarations and vacuous recommendations were the standard rhetorical fare of the early 1940s, repeated *ad nauseam* by the nationalist right and its vocal film community members. References to the commodification of Hungary's traditional culture, raw profit motive, urbanist or cosmopolitan sentiments, and the superficiality of all pre-1940 films reeked of the period's rampant racism and antisemitism. Naturally, nearly all who preached that Hungary must develop a new style of film agreed that these elements must be eliminated for the new style to germinate. Adopting a populist cinematographic vision required filmmakers first to expunge from their repertoires the stereotypes of Hungarian culture – the *Puszta* cowboys, embroidered dress, and oft-repeated musical refrains – and replace them with more realistic portrayals of Hungarian life. Second, Hungarian filmmakers must shed the characteristically 'Jewish' styles of the operetta[10], the formulaic comedy with its 'ring-road witticisms', the farcical melodrama, and the middle-class/peasant synthesis.[11] But knowing what they wanted to eliminate, actually excising it, and then taking the additional step of providing an attractive alternative were entirely different matters.

Figure 15: Might they be Hungarian? This still, from the 1938 movie Márton Keleti-directed *Barbara in America*, represents the type of overwrought costuming which raised the hackles of advocates of greater realism and a new Christian national Hungarian style. Source: MaNDA, Budapest.

Christian vanguard producers did rally to the calls for seriousness of purpose and weightier themes.[12] In their pursuit of an authentic Hungarian culture, they returned to subjects from Hungarian literature and history to combat the 'shallowness' of the Jewish-dominated 1930s.[13] A second option, 'hero' films, also gained traction. 'Nations are built upon the shoulders of their great men', suggested Mihály Fetter, so biographical portraits or stories about 'scientists and scholars, statesmen and great men' were far more nationally edifying than dramas where young ladies fell for company directors.[14] Somber, reverential tributes to Hungary's pantheon, its historic and cultural figures and their achievements, would ensure that Hungarian citizens were properly nationalized, informed about the 'real Hungary'.[15] A third proposal, commonly raised in the early 1940s, was to make so-called 'problem films'. This genus included movies about 'Hungarian life' or supposedly fact-based 'stories of the people', films about the everyday experiences and travails of regular Hungarians.[16] Realistic films exhibiting the 'growing problems associated with the Hungarian land, Hungarian race, and Hungarian workers' presented a historic opportunity to fabricate a legitimate national film style, a 'renewal' true to Hungary's 'national-political interests'.[17] All three of these forms, with their foci on realms of culture insulated from Hungary's Jews, would go hand in hand with the 'artistic, ethical, and productive cleansing' of Hungarian culture audiences supposedly desired.[18]

Certainly, there was tangible evidence of a trend toward more serious film, populist themes, and state-inspired 'persuasive' pictures. Hero films showing people of peasant origin gaining access to and acceptance in aristocratic ruling circles, such as *István Bors* [*Bors István*] and *Dr. István Kovács* [*Dr. Kovács István*], were blockbusters. Pictures featuring 'regular' men, lower middle-class heroes who vanquished exploitative and unpatriotic capitalists such as *The Thirtieth* [*A Harmincadik*], *Andrew* [*András*], and *Changing of the Guard* [*Őrségváltás*] also numbered among the most successful and decorated of the wartime period. Movies about the Hungarian past such as *The Song of Rákóczi* [*Rákóczi Nótája*] and *Devil Rider* [*Ördöglovas*] won huge audiences. And as the war radically altered the geopolitical context and Hungary's borders changed, films idealizing the organic link of Magyardom to Transylvania via its peasant national culture and its history, including *Silenced Bells* [*Elnémult harangok*], *People on the Alps* [*Emberek a havason*], and *One Night in Transylvania* [*Egy éjszaka Erdélyben*], gained critical acclaim and box office success.

But by-and-large, these films were exceptions, and many were problematic in-and-of themselves. Stylistically, few of the new Christian national features challenged the conventions of the 1930s or made more than superficial adjustments to common styles or genres.[19] In his study of the 1939–44 era, István Nemeskürty concluded that the genres of films made were, in aggregate, similar to those made in the 1930s. It was only the casts, characters, and themes that changed. Nemeskürty determined that the 'commonplace comedy' remained the dominant Hungarian film type, accounting for 90 of the 227 films made between 1939 and 1944.[20] There were few of the so-called 'serious' films that addressed the problems of the day, and most of those that did used stark, simple terms. As Hungary purged itself of the Jewish, internationally-trained core that had been responsible for nearly all of its pre-1938 production, movie making gradually fell into the hands of inexperienced professionals. These men and women, if they were not secretly working with the purged Jews, resorted to techniques born of the silent era and reborn in the German *Heimat* style: the repetitive showing of recognizable landscapes, locales, and historical sites. Hungarians naturally added their own twists, stuffing films with peasant costumes, aristocratic pomp, feudal authority figures, dashing insolvent military heroes, and gentry values – the same stereotypical characters and themes to which audiences had become accustomed before the age of sound. The famed director István Szőts retrospectively concluded that, 'Since the 'changing of the guard' occurred in 1939, the quality of films and the spiritual basis did not change; at very most [films] were dolled up with stupid little political catchphrases or nationalist-tinged operetta-Hungarianness'.[21] Beyond that was the matter of character substitution, replacing the urban middle-class stars of the 'cosmopolitan' films of the mid-1930s with captivating naïfs from the villages. '[Gyula] Kabos had disappeared from the commonplace comedy, and so had the old offices and clerks of Budapest', wrote Nemeskürty. 'The world of pretty young ladies from the country took possession of the screen'.[22]

If Hungarian filmmaking had veered in an entirely new direction, as Ferenc Kiss claimed, it was less than obvious to the era's audiences. With a few notable exceptions, attendance figures demonstrated that wartime audiences preferred the same fare they had enjoyed through the 1930s. The longest-playing films in premier theaters in 1941, *Military Hat and Jacket* [*Csákó és kalap*], *The Gyurkovics Boys* [*Gyurkovics fiúk*], *Finally* [*Végre*], *Pista Dankó, One Night in Transylvania,* and *Old Waltz* [*Régi keringő*], reveal this reality.[23] *Military Hat and Jacket,* written by the prolific Jewish screenwriter Károly Nóti, was a farce using the standard love story with a twist. A Hungarian archaeologist, returning from a stint in America, falls in love with a gentry woman. He learns from his butler that Hungarian women do not marry civilians, only men in uniform. The near-sighted academic thus joins the army to win his love's hand. After a series of unsuccessful efforts, a happy ending is achieved. *The Gyurkovics Boys,* adapted from Ferenc Herczeg's novel of the same name by the Jewish-born screenwriter István Békeffy, recounted the lives of the five Gyurkovics brothers, all of whom pursue different careers and loves. It was banned in 1944 due not only to Békeffy's involvement in the film, but because some of its actors were also Jewish.[24]

Figure 16: An officer and a gentleman. Two of the *Gyurkovics Boys,* Milán and Géza, played by László Perényi and László Szilvassy. This 1940 film, one of a number of wartime works by the Jewish-born screenwriter Isvtán Békeffy, was not banned until 1944. Source: MaNDA, Budapest.

Finally was a proto-typical love triangle, in which a woman's broken-down car and a storm cause her to get stuck with two men, one serious and ambitious, the other a devil-may-care romantic. The movie concerns her choice between the men-*cum*-suitors. *One Night in Transylvania,* originally titled *Alteregó* and set in the eighteenth century, was also a love story. In this case, a beautiful Hungarian countess from Transylvania falls in love with the adjutant for the Habsburg co-regent, Joseph II, thinking her besmitten was actually Joseph. Another love story, *Old Waltz* presented a prodigal Hungarian millionaire who at long last returns to Hungary and sets up his daughter with the son of a countess. The twist is that the Countess' son favors an actress and the daughter of the millionaire prefers her intended's driver. In the end, love wins out over class divides and the plans of the parents.

These short descriptions demonstrate that only one of the top six Hungarian films of 1941, *Pista Dankó,* came close to the ideal of the new

Christian national vision. The aforementioned five films were all light fare, films of distraction rather than the prototypical films of persuasion – the didactic, serious films the Christian national leadership desired. Several of the features were written or prepared for the screen by Jews. Even *Dankó,* touted as one of the first manifestations of the new Christian national film style, was a problematic example, containing many of the characteristics nationalists denounced as Jewish and phony. This Pál Jávor film did champion the working classes, to which his character, Pista Dankó, supposedly belonged, and it likewise paid homage to the upper middle classes, the aristocracy, and effete army officers. That said, the picture is best described as a love quadrangle whose corners were the famed gypsy violinist Pista Dankó; his jealous gypsy suitor; the blonde-haired, blue-eyed Hungarian he truly desires; and his beloved violin. The movie is far from the standard musical love story, however, and its broad concept of nation makes it a fascinating subject. It evokes Hungarian racism toward gypsies, but ends up validating the marriage between Dankó, a gypsy musician, and Ilonka Jáky, the Hungarian female paragon. The movie may have cast 'gypsies' as devious ingrates who pathologically lie and cheat, but the producers also presented Dankó's gypsy music as legitimately Hungarian and Dankó as a dyed-in-the-wool patriot.

This was hardly the new type of Hungarian national motion picture desired by the right. Rather than shed the worn out comic-opera form and the knee-slapping, pub-frequenting, gypsy-music-singing stereotypes of old, it exploited them, making more money than any film which came before it.[25] Purposely held back by Hunnia so it could be released with great fanfare as the 200th Hungarian sound film, the Ministry of Religion and Education named it the top film of 1940 even before it had opened in Hungarian theaters.[26] Hungarian film authorities entered *Dankó* in the 1940 Venice Film Biennial, and touted it as 'pure [in] Hungarian taste, [and in] Hungarian atmosphere'.[27]

Dankó, in other words, symbolized the disjunction between rhetoric and practice. The film community may have concurred that Hungary needed a new Christian national, populist, racially-pure aesthetic conveyed through film, and they may have preached that what the masses wanted were their own stories, their own problems, projected countrywide. Ticket sales and the actual behavior of the film industry indicated

Figure 17: The Gypsy Hungarian. Despairing of lost love but buoyed by the gypsy music, Pál Jávor plays the virtuoso Pista Dankó, who is shown here admiring his second love, his violin. Source: MaNDA, Budapest.

otherwise. Spectators in Hungary and abroad preferred recognizable stereotypes and styles, and Hungarian elites were afraid to challenge the filmmaking status quo. Gypsy music, love stories, and comedy remained the standards of wartime Hungarian film.

Mixed successes like *Dankó* mollified few. People of all political stripes – whether parliamentarians, film critics, regulators, or even producers – endlessly complained about the deterioration of Hungarian film. In September 1940, ONFB President Wlassics took the film community to task:

> A serious sense of vocation must permeate all activities of the state-directed Hungarian film production. In this time of national renewal, it is not acceptable to be guided by the public's more base instincts or to generate the widest sales.[28]

Critics and producers agreed that filmmakers' continued reliance on 'gypsy music', the 'stumbling-bumbling peasant' and 'volkish attire' to identify

Figure 18: Traditional Hungarian dance scene from *The Flower of the Tisza* (1938).

Figure 19: The *puszta* cowboy in *The Yellow Rose* (1939).

Figure 20: The Transylvanian peasant in *Bence Uz,* a 1938–9 smash. Source: MaNDA, Budapest.

their films as Hungarian was indicative of the absence of an inherent sense of Hungarianness in film production.[29] The clamor for 'serious' movies grew shriller. The radical right member of the lower house of Parliament, Zoltán Meskó, groused in late 1941 that the displacement of 'historical works portraying Hungarian heroes [and] symbolic works' by 'love triangles' threatened Hungarian self-consciousness and left Hungarians dependent on 'foreign-subject movies…the heroes of foreign nations and their virtues'. 'Good *völkisch* [*népies*] and historical works are necessary in the villages', he concluded, but filmmakers were failing to provide the people what they needed.[30]

The litany of grievances mounted in 1942. On nearly a weekly basis, articles appeared in *Magyar Film* and other press outlets restating the expectation that a Christian and national-minded film production would develop. And as if there were a script, these same articles would bemoan the fact that no such production had yet evolved, that Hungary seemed unable to exorcise the 'Jewish spirit' from movie-making.[31] While these criticisms became less frequent in late 1942, the pace accelerated again in 1943. At a party early in the year, the fascist-leaning actor Antal Páger reportedly raged that 'the Hungarian film industry will continue to produce crap until the leadership of Hungarian film production is taken out of the hands of those elements expelled from Berlin'.[32] 'In Hungarian films one cannot find Christian morals and deep Hungarian culture', grumbled parliamentarian József Közi-Horváth during his questioning of the Minister of Religion and Education, Jenő Szinyei Merse in December 1943. Quoting the Catholic Action movement's denunciation of Hungarian film and other sources, Közi-Horváth charged that instead of imparting Christian thought and nation-serving propaganda, Hungarian films were destroying the old countryside dynamic with empty 'frivolity', brain-washing the village women of

Figure 21: A man of the non-Jewish people. Publicity still of the popular, talented, and antisemitic Antal Páger. Source: MaNDA, Budapest.

Hungary, 'waves' of whom now spent their Saturdays getting their hair coifed to match their favorite starlets rather than tending to their traditional duties.[33] Whether one believed the film industry was purposely damaging national values, that it had resorted to a strategy of feeding audiences a diet of the 'most worthless, most indistinctive hodge-podge' while hoping the moviegoer would unconsciously 'get' the charm of the 'national character',[34] or that it had failed to 'discover the land...along with the values of the people', few film or political elites seemed satisfied with the direction of their industry.[35]

Film Europe and Film Stasis: International Impediments to Hungarian Christian National Film

In a 1939 article, the critic Sándor Eckhardt characterized Hungarian film as bursting with stock characters and musical monotony. Unlike other critics with similar objections, Eckhardt did not blame filmmakers for misunderstanding audiences' needs. Rather, he believed Hungarian producers made patterned films in response to the desires of foreign political and cultural elites to advance their own nation-building programs. It was the need for the other, the exotic, he wrote, that international film audiences so readily consumed particular images of Hungary, especially the Puszta cowboy, the fabulous culture and society of the nineteenth-century aristocracy, and the gypsy girl and her music.[36] Eckhardt was certainly correct in his analysis. Hungarian filmmakers resorted to these forms so that their products might sell outside of Hungary. Once Hungarian films broke through in Yugoslavia, there was increasing pressure to continue shipping recognizable visual and musical forms abroad, the very same dances, customs, landscapes, monuments, and characters that critics at home found so exasperating. Just as others lambasted operettas, gypsy music, and repetitive landscapes, János Bingert, the insightful Hunnia Director, called for 'picturesque musicals', believing they offered Hungary the best potential for export.[37] What Bingert grasped and what Eckhardt and others failed to acknowledge was that foreign audiences generally welcomed these representations for much the same reasons Hungarian audiences did. Film inspired the process of nation-building Eckhardt pinpointed occurring outside of Hungary inside

Hungary as well. This explains why Hungarian producers continued to churn out the very images and film types they claimed to despise, and it explains why Hungarian censors heartily approved of them. Images of cultural distinction declared the singularity of Hungarian culture and thus its value at home and especially abroad, even if that singularity took the form of well-worn stereotypes.

Pressure to remain true to the stereotypical tropes, styles, and imagery of Hungary's earlier years of film production also came, somewhat paradoxically, from the Nazis. In order to provide some ideological and cultural backbone to their program of ridding the continent of American film, the Nazis began a 'Europe for the Europeans' campaign, concomitant with the reinvigoration of the International Film Chamber in 1941.[38] Rooted in the 'Film Europe' concept that had arisen in the 1920s, the premise of this campaign was that Hollywood's cultural products were perverse and its economic practices oppressive.[39] Hollywood features denied Europeans their own marketplaces and by taking up important cultural space prevented native European cultures from flourishing. Thus, ejecting Hollywood from Europe was akin to removing chains, emancipating the cultures of the small states from the restraints of unfair competition.[40] Once unfettered, quality, crafted European products would come to market.

The program of the Nazis – using the IFK to force American film from Europe and consolidate the European film economy – would have an 'unmistakable cultural-political influence', wrote IFK General Secretary Karl Melzer. '"Indirectly" it would create a consciousness of European film…conceding to film one of the valued places in the European spiritual and cultural life'.[41] The more direct effects of the Nazi program would be thematic changes and greater regimentation of filmmaking processes. Melzer told his IFK colleagues that Europe was a 'house which had 100 million places' whose films, by means of their subject matter, must 'breathe of the *spirit of the European family*'.[42] He recommended they make 'serious problem films' reflecting the 'present mentality', the same formula many in Hungary's film industry advocated as the most direct route to a national film style.[43] The best way to actualize these movies was to film on the basis of pre-filed scripts, avoiding seat-of-the-pants productions. A well-written, pre-reviewed script would give rise to a more professional, more artistic, and more worthwhile product. This rise in quality, in turn, was the 'overall solution for European film'.[44] 'If serious, critical, and optimistic artists, technical professionals, businessmen, and politicians all aware of their responsibilities are active in European film, then the influence of American films…[on] revolutionary Europe will quite quickly cease'.[45]

Many Europeans, especially those in neutral countries who had joined the IFK, saw through Melzer's smoke and mirrors and dismissed the concept of 'European film'. They recognized that the Nazis were simply re-clothing their national interests in the garb of 'Europe'.[46] Some of Hungary's film officials, however, played along. The focus on quality craftsmanship, with its stress on the art of production rather than its business side, meshed nicely with Hungary's own crusade against residual 'Jewish influences'. Many in the Hungarian film community also found Melzer's recommendation that European film concentrate on everyday issues alluring. But nothing was more hypnotic than the prospect of permanently installing themselves as suppliers of film throughout Europe. Hungarian officials proclaimed their selfless willingness to produce European film for two reasons. They would insure the survival of smaller filmmaking countries and supply European theaters the materials they needed to guarantee continuous cinema programs.[47]

What then, in practice, did it mean for Hungary to make European films? In theory, Hungary needed to finally do what Interior Minister Jószef Széll had promised Joseph Goebbels in 1933: fashion recognizable national films highlighting the uniqueness of Hungarian life. This should have posed few problems for Hungary's film elites in the early 1940s and ought to have propelled the advance of new Christian national film forms. Distinctive histories, realist films with social themes, populist films, films about Hungary's great men and more, constituted the stuff of Christian nationalist dreams, perfect for the new European film market.[48] Creating these products, however, was not easy. In fact, the idea of 'European film' brought contrary pressures to bear, submarining Hungary's attempts to fashion new national products. Works too national in character, such as historical or biographical films, might be unexportable. Hungarian social problem films could be too topical and perhaps controversial, or at least too specific to be understood or appreciated by foreign audiences. Nationalistic films with obvious revisionist sentiment were unlikely to be purchased for sale abroad. So, instead of encouraging experimentation, the

European film idea, like the desire for exports in general, pulled Hungarian filmmakers back toward familiar stereotypes and cinema forms.[49]

That German authorities thought of European film as no more than a mélange of niche national films that they controlled became transparent in correspondence between the IFK General Secretary Melzer and Hunnia Director Bingert. Melzer wrote in October 1942 that Hungary's contribution to the realization of the new 'European film project' should be the production of three to four modern Hungarian-themed movies. These films, Melzer believed, would be easier to sell in different lands than historical films and they would also be less costly to make. They should contain 'typical Hungarian subject matter...'[50] Germany's European film project thus perpetuated Hungary's reliance on hackneyed tropes and stereotypes, the typical Hungarian characters, milieus, music, and humor, developed, in many cases, by the Jewish film professionals the Nazis so despised.

Bureaucracies and Backwardness

The Hungarian film community's failure to forge a Christian national style or aesthetic is understandable in hindsight. Achieving a consensus about what the nation was, rather than what it was not, was perhaps impossible, an attempt to define a unity that could not sustain itself. There were far too many centrifugal forces pulling film producers away from the condensed national center and toward other conceptual bodies. Even as Hungary's film establishment spoke of the need for a national style, its members also recognized the need to maintain the health of the industry, to produce exportable films, and to satisfy other contradictory economic or cultural-political imperatives. They called for reform but instead created additional layers of bureaucracy. They insisted on a new generation of innovative professionals, but either neglected to build the institutions to train the young filmmakers or instructed them in the old forms. They called for 'serious' problem films but then retreated, afraid of the disruptive effects these films might have on Hungary's fragile social and political hierarchies. They demanded historical films, films about great Hungarian men, but found neither suitable subjects nor funds to pay for such grand creations. They called for ideological films while willfully ignoring mass media's need to make money to survive. In short, Hungary's film establishment, although purged of most Jews and in theory more internally cohesive than at any period in its history, remained far too divided against itself, unsure of its priorities, and too backward to reach a lasting agreement about a national film style with nationalist potential.

Case in point was the country's development of its film professionals. Ministry of Religion and Education discussions concerning the creation of a state-funded Film Academy pre-dated 1937, yet through 1942 the Ministry still had not acted. An academy responsible for artistic and technical instruction remained a vision, despite persistent complaints of talent shortages and the consequential continued reliance on Jewish expertise. Prior to the war, Hungary's sole state-recognized film school was run by the Jewish director Béla Gaál, but the Jewish Laws forced its closure in 1939. The country thus went nearly two years without any organized professional training program. In the interim, select groups of Hungarian film professionals did internships in Italy or Germany. Yet these options were limited and catered primarily to established film professionals and technical workers, especially actors, directors, cameramen, and sound professionals. Hungary's private film community feebly attempted to address the training void in 1941, offering episodic lectures and seminars. For the average Hungarian film professional, such as a screenwriter, set designer, or theater manager, there were few choices. Finally in 1943, a frustrated István Erdélyi took action. Opening the Kárpát Film Academy, he immediately began offering classes and organizing events for Hungary's film workers.

Months later, the Ministry of Religion and Education at last announced plans to open a Hungarian Film Academy in late 1943 or early 1944, one which should have displaced Kárpát. This state-backed film school was to offer a three-year training program for all individuals engaged in exhibition, direction, trade, and the technical side of film production. During the first year, all students would receive instruction in theoretical aspects of filmmaking and exhibition, in order to groom more 'cultivated experts' and raise the aesthetic bar. Courses on Hungarian language, literature history, and the Hungarian fine arts would provide all film professionals with a common nationalist background. From there, depending on their areas of interest, students would advance through the basics of film practice, ranging from beginning editing to concepts of cosmetics to the economics of film. In the second and third years, studies would tend toward the practical aspects of the profession. Officials were optimistic that the Film Academy would birth a new generation of experts. They were especially hopeful that the Academy would solve Hungary's 'star problem', whereby the small circle of talented actors were demanding ballooning salaries and causing filmmaking costs in Hungary to skyrocket.[51] Film Academy proponents told themselves professionalization would be welcomed by a public yearning for 'new artists and new artistic performances…Hungarian creations which, cleansed of foreign mentalities and foreign elements, would strive unswervingly toward perfection'.[52] The Film Academy, in other words, would train a new cohort untouched by the money-grubbing trade practices and superficial comedic/operatic motion picture styles associated with Jews.

Alas, a three-year program begun in early 1944 would have graduated its first class in 1947, had the Hungarian government actually established a film academy. It did not, and the education of Hungary's wartime professionals remained largely in the hands of István Erdélyi, whose curriculum was not designed to usher in an era of new thinking.[53] Instead of purging foreign mentalities and elements, Erdélyi embraced them. Kárpát study groups watched Russian, French, English, and American films, purposely avoiding German products.[54] A 1943 Kárpát-sponsored series on the practicalities of filmmaking, acting, sound techniques, and other topics given by well-known Hungarian film directors, company chiefs, screenwriters, and university directors received extensive coverage in *Magyar Film*. Guest speakers and those invited to seminars not only included the most desirable starlets of the day, but also Jewish technical experts, out-of work Jewish actors, and other prominent 'non-Aryan' former members of the film establishment.[55] If it did not actually hinder the development of alternative thinking among filmmakers, Erdélyi's academy did little to encourage the maturation of a new aesthetic attractive to racial nationalists.

Bureaucratic developments and shifts in the political winds likewise impeded the growth of new national forms, although on occasion they could actually have the salutary effect of encouraging them. In 1942, a series of upheavals altered the politics of film in Hungary. First, Ferenc Kiss stepped down from his positions as Film Chamber President and government Film Commissioner. Tendered in late December 1941, Kiss' resignation went into effect officially in March 1942. He was replaced by the slightly less radical and entirely ineffective András Cziffra, who, as an 'invisible government Commissioner' insured that the Film Chamber remained dormant and did little more than publish *Magyar Film*.[56] The second and more earth-shaking development was a full-fledged change in government, with Miklós Kállay, an advocate of a more neutral foreign policy and a more moderate, less antisemitic domestic program, replacing László Bárdossy as Prime Minister.[57] Along with Kállay came a gradual change in the political order. In April 1942, Kállay named István Antal to a newly contrived political post, Minister of National Security and Propaganda, a sop to the radical right. He charged Antal's ministry with encouraging national unity, patriotism, and Christian morality through mass media. To even the political balance, Kállay helped engineer the July 1942 transfer of Bálint Hóman, ending his long tenure as Minister of Religion and Education and moving him to a much less powerful position as head of the Pál Teleki Scientific Institute. In Hóman's place Kállay installed the more pliant, less ideological, and less active Jenő Szinyei Merse.

The wake generated by the arrival of Kállay, Antal, and Szinyei Merse rippled through the movie business, creating both space for experimentation and new, more constraining boundaries. Initially, excitement reverberated through the film world, which hoped that the shake-out in officialdom might invigorate Hungarian national film production and even promote its quest for change in form and content. Antal's office gave him oversight of a variety of media, and shoring up government positions in radio and film was his primary responsibility. He recognized how the government's own film politics undermined the goal of realizing an organic national style. Hungary had developed a fly-by-night film production system, whereby the new Christian companies the government promoted resorted to trendy superficiality to appeal to domestic or foreign audiences to insure their survival. If they did not, they disappeared after one failed film.[58] Antal swore to end this counterproductive behavior and his prescriptions seemed to herald increased state involvement and direct investment in feature production.

Not all of the film community greeted Antal's appointment positively. In terms of antisemitism, Antal was a zealot.[59] The creation of his office, however, placed a check on the activities of István Kultsár and was a death sentence for the already moribund Film Chamber, as it usurped responsibility for encouraging filmmaking in accordance with Christian morals and Hungarian national ideals. Antal's appointment was also symptomatic of the continued backwardness of Hungarian governance. Instead of streamlining supervision of all film production, the formation of the National Security and Propaganda Ministry added yet another level of bureaucracy, and an under-funded one at that.[60] With limited cash and personnel, Antal found himself compelled to make short historical culture films to achieve his propaganda goals.[61] This was a great disappointment to those who expected Antal to command resources on a scale similar to Joseph Goebbels.

While Antal himself preferred to keep his distance from the feature film business, several of his proxies aggressively asserted themselves in that realm. From mid-1942 on, the Propaganda Ministry had a designate on the Hunnia board and fairly regular representation on the ONFB and OMB. Beginning in 1943, the person who held these positions and served as chief of film operations in Antal's Ministry was Hunnia's Assistant Managing Director Sándor Nagy. Nagy, one of the more 'terroristic' advocates of Christian nationalism in the industry, was aided by another Antal assistant, György Patkós.[62] Both played significant roles in scripting films at Hunnia. In his official capacity in the Propaganda Ministry, Patkós wrote several films as well. While government film bureaucrats had long been involved in the re-writing of scripts and in 'suggesting' what private producers should make, when Patkós initiated screenplays himself it was an unprecedented step.

One of Patkós' films, *To the Fourth Generation* [*Negyediziglen*], executed with the financial support of the Army and the Prime Minister's office, deserves attention in deference to its audacious, almost laughable propagandistic content.[63] Written to promote patriotism, its unabashed anti-Bolshevik, anti-Soviet message and imagery included casting the Soviets as ugly, scarred baby-killers who force their women to fight. This apparently appealed to Hungarian audiences, even if they found the images overwrought. The movie's portrayal of Hungarians as magnanimous, trusting soldiers and as simple farmers who read the Bible, owned their own cows, and had perfect porcelain-skinned, impeccably-mannered, intellectually-gifted children must have struck a chord as well. *To the Fourth Generation* was one of Hunnia's most profitable films in 1942, and even a popular hit at that year's Venice Biennial.[64]

Figure 22: The ugly enemy. Still of a Soviet soldier from *To the Fourth Generation.* Source: Film copy purchased by author from the Hungarian Film Archive (MFA), now a component of MaNDA, Budapest.

Thus, while Antal himself paid only sporadic attention to features, his Ministry did exert influence upon motion picture production.[65] The example of *To the Fourth Generation* persuaded some that the resources of the state could be accessed by the film industry if the industry could devise a recognizable national or nationalist style. In fact, nearly coincident with the creation of the Propaganda Ministry, filmmakers made their first legitimate attempts to create a new set of motion pictures. In 1942, work was underway or had just been completed on a variety of what film officials hoped would be characteristically Hungarian films. These included national propaganda pieces (such as *To the Fourth Generation*), 'problem films' and other movies addressing the so-called everyday life of the common Hungarian, and historical features. Even with the bureaucratic mayhem and the emasculation of the more radical elements in the Film Chamber, Hungary seemed on the verge of producing a critical mass of films indicative of a robust Christian national style. To understand why this never occurred requires a return to early 1942 to examine the content and fate of a number of the top-selling and most influential wartime motion pictures.

Censorship and the End of the Populist Experiment

One of the first, most successful, and politically important of the wartime populist problem films was *Dr. István Kovács*. Directed by Viktor Bánky, the film debuted in April 1942 and spent a record 119 days playing in three of Budapest's top premier theaters, the Nemzeti Apolló, the Atrium, and the Szittya.[66] In this smash hit, Dr. Kovács, a famous historian, recounts lessons from his career. The son of peasants from northern Hungary, Kovács attended university and became an 'honorable professor'. He was to marry Ada Tatár, the daughter of a parvenu lawyer. Kovács and his family meet Ada's parents, who react with disdain, offended by the Kovács' humble origins. The engagement is broken off, and Kovács instead marries his childhood sweetheart, Ágnes Balog, who like him was a village peasant.

Kovács' marriage and his lectures about Hungary's Eastern-based, anti-urban cultural origins, divide opinions. Students and acquaintances take up sides in support of or opposition to what Kovács represents. Because of the lesser pedigree of his wife Ágnes, the town's gentry shun her. In fact, the younger brother of Kovács' former fiancée Ada insults Ágnes, and Kovács responds by slapping the boy's face. A row develops between the Kovács and Tatár families, but the university takes Kovács' side. When a German-named friend of Tibor Tatár ups the ante by charging that Kovács means to incite rebellion through his lectures, Kovács resigns his teaching post. In the end, Hungary's triumvirate of authority, represented by a government bureaucrat, an army official, and a cleric, support Kovács and his theories about Hungarian national purity and renewal. Urged on by his peasant father, he returns to the university and finds vindication. At his return dinner, the Minister of Education kisses the hand of his wife, symbolizing that in legitimate Hungarian circles of power, origin makes no difference. This facsimile of German *Blut und Boden* ideology proclaims the centrality of the working classes to the nation, and that the 'deep Hungarian' roots of the nation were in its peasantry.[67]

The reactions the film provoked provide a wonderful example of how, once again, conflicting imperatives clashed and prevented the further coalescing of a new Hungarian film style around a populist core. Initially, the Censorship Committee voted five to one to refuse the film an exhibition license. The censors worried about the 'streak of national-colored democracy' that ran through the picture, the 'demands for equality' from below and the 'social-political anxiety' the film would elicit. They concluded that the film was too provocative, and that it would undermine the balance necessary to sustain Hungarian society.[68] In essence, they concluded that a picture that actually silenced and colonized the peasantry, speaking for it through its intellectual tribune, Kovács, rather than allowing peasants to speak for themselves, was still too populist, that the ideology it endorsed was too close to democracy.

In response to the apprehensions of the gentry censors, Bálint Hóman, still serving as Minister of Religion and Education, reviewed the film and agreed that it must be reworked. At the MFI studio, director Viktor Bánky cut numerous scenes and inserted an entirely new prologue designed to mute the theme of class conflict between the peasantry and the gentry/aristocracy. The film now began with Dr. Kovács explaining that the people, Jew and bourgeois, who discriminated against him because of his origins were declining in number and had no place in the new Hungary. These changes placated the oligarchs who crowded the Hungarian bureaucracy. They appeared to deflect blame for social strife from the gentry and aristocracy to the snobbish Jewish bourgeoisie.

Nevertheless, once the film hit premier theaters, it brought in record crowds and had the unintended consequences the censors originally feared. Some critics proclaimed it the first film to be representative of the new national style, the first with the courage to 'lash out' at the problems of Hungarian society.[69] The film that Hungarian authorities nearly forbade became the top drawing feature of 1942, attracting new audiences unaccustomed to attending movies.[70] The populist writer Géza Féja's review of *Dr. Kovács* in the government mouthpiece *Magyarország* explained the phenomenon:

> Generally a different type of audience is going to see the film *Dr. István Kovács* than to the [other] popular film sensations. Most of the audience is in shabby clothing. Some of them even bring with them the smell of the workshop, factory, or grocery. There is no doubt that mostly common men are going to see this film. Their thunderous applause bursts out every day in the darkened hall....Please look at the eyes and faces after the performance. They bear the traces of a dramatic catharsis. People receive something from this film that they have sought elsewhere in vain[71]

This was precisely the Damocles dilemma that so frightened Hungary's censors. To build a stable base for the Christian national film industry, new working-class audiences were mandatory. But appealing to these spectators meant addressing potentially destabilizing social themes. Newspapers wrote that the film produced a 'hatred of the gentry' and reported on middle-class panic that the film might spur the peasantry to 'think too much of themselves'.[72] These responses must have shaken the censors and other government bureaucrats, persuading them that they could not completely control the messages audiences received and that themes of unity could easily be interpreted to support notions of difference. At the same time, others, especially the right, warmly welcomed the film, pleased that Hungary's 'thought' and its 'social truths' were at last translated into motion picture form. *Magyar Film* lauded *Dr. Kovács* as proof of the might of Hungary's new national film production, a signature of mass media's power to reflect and shape public opinion.[73]

It was exactly this power that alarmed Hungarian authorities and prevented the emergence of a true consensus in support of a populist identity for the Hungarian nation. As a result of the class-based reaction to *Dr. Kovács* and in fear of the domestic radical right Arrow Cross movement, Hungary's film institutions narrowed the populist message in the films they made and allowed through 1944. The behind-the-scenes revisions of *The Thirtieth* [*A Harmincadik*], a film directed by László Cserépy, provides evidence of the degree to which state authorities intervened. *The Thirtieth*, which premiered in October 1942 only a few months after *Dr. Kovács*, appeared to be the story of workers struggling against a capitalist mine owner. At a mine settlement on the outskirts of Pest, the workers organize to form a school on the premises. They raise tuition and hire a teacher. The teacher, played by Antal Páger, the ubiquitous face of Hungarian rightist populism, fights to set up the school, only to see his efforts wasted when the mine owner's lackeys confiscate his desks and supplies and convert the school room into a garage. The feckless mine manager, citing legislation and bowing to pressure from the owner, says that there may only be a school if there are thirty pupils. The teacher and his loyal mining friends do not give up, and eventually, the problem resolves itself when new families move to the community. The number of students passes thirty, and with the support of the mine manager, the school is permitted to function.

Prime facie, this film presented the same class conundrum *Dr. Kovács* had. The actor Róbert Bánky claimed in postwar criminal testimony that state authorities were going to prohibit the picture, risking HP 300,000 of production investment. The work-around was to scapegoat the mine owner, blaming him alone, not capitalism in general, for the class strife. In the original script written by Márton Kerecsendi Kiss, there were scant references to the mine owner. He appeared briefly in only one scene. To placate the censors, the mine owner's Jewishness was made more apparent, and a new scene was inserted making him appear as the only person actively exploiting his workers. To further highlight his treachery, the mine owner was portrayed as callously enjoying hindering the schooling of his employees' children.[74] Other minor revisions included transforming the mine manager, Mr. Pongrácz. In the first version, Pongrácz 'was not a bad person and originally was not opposed to the school. [He was] a little helpless', bent to the will of capitalism by the invisible hand.[75] Pongrácz's character was made slightly more receptive to the wishes of the workers and in the final version he becomes empowered, willing to declare his loyalties to his racial/Christian working-class kin, siding against his recalcitrant Jewish boss.

By the waning months of 1942, it was clear to government film bureaucrats that unadulterated populism, because of its many similarities to socialism and democracy, was a real threat to their regime. This helps explain why populist themes, although present in many Hungarian wartime films, and social 'problem films' dealing with inequality, were generally stillborn. Despite endless chatter from the right, calling for honest, realistic films dealing with the problems of the 'Hungarian people', films could only address these issues if they were packaged in certain ways. Land reform remained a taboo subject, as did true democracy. Class divides could only be discussed if unity triumphed or if the divides were presented as artificial, a function of Jewish manipulation of the system. On the rare occasion when filmmakers attempted to package their materials differently, they were stymied, regardless of their professional standing. In late 1942, the doyen of wartime directors, Viktor Bánky, teamed with István Erdélyi to make the film *Your Loving Son, Peter* [*Szerető fia, Péter*]. The movie reflected badly on the old gentry. Its plot revolved around the deathbed confession of an elderly wife of an estate farm hand to her son. She reveals his illegitimate origins, that he is the son of the estate owner, Count Pálossy. The Count takes the news well, happy to learn that he is the father of a hard-working, honest man. The problem is that Péter is more admirable than the Count's legitimate offspring and heir, Béla. A distasteful, spoiled, sickly, and aimless member of the landed aristocracy, Béla has nothing more productive to do than scheme. He decides to spoil Péter's life by seducing Péter's fiancée. Just as he is about to make his conquest at a Mayerling-like hunting lodge, he crashes his car, a victim of his own drunkenness. He is found by Péter, who carries the expiring Béla home and selflessly offers to donate blood to save his half-brother's pitiful life.

The movie premiered on Christmas Day, 1942, and played for nearly three weeks in Budapest's top theaters before the Censorship Committee suddenly revoked its license. The actual censorship decision for this film no longer exists, but circumstantial evidence indicates that the Interior Minister and the National Security and Propaganda Minister disliked the film.[76] István Erdélyi, the film's producer, claimed that the reason the OMB revoked the film's license was that the film challenged class divides, that it was too anti-landowner and pro-peasant. Yet that is not all. Erdélyi asserted that he was informed that if a sufficiently antisemitic prologue were added, similar to what had been done in *Dr. Kovács.* the film would be immediately re-approved for exhibition.[77] Apparently, the filmmakers refused, believing that there was no easy way to transfer blame to the Jews, and *Your Loving Son, Peter* remained shelved for the duration of the war.

The recall of *Your Loving Son, Peter* jolted the film world. It particularly shocked the money men who backed motion pictures. Many now publicly feared that any motion picture could be yanked from theaters mid-showing, jeopardizing their investments. Conversely, the controversy surrounding *Your Loving Son, Peter* demonstrated the lengths those in high offices were willing to go to make sure that class conflict was conflated with or at least mitigated by antisemitism. Hunnia's János Bingert interpreted the ban of *Your Loving Son* differently. He viewed it as the death knell for populist movies. In response to a proposal submitted to Hunnia concerning a populist subject that he smothered, Bingert justified his decision by claiming the genre had fallen out of favor:

> …since the government decided to forbid the *Your Loving Son, Peter* peasant drama, I believe you understand that now such filmmakers who have politically related or socially discordant

> film themes might not dare to stretch and stick with them. The cowardice, however, I believe, is justified, because, as I see it, from the government position, they do not want to see such types of subjects put on film. In consequence, one cannot speak of making a film about this or a similar motion picture idea.[78]

If Bingert, perhaps Hungary's most power film personality, a man who found starring roles for women he fancied, believed the government did not want populist film, the government did not want populist film.[79] The age of the populist problem film was very short indeed.

Antisemitism in Hungarian Film: The *Changing of the Guard*

A current of antisemitism ran through many of the films of the wartime period, but blatantly antisemitic themes such as that in *The Thirtieth* were uncommon. However, in the chaotic summer of 1942, the bureaucratic upheaval owing to the creation of the Propaganda Ministry and the replacement of the Minister of Religion and Education opened space for the premier of Hungary's analogue for *Jud Süß*. Written by Miklós Tóth, directed by Viktor Bánky and produced in the MFI studio by the Pannonia Film Company, *Changing of the Guard* [*Őrségváltás*] is among the most important and influential films made during the wartime period. One of the top performing pictures of 1942, its run in premier theaters lasted 91 days, making it an 'extraordinary success'.[80] *Magyar Film* touted it as 'the truth' about the Jewish 'occupation' of Hungary, a contemporary problem film whose time had come.[81]

The film is staunchly antisemitic, the story of Christian engineer Péter Takács, played by Antal Páger, and his struggles to overcome the machinations of the Jewish management of a large industrial firm. Set in the late-1930s, the film presents Takács as a neglected talent, ignored by Jewish company directors who reward not the deserving, but other Jews. The film includes references to the anti-Jewish positions of Prime Ministers Gyula Gömbös and Béla Imrédy and implies that the entire country expects government reforms designed to rectify the power imbalances which favor Jews. Of course, the Jewish laws do come into force, requiring the removal of all Jewish upper managers. In order to retain behind-the-scenes control of the company, these men – nearly all of whom are caricatured by long-nosed, bespectacled, poor-postured actors – appoint puppets, strawmen. The primary shared trait of the strawmen is that they are stupid, easily bought off by the duplicitous Jews. The film's director, Viktor Bánky, and its cinematographer and cameraman, Barnabás Hegyi, dramatize the behind-the-scenes control literally, placing silhouettes of Jewish puppeteers behind every Christians they appoint and manipulate.

The Jewish directors begin funneling the firm's assets to the London bank account of the former chairman who had stepped down in deceptive compliance with the Jewish laws. The Christian engineer Takács, whose only real interest is to produce materials for the good of the nation, observes the destruction of the company, comprehends the plot, and decides to take action in defense of Hungary. He succeeds, to an extent, in educating the good-natured but dull Christian gentleman the Jewish leadership chose to head the company (by persuading him to read István Széchenyi's *Credit*). His pupil dies suspiciously when a police investigation begins turning up evidence of the Jewish plot to pilfer the company. Eventually, Takács triumphs, selected by a government commission to be the firm's technical director. With his victory the Jews are vanquished, symbolizing that Hungary itself, not just its industry, is returned to its rightful Christian owners.

Similar to other 'problem films' such as *Dr. István Kovács* and *The Thirtieth*, the government's treatment of *Changing of the Guard*, from its conception through its distribution, suggests that substantial divisions existed regarding the film's message and its efficacy. Viktor Bánky claimed in postwar testimony that he was reluctant to direct the film. Only after large sums of money changed hands and ONFB members made personal appeals to convince him that the feature was to be 'important and desirable' did Bánky agree to direct. He also claimed to have tried to diminish the film's antisemitic message, preferring to concentrate on class conflict. The censors, he charged, cut some very important scenes of the masses that would have changed the tenor of the picture.[82]

While there is ample evidence that most government film bureaucrats wished class conflict to be shown only as a symptom of Jewish capitalism, the overt antisemitism of *Changing of the Guard* worried some of the more moderate members of the government. After the war, ONFB deputy chief László Balogh recalled that extensive discussions occurred within the Hungarian government, outside of the normal censorship channels, concerning how the movie should be made and whether to allow its exhibition. ONFB head Wlassics, the Interior Ministry, and the Propaganda Ministry all had input into the work. Balogh remarked that Interior Minister Keresztes-Fischer wanted the word Jew to be struck from the film. He opposed any 'abuse of Jews'. Balogh remembered that once the film was made, the OMB was quite anxious about its potential to lend support to Hungarian extremist groups, and they approved it with reservations. ONFB members shared enough of these reservations that Wlassics arranged a special preview for members of the Prime Minister's Council. This group, which included all of the top members of the Kállay government, was bitterly divided. The Chief of the General Staff, General Ferenc Szombathelyi, came out against the film, worried that it would 'upset domestic tranquility'. Keresztes-Fischer suspended the picture's permit and entrusted Propaganda Minister István Antal with re-working the film. Antal's alterations apparently satisfied his colleagues, and *Changing of the Guard* debuted in August 1942.[83]

When crowds flocked to *Changing of the Guard,* just as they had to *Dr. Kovács, The Thirtieth,* and a handful of other features with slightly less transparent antisemitic or populist undercurrents, a consensus must have developed among Hungary's upper echelons. Hungary's film bureaucrats rejected pure populism as a basis for a new national style in late-1942 and early 1943, despite, and in fact because of its box-office draw. They also began to exclude obvious antisemitic themes after the *Changing of the Guard* experiment. As Bánky was completing work on the first version of *Changing of the Guard* in the MFI studios, Arzén Cserépy, a Hungarian director who had long worked in Germany, wrote to Hunnia chief János Bingert. Cserépy communicated his wish to make a 'Jewish-subject film' based on the neo-conservative Miklós Bartha's turn-of-the-century sociographic study *On the Kazár Lands* [*Kazár földön*], promising that it would be the real Hungarian *Jud Süß.* According to the film historian Tibor Sándor, Bingert twice turned down Cserépy because he could already detect a palpable change in the winds.[84] György Patkós and Sándor Nagy teamed up to write a film about the Tiszaeszlár blood libel trial, but opposition in Hunnia prevented that script from reaching the studio as well.[85] During the Kállay government interregnum, inflammatory antisemitism would not become a hallmark of government policy nor of the national film style.[86] Passive antisemitism, meaning representations of Hungary as a Christian nation and associations of exploitative capitalism with Jewishness, remained acceptable, if not encouraged. Anything more emotive risked stirring up the masses, supporting the radical agenda of the far-right Arrow Cross, and giving in to German demands. Through March 1944, most of those who determined Hungarian film politics had little desire to do any of those things.

The People of Hungary? The *People on the Alps*

There was one 'problem film' made in 1941–2 that was well-received by the entire film community, from regulators, critics, and movie-goers in Budapest to those in the countryside. Interestingly, it was a film which, while it painted a clear picture of a non-Jewish Hungary, completely sidestepped antisemitism and largely avoided flogging the aristocracy or the gentry. The film was István Szőts' *People on the Alps* [*Emberek a havason*], by most accounts the best and most innovative Hungarian film of the entire sound period through 1944. Based on a József Nyirő novella and filmed in the mountains of Transylvania, the picture was Szőts' first. It was one of wartime Hungary's top-grossing pictures and it collected awards at film festivals throughout Europe in 1942.[87] Hungarian authorities, in fact, were convinced that only Nazi back-room arm-twisting, which dictated that either a German or Italian film win the top prize at the Venice film festival, prevented *People on the Alps* from taking home the highest honor.[88] Italian newspapers declared *People* a new kind of realist Hungarian film style, an assessment repeated by film historians.[89] Even the Nazi *Film-Kurier* admitted the film was a 'Heimat-like picture' that was 'particularly worthwhile'.[90] *Magyar Film* touted the picture as 'strange, different, and new', distinct from nearly all other Hungarian works of the period.[91] Hungarian film and cultural figures lavishly praised Szőts' opus. In parliamentary discussions of film matters in November 1942, both the new head of the OMME, Jenő Padányi-Gulyás, and the radical right journalist György Oláh lauded *People* as artistically accomplished and politically powerful, especially due to the desirable national geography it promoted, which focused on the returned lands of Transylvania.[92]

Figure 23–4: 'True' Hungary. Stills from *People on the Alps*. Source: MaNDA, Budapest.

What was so novel, so special about *People on the Alps?* The film told the devastating tale of Gergő Csutak, a Transylvanian lumberjack recruited by a timber company manager to move to the high Carpathians with the promise of good pay. After the family relocates to the picturesque snow-covered mountains, the manager rapes Csutak's attractive wife, Anna, while her husband fells trees. Shocked, crying, and bleeding, Anna escapes with her son, running blindly from her cabin, which catches fire when a candle she knocks over ignites the structure. Her flight ends tragically, as she slips and falls into a frozen crevasse. Although she and her uninjured son are rescued and she recovers briefly, Anna eventually dies of complications from her injury. Without the funds to send her body home for a proper burial, a grief-stricken Csutak takes his wife's body with him on a train, pretending she is alive. Only the good will of the conductor and of fellow passengers allows Csutak to succeed. He returns to the mountains to exact revenge, killing the manager who savaged his wife. He is arrested, but escapes in order to spend Christmas with his motherless young son. Pursued by the police, Csutak is shot dead near his camp where his fellow woodcutters find and protect his body. They collect a reward for apprehending a fugitive, which they use to buy clothes and boots for Csutak's now orphaned child.

A number of characteristics made this work stand out. First, nearly every cast member was a neophyte, and most were quite good. This alone distinguished the movie. Second, the cinematography was commendable. While not radically innovative, the vistas were stunning and the camerawork included a number of firsts. It was the first Hungarian sound feature to film the majority of its scenes outside of a studio and it was the first sound film to show partial nudity, an addition which made the rape scene much more poignant. Third, the film had a realistic quality to it that far exceeded anything else produced between 1931 and 1944. *People* convincingly communicated the sense of life's unexplainable vagaries, the harshness and poverty of working-class existence, the pain of losing a loved one, and the reality of violence. Unlike most of the period's films, *People* dealt with grey areas, such as justifiable revenge and the question of legitimate justice. Yet it also unmistakably declared Transylvania as Hungarian, performing such an important revisionist task that Hungary's authorities and Hunnia officials singled out the film as one of 'national value' before production had even begun.[93]

Figure 25: The hardened Hungarian. The reality of the Transylvanian lumberman's life, portrayed in *People on the Alps.* Source: MaNDA, Budapest.

Like the other 'problem films' of the early 1940s, a heavy anti-modern, peasantist, Christian undercurrent ran through *People.* In this case Catholic symbolism appeared incessantly. The problem was that the movie's anti-capitalist mentality – although without the trappings of antisemitism – was the same sort of worker/peasant collectivist thought that eventually ran afoul of film authorities.[94] Just as the serious problem film seemed on the verge of blazing a new path for Hungarian film in late 1942, the genre fell out of favor. The need for domestic tranquility and unity now trumped the imperative to expose the plight of the average Hungarian. Hungary's governing oligarchs, despite the box office success of *People,* determined they could no longer permit movies of this ilk, pictures which risked sowing further social dissension. By early 1943, after the devastation of Hungary's armies outside of Stalingrad, all problem films, whether they lauded the 'deep magyar peasant', critically sketched the gentry class, touched on class inequities, or highlighted antisemitic themes 'had hit a bad patch'.[95] Jenő Padányi-Gulyás, the OMME President, outlined the change in thinking among the Hungarian film elite:

> Allow me in advance to say that the producers of such themed films [political propaganda and 'problem' films] should make them with the best intentions and make them well. Not long ago that was exactly the intention of the highest leaders, that Hungarian filmmakers not pay homage to simple Muses, rather they should intensively interrogate Hungarian society's vital questions. Nevertheless, perhaps it would be better if the preparation of films dealing with these subjects were [now] put aside until after the war.[96]

Policies and priorities shifted while *People on the Alps* played in Hungary's theaters. The most important matters of the day had become too popular, too threatening, too uncontrollable for Hungary's film establishment to allow them, despite the fact that spectators were climbing over one another to see films that addressed these subjects. There could be no more trenchant proof of the inconsistencies internal to the Hungarian wartime concept of national film. Hungary had succeeded in developing a genre of serious topical films that sold tickets and even attracted audiences unaccustomed to movie-going. This popularity, this stirring of emotion, was precisely why those who regulated film, the literal caretakers of social stability, had little interest in seeing them made. Thus, they effectively squashed the genre.

Great Men and History

A second genre, the historical film, also offered the prospect of uniquely shaping Hungarian national identity and fortifying patriotic pride. This category, which consisted of 'great man' films, historical pictures featuring the exploits of Hungary's lesser heroes of the past, and features about seminal events in Hungarian history, received added attention with the outbreak of war. The first Hungarian experiment with historical film was the dramatic biographic about the Viennese Hungarian doctor, Ignác Semmelweis, and his struggle to eliminate post-partum fever. Semmelweis, who used antiseptics to clean medical equipment and promoted sanitary operating conditions, appeared to be a perfect Hungarian facsimile of the German medical biopic *Robert Koch.* Scripted in 1939 by a team of writers and directed by the famed André de Tóth, who returned from Germany specifically to complete this project, the work was titled simply *Semmelweis*.

The Mester Film Company released the picture in early 1940, and it flopped, losing approximately HP 30,000 before reprising late in the war.[97] German authorities reported that the goal of *Semmelweis* was to cast a new type of Hungarian hero, a modern scientist who contributed to the advancement of Europe. German sources attributed the film's failure to Hungarian historical ignorance. Hungarian audiences did not know who Semmelweis was, asserting that is 'too German-sounding' name caused some viewers to avoid the movie, accounting for its disastrous run.[98]

Figure 26: A great Hungarian addresses his nation but no one listens. With his students sitting in rapt attention on the edge of their seats, Ignác Semmelweis lectures about germs and antiseptics. Unfortunately for the film's makers, Hungarian audiences did not show similar interest. Source: MaNDA/MFA Photo Collection.

Scarred by the *Semmelweis* experience, Hungarian filmmakers were slow to return to the historical great man genre. One might include *Pista Dankó* in this category, but he was hardly a recognized national hero, and the film made about him was a musical melodrama, not a historical work. Between 1940 and 1942, however, pressure rose to try again. Ufa representatives communicated their interest in importing works about Hungarian history, whether culture films or features, into Germany.[99] Journalists, politicians, and Propaganda Minister István Antal proclaimed the need to see the past on the silver screen. Accompanying these incessant pronouncements were numerous drafts of historical dramas submitted to, and almost universally rejected by, the ONFB. For example, the ONFB turned down a proposal for a feature about the 133 days of Bolshevik rule under the Béla Kun regime in 1919 despite the film's patently anti-Bolshevik slant. ONFB head Wlassics wrote that even though the film's writer had the support of the Army and the Foreign Ministry, 'one would not be free to make this sort of film'.[100] Wlassics' committee denied a number of other historical film proposals based on similar obtuse anxieties about inciting the masses. Lack of adequate Christian funding torpedoed many other screenplays.

When it became clear that social problem films would not evolve into the foundation of a new national style, Hungary's filmmakers again looked long and hard at historical films, encouraged by the Kállay government.[101] The difficulty facing Hungary, believed ONFB leaders, Hungary's studio heads, and top producers, was that the only way to make a successful historical film was to mimic the massive *Madame du Barry*-like costume drama. History on its own would not attract large enough audiences to make a profit. Extravagance and big showy productions with incredible sets and hundreds of extras were the tried-and-true ingredients for the blockbuster historical dramas of the 1920s and 1930s, and there were few in Hungary's film establishment capable of challenging convention. This meant that if Hungary were to wade into the genre of historical films, it must be willing to invest unprecedented amounts. And since capital was still short due to restrictions placed on Jewish activities in the film industry, not to mention the financial drains of war, a big-money production in the early 1940s was out of the question.

Using the excuse that it wished to avoid all Jewish capital, Hunnia began producing its own films in the 1941–2 film year. Prior to mid-1941,

Hunnia itself had not initiated productions, written scripts, nor gathered casts. It had only provided studios, technical experts, and film laboratories to independent companies, while also bearing some of the production costs and capital outlays. Once Hunnia ventured into production, it meant that larger accounts could be tapped, the kind of sums necessary for glorious historical works.[102] In late 1942, just months into his tenure, Jenő Szinyei Merse addressed both houses of Parliament concerning his cultural-political program. In these two speeches, Szinyei Merse promised to aid the 'cultural endeavors of our nationality'. To achieve this goal, he specifically called for films to be made 'about our national greatness, about the historical times when our nation arose'.[103] He promised stipends and awards to those who could write such movies, indicating perhaps that the state was ready to help underwrite the enormous cost of a historical opus. By late 1942, Hunnia executives were salivating over the prospect of direct subsidies.

In the first months of 1943, Hunnia employees, in cooperation with the ONFB and the Propaganda Ministry, developed outlines for a series of great man films. Initially, they conceived of three major historical films, focusing on a few of modern Hungary's founding fathers: Prince Ferenc Rákóczi, Count István Széchenyi, and Lajos Kossuth. The first project was to be a massive historical costume drama about the anti-Habsburg rebellion of Prince Rákóczi of Transylvania in the early 1700s. Using a superstructure of romance girded by music, the film's contemporary purpose was to put 'Hungarian freedom, Hungarian communal feeling, Hungarian fighting spirit, and love of country' on display.[104]

Everything was done to achieve that end. Hunnia hired one of the wartime period's finest writer/directors, József Daróczy and signed a top screenwriter, Miklós Asztalos, to co-write the script. To compensate for a textile shortage, the Ministry of Religion and Education promised to secure costumes from the wardrobes of the National Theater and National Opera. The Propaganda Ministry reviewed all of what Asztalos wrote, insuring that the film included 'national feeling', 'patriotism', and 'self sacrifice'.[105] *Magyar Film* called *The Song of Rákóczi* [*Rákóczi Nótája*] Hungary's 'first historical film' and bragged that it was prepared according to 'American standards'.[106] Hunnia designers built sets of never-before-seen-in-Hungary complexity. On-location filming was done in Kassa [Košice], Munkács, and

Sárospatak. The film starred one of Hungary's most popular actresses, Klári Tolnay. After its release in November 1943, the picture became an immediate success, yielding box office receipts equal to *Dankó Pista* but in fewer than four months. Yet the HP 380,000 that *The Song of Rákóczi* earned barely covered half of the film's HP 714,000 production costs.[107] In fact, this income was only slightly more than Hunnia itself had invested in the endeavor.

Even before the release of *The Song of Rákóczi,* naysayers predicted that the genre would be stillborn. The structural problems that had prevented Hungary from producing historical costume dramas in the past remained. Hungary simply could not risk losing HP 2–300,000 on every film it made, argued *Magyar Film* editor Géza Matolay, who months earlier had advocated the same dramas. He recognized that Hungarian historical subjects were virtually unknown elsewhere in Europe and thus producers could not expect to recoup domestic losses on sales abroad. Neither Germany nor Italy, he asserted, had interest in Hungarian historical film. France, Spain, Sweden, Portugal, Belgium, and Denmark paid little attention to Hungarian history, and Hungary's neighbors, excepting Bulgaria, had no desire to screen a Hungarian view of history.[108] János Vaszary put it more

Figure 27: Klári Tolnay, star of *The Song of Rákóczi.*

bluntly, pronouncing the historical genre dead before *The Song of Rákóczi* debuted. 'The time of Hungarian history films', the director/screenwriter foretold, 'has passed'.[109]

In fact, Matolay and Vaszary were correct. The age of Hungarian historical films was over before it began. An entire slate of these films was proposed for the 1943/44 production year. Excluding *The Song of Rákóczi*, only three of the nine, *Devil Rider*, *Éva Szováthy* [*Szováthy Éva*] and *Wildfire* [*Futótűz*], were actually completed. *Devil Rider*, released on the day German troops occupied Hungary, hardly deserved the label 'historical' film. The picture concerned the actual love affair and eventual marriage of the Hungarian equestrian hero, Count Móricz Sándor, and Leontine Metternich, one of Prince Metternich's daughters. Sándor battles outlaws and Prince Metternich's opposition to win Leontine's hand. *Éva Szováthy*, likewise a love story, was loosely based on the 1939 German Terra Film hit, *Mária Ilona*. Historical only in the sense that it referenced 1848,

Figure 28: Still from *Devil Rider*, one of the few successful historical costume dramas made in wartime Hungary. Source: MaNDA, Budapest.

it was a drama tying together love, revenge, and suicide, whereby a Viennese beauty chooses her Hungarian paramour rather than loyalty to the Empire. The plot of *Wildfire* too revolved around love, but this time a famous actress in Kolozsvár in the 1820s makes the honorable decision and does not marry the dashing, but semi-engaged man who falls in love with her. Like *Éva Szováthy* and *Devil Rider*, authorities designated *Wildfire* historical only because it was a costume drama situated in the past. Audiences were not expected to take pride in nation as a result of the historical content of the work.

Plans for films that addressed more meaty historical subjects went unrealized. Neither of the remaining great man features ever came to fruition. Authorities rejected a proposal for transforming a culture film about István Széchenyi into a historical feature because the script, written by a university professor and a museum curator, was so miserable.[110] A sketch for an enormous drama about the leader of Hungary's 1848–9 revolution, Lajos Kossuth, made it a step beyond the Széchenyi outline, but not much further. Unprecedented in size and scope, Hunnia projected the Kossuth feature would cost nearly HP 1.2 million. The Propaganda Ministry pledged to pay half of the film's cost and drafted one of Hungary's top talents, Lajos Zilahy, to write the screenplay in early 1944.[111] Filming was scheduled for May 1944, but it appears that German occupation authorities were determined to save the Hungarian film community money. The film never went into production. *The Song of Rákóczi*, alas, was the first and last of Hungary's true historical dramas.

Conclusion

World War II Hungary was the ideal time and place for a national film style to emerge. Wartime conditions created an existential need for an emotive, binding national unity. The war and German actions had relieved Hungarian cinema of competition from the United States, Great Britain, Poland, Czechoslovakia, Austria, and, for a time, France. Territorial revision meant that Hungary's actual borders now corresponded more closely with those imagined by the cultural and political elites. Racist, 'monochrome nationalist' sentiments were more pervasive than ever, causing an upwelling of support in the film industry to follow an explicitly Christian national course of development.[112] Legislative restrictions on the definition of who belonged to the Hungarian nation were the strictest they had been in the twentieth century. There is no doubt that Hungary succeeded in manufacturing a collection of nationalistic works, meaning works that promoted loyalty to and love for the presumed nation. However, with few exceptions, this collection did not amount to a body of work substantially different in form or content from Hungary's pre-1939 cinema heritage. Neither did it represent some sort of coherent aesthetic and culture distinct from that of other national groups.[113] According to the contemporary standards of Hungary's film elites and nationalists, Hungary failed to demarcate itself. It found neither a semblance of wholeness, nor a consensus vision of a usable past transferable to celluloid. Analysis of films and scripts censored, abandoned, or rejected, make these conclusions ever more evident.

Hungary's attempts to conceive of a distinct national culture and to project it on screen were hampered by its own failure to unify and consolidate its film industry. The wartime government neglected to provide training for new film professionals. It added new layers of bureaucracy to film regulation and creation, such as the Propaganda Ministry, rather than streamlining existing ones. It left this new ministry short of funds, limiting its activities from the outset. It adopted superficial reforms such as Hungarianizing all movie theater names (e.g. the City became the Szittya) and the universal terms filmmakers used (mandating the use of *vágó* rather than editor/cutter) in lieu of more comprehensive steps.[114] It pursued strategies designed to maximize export potential, succumbing to the pressure to produce films based on a store of stereotypes. Thus the products of wartime filmmakers were often stale renditions of the same forms produced in the 1930s or adaptations of international trends, such as crime, spy, or science fiction films.

'No homeland', Theodor Adorno tells us, 'can survive being processed by the films which celebrate it…[films] which thereby turn the [homeland's] unique character…into interchangeable sameness'.[115] Yet Hungary never even reached the point of defining what its new, unique character and culture were. As a result, its film establishment bumbled through trysts with problem films and historical films in a series of failed attempts to capture that elusive national difference and imagined consensus. By 1943, some suggested giving up the search, or at least admitting that Hungarian national film should be transnational. In a film magazine interview, János Vaszary, one of Hungary's most respected film professionals, remarked: 'I don't believe Hungarian culture should exist in and of itself, being only useful domestically. Thus, it would do our culture more use [to have] exportable, gleeful film [as opposed to historical, propaganda, or didactic film]'.[116]

Bureaucratic shortcomings and the desire to project Hungarian culture abroad were just the tip of the iceberg crushing the development of a national film style. The most important impediments blocking its emergence were the political contradictions embedded in Hungarian society. The divisions between urban and rural, middle class and worker, gentry and peasant were so problematic that in order not to offend or inflame, films often were made in a classless 'vacuum' with 'heroes who did not belong to a profession'.[117] So great was upper officialdom's fear of the masses, of democracy, of radical antisemitism, and of loss of message control that it largely abandoned the most promising national film style, the social problem film. And the historical drama, the other genre movie makers and cultural commentators believed might lead toward a national style, was sunk by a profound lack of imagination and capital. In sum, these failures contributed to the relative stability in genres and representative forms from the early years of sound through 1944. Hungary's government bureaucrats had come face-to-face with a dilemma Homi Bhabha would describe some forty-five years later. In declaring the supremacy of a particular notion of culture, the authority of those asserting (or resisting) supremacy is simultaneously undermined. When cultural difference is enunciated, Bhabha suggests, it is generally split between claims of eternal certitude – of tradition and ancient community – and change – 'new cultural demands, meanings, strategies in the political present'.[118] Afraid of undermining their own claims to authority by acknowledging other voices, Hungary's governors could not bring themselves to endorse any new type of national cultural form, even if those forms were profitable and popular.[119] In cinema, Hungary's nationalizing state bumped against its intrinsic limits, and its 'imaginative community' failed to find an imagined consensus.[120]

'Terrified by a new war', concludes István Nemeskürty, 'Hungarian film production ... took refuge in the remotest past of film history in its search for stories and inspiration .... The majority of the wartime directors ... were ignorant of the very primary rules of the profession, and easily confused by any intricate story…'[121] Nemeskürtky's disparaging of the content, quality, and complexity of Hungarian wartime features unfairly blames producers alone for the banality of the period. Hungary's filmmakers were constrained by shortages of raw film and crippling limits on their expenditures and studio time. They were restricted by two layers of official censorship, the ONFB pre-production and the OMB post-production, subject to meddling by the Propaganda Ministry and constricted by self-censorship. All severely limited the range of topics they could address and the revenues they could collect. Filmmakers may also have been hamstrung, or guided, by what they believed Hungarian audiences desired. Despite all protests to the contrary, Hungarians had to produce for their market. If there was a decline in the quality of Hungarian movies, suggested György Dobos, the owner of the Délibáb Film Company, it was because Hungarian audiences were still willing to buy tickets to bad films. Hungarian directors, producers, and writers of the period, he implied, should not be dressed down for merely responding to what they understood viewers wanted.[122]

Hungarian movie-goers craved precisely what cinema patrons the world over wanted, especially during the war – entertainment. Some preferred hero films, others did not. Some wanted films about people resembling themselves; others had no desire to relive their day-to-day experiences in the confines of a theater. Most sought a few simple hours of escape. Eighty percent of Hungarian movies, perhaps more since the war had begun, concluded with a happy ending, estimated the editor of *Movie News* [*Mozi Ujság*], Mrs. István Szentpály. This was true not only because Hungarian films aimed to imitate American ones, but because Hungarian audiences demanded it be so.[123] Whether you make serious films or comedy, wrote a crass cinema owner in Nagyvárad [Oradea], 'make sure you make them with women'.[124] 'More comedy, light, etc. films will have to be produced to replace the propaganda or serious-theme films', remarked Nándor Jenes after his return from Italy in November 1942.[125] A 1943 survey of cinemagoers by the Propaganda Ministry revealed a huge majority of those polled (82 percent) claimed to want to see something other than that which they were being offered. However, when asked what specific type of movie they would like to see more frequently, only 2.5 percent of the respondents felt Hungary needed more historical films. Less than 4 percent felt more films about the everyday life of the common man should be produced. Yet nearly 12 percent believed

Hungary needed more comedies although comedies were already Hungary's most common genre.[126] A random survey of cinema owners conducted by a *Magyar Film* writer in September 1944 yielded similar results. When asked what his audiences wanted, a Budapest cinema manager answered bluntly, 'Give us Hungarian comedies!'[127]

In this context, Hungary's retreat from propaganda films, problem films, and historical films is more understandable. Not only did Hungary's governing authorities shy away from a national film style built upon topical issues, history, and propaganda, neither did Hungary's audiences want them. That did not prevent certain segments of the film industry from continuing to pursue the quest. Imre Bencsik, for example, the founder of a small Eisenstein-influenced art film movement, offered an imaginative solution. In arguing for the creation of a national style, Bencsik concluded that film was a work of art capable of reflecting the eternal tastes, customs, and moral qualities of a people; that it 'conceals in itself the notion of national art....The national film style', he continued, '[was] not expressed in externalities, rather the film, as art, must hide it in itself'. Instead of calling a film Hungarian, because 'cowboys eat goulash on the Hortobágy plain for two hours', Bencsik proposed that collective Hungarian characteristics could become manifest through the rhythm of the editing:

> We must implement the ancient Hungarian rhythm into the rhythm of the film. ...As foreign examples I can only speak of Slavic films, which have some sort of monotone rhythm, [with] pictures presented in a uniform duration of somewhat too great length.
>
> The Hungarian rhythm guides all of the manifestations of our lives, it immediately directs our pronunciation: the stress on the first syllable, the others unstressed; in addition, for example in the *csárdás*, the 'stress' on the first step to the right, the second unstressed, and this repeated to the left. We find this Hungarian rhythm in the meter of our verse, in the ancient eighth, in music, and in all our temporal arts. We must achieve this in the rhythm of film editing. We must begin scenes with longer, expressive, 'stressed' pictures, which are followed by shorter, quicker...'unstressed' pictures. A stressed picture then moves the action forward again, followed again by unstressed pictures, etc.

> Consequently, our films will not only be garbed in 'peasant undergarments' but the typical Hungarian films will be girded in all sorts of private ways.[128]

Fortunately or unfortunately for Hungary, it lost its chance to change its cultural undergarments. When German troops arrived in March 1944, Hungary forfeited remaining control over the destiny of its film industry, whose quest for crowning glory, a national style, came to an abrupt end.

Notes

1. Propaganda Minister I. Antal, speech given in Eger, 10 February 1943. *NARA* – RG 262, Records of FBIS, Transcripts of Monitored Foreign Broadcasts, 1941–6, Budapest 1941–6, Box 153, File February 1943. Budapest short wave 10 February 1943, Record 48716.
2. Ferenc Kiss, 'A magyar film 1940-ben', *MF* I/45 (23 December 1939), 1.
3. Imre Huzly, 'Az új népi film', *MF* I/18 (17 June 1939), 13.
4. Miklós Lackó, 'Budapest during the Interwar Years', in A. Gerő and J. Poór, (eds), *Budapest. A History from Its Beginnings to 1998,* trans. J. Zinner, C. Eby, & N. Arató (Highland Lakes, NJ, 1997), 184.
5. Manchin, 'Fables of Modernity', 282–3.
6. In Jenő Király's opinion, the arrival of the 'vamp' and Hungarian *film noir* symbolized a turn away from the more optimistic styles of the 1930s, evidence of a more fear-filled, anxious time, but he does not attribute this to Hungary's increasingly radical Christian nationalism. J. Király, *Karády mítosza és mágiája* (Budapest, 1989). Györgyi Vajdovich also sees a turn towards *Blut und Boden* film reflective of the 'depressing, melancholy atmosphere' of the wartime period. Vajdovich, 'Művészetek: Filmtörténete. A magyar film 1918-tól 1945-ig', in *Dunánál. A magyarok a 20. században*, http://mek.oszk.hu/01900/01906/html/index4.html and Vajdovich, 'A magyar film 1939 és 1945 között', *Metropolis* 17/2 (2013), http://metropolis.org.hu/?pid=16&aid=480 Vajdovich draws from István Karcsai Kulcsár's earlier works, including, '"Őrségváltás" után. "Népi film" 1939–1944', *Filmkultúra* XXI/11 (1985), 14–24. See also Tibor Sándor, 'Urak és parasztok – viták és mitoszok', *Budapesti negyed* 5/2–3 [16–17] (1997), http://epa.oszk.hu/00000/00003/00014/sandor.htm.
7. Hunnia's June 1942 public hearing address was emblematic. Market expansion, claimed Hunnia's directors, created 'a need for Hungarian film…in which we seek a Christian and national spirit'. *BFL* – Hunnia Filmgyár Rt. – File # Cg29830-3949. 'Közgyülés', 24 June 1942.

8. József Daróczy, 'Hogyan és milyen legyen a megújhodó magyar filmgyártás?' *MF* I/3 (4 March 1939), 4–6; Antal Güttler, 'Legyen "új" az új filmgyártás!' *MF* I/5 (18 March 1939), 2–3; Kádár Miklós, 'Néhány szó a magyar filmtémákról', *MF* I/14 (20 May 1939), 5.
9. Ádám Szücs, 'Van-e magyar filmstílus?' *MF* II/29 (20 July 1940), 4–5.
10. On the operetta and its Jewish/democratic character, see Péter Hanák, 'The Vienna-Budapest Operetta', in *The Garden and the Workshop* (Princeton, NJ, 1998), 135–46, esp. 145. On subversive cabaret and its 'ring road witticisms' that were the sources of many films, see Tibor Bános, *A Pesti Kabaré 100 éve* (Budapest, 2008).
11. Miklós Kádár, 'Javítanivalók a magyar film', *MF* I/28 (26 August 1939), 6.
12. This corresponded with developments in other world cinemas. Hollywood, for example, began making more social problem films, hero films, and historical dramas in the 1930s. It stuck with some of the genres during the war.
13. This vacuous rhetoric ignored the fact that 1930s' films had regularly drawn from literary and historical classics.
14. Mihály Fetter, 'Egységes magyar filmgyártást!' *MF* I/26 (12 August 1939), 16.
15. Szücs, 'Van-e magyar filmstílus?' 5. Szücs and many other conservative right and fascist cultural figures made similar recommendations. See Sándor, *Őrségváltás után,* 165.
16. 'Szabó Dezső előadása a filmről', *MF* III/23 (7 June 1941), 6.
17. János Dáloky, *Igy készül a magyar film. A magyar film múltja és jelene, kulisszatitkai és problémái. A könyv többszáz érdekes képpel világítja meg a magyar filmek születését* (Budapest, 1942), 184–5. ONFB chief Wlassics introduced the text. See also Dáloky, 'Die neue ungarische Filmakademie', *Interfilm* 7 (23 November 1943), 109.
18. Dáloky, *Igy készül a magyar film,* 185.
19. Possible exceptions were early Hungarian variants of neorealism featuring Katalin Karády. These films, while popular, were not welcomed by rightist nationalists, who saw them as aping tradition. Additionally, they despised Karády, her politics, and the fact that Jews were engaged in writing her films.
20. Nemeskürty, *Word and Image,* 118. Manchin, in contrast, argues that there was a fundamental difference in the nature of the comedy. Manchin, 'Fables of Modernity', 282ff.
21. István Szőts, *Szilánkok és gyaluforgácsok* (Budapest, 1999), 29.
22. Nemeskürty, *Word and Image,* 118.
23. Statistics from 'Több mint 1000 napot futott premierben 39 magyar film az elmult évben', *MF* IV/2 (12 January 1942), 2–3.
24. *BFL* – MFSzSz – XVII 647/1 – 287 B, Erdélyi István. 'A Kárpát Film betiltott filmjei 1944 évi január és április hó közötti időben'.

25. *MOL-Ó* – Hunnia, Z1123-r1-d2, 1940–2, 72. 8 October 1941 Jgykv – Hunnia Igazg-i ülésről, minutes 5. Isvtán Langer calculated that *Dankó*, by the end of 1942, had yielded a net profit of nearly HP 218,000, at the time a record. Langer, 'Fejezetek', 221.
26. 'A filmverseny díjai', *MF* II/27 (6 July 1940), 4.
27. 'A magyar film nagy hete', *MF* II/37 (14 September 1940), 1.
28. Gy. Wlassics, Jr., 'Komoly hivatástudatot várunk a magyar filmek alkotóitól', *MF* II/38 (21 September 1940), 1.
29. I. György, 'A magyar film nyomában', *MF* III/25 (21 June 1941), 2. György's suggestion for making films more Hungarian was to infuse them with the aforementioned stereotypes and the 'aroma, taste, color, and intimacy that seduces all foreigners arriving in Hungary'. So much for practical advice.
30. 'A kultuszminiszter válasza a magyar filmet ért támadásokra a Képviselőházban', *MF* II/46 (17 November 1941), 2.
31. For the first three months of 1942, for example, consider this selection of articles: 'Annyi stróman vállalat alakult, ahány zsidó filmest elbocsátottak', *Magyarság* (4 January 1942); J. Szakály, 'Eszmény és lelkiség a keresztény filmművészetében', *MF* IV/1 (5 January 1942), 3–4; J. Kodolányi, 'A magyar szellem árulói', *Református Élet* (10 January 1942); 'A holnap filmje', *A Holnap* (13 January 1943); vitéz Szegedi László, 'A magyar filmművészet jövője', *Új Európa* (1 February 1942); Á. Pacséry, 'Komámasszony, itt az olló', *MF* IV/6 (9 February 1942), 4, 6; 'Filmszakmai örjárát', *A Nép* (5 March and 12 March 1943), two part series; K. Lajthay, 'A magyar film fejlődése', *MF* IV/10 (10 March 1942), 2; 'Bolváry Géza: Az én oldalom', *Film, Szinház, Irodalom* (13 March 1942); 'Mit vár a filmre?' *Szabadság* (13 March 1942); and J. Szakály, 'A feltámadás útja', *MF* IV/13 (31 March 1942), 3–4.
32. 'Ezt irja az Olvasó', *Magyar Nemzet* VI/6 (9 January 1943), 6. Páger was of course referencing the Jewish Hungarians expelled in 1932–3.
33. 'A magyar film a törvényhozás színe előtt', *MF* V/51 (22 December 1943), 1–4, esp. 2–3.
34. Pál Barabás, 'Amig egy film eljut odaíg…Milyen legyen a magyar film?' *Film Hiradó* II/15 (13–20 April 1944).
35. Szőts, quoted in 'A magyar filmhez fel kell fedezni a magyar tájat', *Film Hiradó* II/18 (4–11 April 1944).
36. Sándor Eckhardt, 'A Magyarság külföldi arcképe', in Gy. Szekfű (ed), *Mi a magyar?* (Budapest, 1939), 127–30.
37. János Bingert, 'Kvalitás és export', *MF* II/51 (21 December 1940), 3.
38. Martin, '"European Cinema for Europe!"' in *Cinema and the Swastika*, 28–32.
39. Kristin Thompson, 'The Rise and Fall of Film Europe', in A. Higson and R. Maltby (eds), *'Film Europe' and 'Film America': Cinema, Commerce and Cultural Exchange, 1920–39* (Exeter, UK, 1999), 56–81.

40. Goebbels outlined these ideas in his July 1941 plenary address to the IFK in Berlin, quoted in 'Der Film im geistigen Gesamtgut Europas', *Völkischer Beobachter* 203 (22 July 1941); and 'Der Film in neuen Europa' *Deutsche Allgemeine Zeitung*', 347 (22 July 1941).
41. Karl Melzer 'Europäische Filmversorgung im zeichen der Qualität', *Interfilm* 3 (November 1942), 24.
42. Ibid., 26–7. Italics in original.
43. Ibid., 25.
44. Ibid., 26–9.
45. Ibid., 29.
46. For a discussion of the Nazi film policy recasting its 'national' interests as 'European' interests, see V. de Grazia, 'European cinema and the idea of Europe, 1925–95', in G. Nowell-Smith & S. Ricci (eds), *Hollywood and Europe: Economics, Culture, National Identity: 1945–95* (London, 1998), 21–6.
47. Julius Wlassics, 'Der ungarische Film', *Interfilm* 4 (1943), 25.
48. Mihály Fetter, 'Az új európai filmgyártás', *MF* IV/13 (31 March 1942), 3–4.
49. Andrew Higson and Richard Maltby conclude that the same dynamic occurred in the 1920s, with the first iteration of 'Film Europe'. See '"Film Europe" and "Film America": An Introduction', in Higson and Maltby (eds), *'Film Europe' and 'Film America'*, 56–81.
50. *'typisch ungarische Stoffe…' MOL-Ó* – Hunnia, Z1124-r1-d23. Nemzetközi Filmkamara, 1941–3. Melzer to Bingert Mel/Rö, 22 October 1942.
51. On star culture, see Brigitta Skaper, 'Magyar filmsztárok a két világháború közötti Magyarországon', *Médiakutató* (Fall 2008), http://www.mediakutato.hu/cikk/2008_03_osz/08_regi_magyar_filmsztarok. On the star problem and wages, see note 52.
52. János Dáloky, 'Die neue ungarische Filmakademie', 108–9.
53. One could also work privately with the professor of film Miklós Kispéter at the University of Budapest, but otherwise options were quite limited. Nemeskürty, *A képpé varázsolt idő*, 640.
54. Nemeskürty, *A magyar film 1939–1944*, 111–12; Nemeskürty, *A képpé varázsolt idő*, 610–11. Nemeskürty also says that Erdélyi assembled the best in the business to secretly watch forbidden films.
55. *BFL* – MFSzSz, XVII 647/1 – 287 B, Erdélyi István. Testimony of István Erdélyi, 28 June 1945. 'II. Jgykv', 8. Jgykv felvéve a NB által a MMSzF-ához kiküldött 289/b. szamú IB, 28 June 1945. That Jews served as instructors in Erdélyi's Kárpát Academy was confirmed by United States diplomatic sources. Burton Y. Berry, the American Consul General in Istanbul, reported in early 1943 that Hungarian Jews continued to play leading technical and teaching roles in the Hungarian motion picture industry. See *NARA* – RG59 [State], M-1206, Roll 15, 864.4061/178 Race Problems. Berry to Sec State, No.127 (R-121).

56. Sándor, *Őrségváltás után*, 168–9.
57. Nicholas Kállay, *Hungarian Premier* (New York, 1954), 10, 14; Loránd Tilkovszky, 'The Late Interwar Years and World War II', in P. F. Sugar, et al. (eds), *A History of Hungary*, 349.
58. 'A magyar filmgyártás ügye a képviselőház előtt', *MF* IV/47 (23 November 1942), 2.
59. Antal's antisemitic predilictions were well-known, and are attested to in various People's Courts' document collections. See especially *BFL* – Dr. Nagy Sándor – Nb 3482/1946; Bánky [Bánki] Gyula Viktor, Nb. 2540/1945; and Dr. Haász Aladár, Nb. 4580/1946.
60. Further evidence of Hungarian bureaucratic complexity is that the average film needed permissions from six different government offices prior to production. *BFL* – Hausz Mária Filmkölcsönző és Filmgyártó – File # Cg41186.
61. Sándor, *Őrségváltás után*, 180–1, 186. Sándor claims Antal resorted to historical propaganda shorts because he felt he could get the most bang for his *pengő* using this media.
62. *BFL* – Dr. Nagy Sándor – Nb 3482/1946, Testimony of Béla Pasztor, 9 August 1945 & 12 September 1945; Sándor Gárdonyi, 4 May 1946; András Bródy, 14 September 1945; and Mrs. Árpád Sebes.
63. For a detailed analysis of this film, see David Frey, '"Why We Fight" Hungarian Style. War, Civil War, and the Red Menace in Hungarian Wartime Feature Film'. *KinoKultura* Special Edition (February 2008). http://www.kinokultura.com/specials/7/frey.shtml
64. The film was well attended, but critically panned at the Biennial. *MOL-Ó* – Hunnia, Z1123-r2-d5. Igazg-i iratok, 1940–3. Undated [1944] report on Hunnia titled 'Hunnia Filmgyár Rt'. Concerning *To the Fourth Generation's* success in Venice, see *MOL-Ó* – Hunnia, Z1123-r1-d2, 1940–2, 71. 23 September 1942 Jgykv – Hunnia Igazg-i ülésről, minutes 2.
65. Sándor estimates that the number of films that could count as propaganda, meaning those either directly funded by a department of the government or written/directed by someone in the Propaganda Ministry did not surpass ten through 1943. Sándor, *Őrségváltás után*, 195.
66. *Interfilm* Sonderheft 1 (Spielfilm-Produktion 1942), 50–1.
67. John Cunningham, 'From Arcadia to Collective Farm and Beyond. The Rural in Hungarian Cinema', in C. Fowler & G. Helfield (eds), *Representing the Rural. Space, Place and Identity in Films about the Land*, (Detroit, MI, 2006), 297; Györgyi Vajdovich, 'Ideologiai üzenet az 1939–1944 közötti magyar filmekben. *Metropolis* 17/2 (2013). http://metropolis.org.hu/?pid=16&aid=484
68. *BFL* –László Balogh, Nb. 1699/1945, 565, 567–8. 'Másolat. A Dr. Kovács István' c. játékfilmnek f. évi december hó 4.-én az OMB-ban megvizsgálásáról…
69. István Somody, 'Az első lépés a magyar film útján', *Magyarság* (12 April 1942).

70. The film played in Hungarian premier theaters for seven days longer than the number two film, *Gentry Nest. Interfilm* Sonderheft 1 (Spielfilm-Produktion 1942), 48–55.
71. Géza Féja, 'Dr. Kovács István', *Esti Magyarország* (25 April 1942), 3, translated in T. Sándor, 'Popular Film in Hungary: History, Politics, Film (A Workshop Study), 1938–1944', *MozgóKépTár. Magyar filmtörténeti sorozat,* CD, vol. 1: Játékfilmek a kezdetektől 1944-ig. Magyar Filmintézet, 1996.
72. Sándor, 'Popular Film in Hungary'. MKT 1, op. cit.
73. 'A sajtó és az új magyar népi film', *MF* IV/18 (5 May 1942), 1.
74. *BFL* - Robert Bánky [Bánki] (a.k.a. Robert von Bánky), 17670/1949. Bánki to Törvényszék 28 December 1949, 29–34. Robert, unrelated to the film's director, Viktor Bánky, was an actor who had a minor part in the film but a major part in the film's financing. He was a partner in Iris Film, the company which produced *The Thirtieth.* The film's director claimed the re-write was coerced, and although it was not his intention, he knew he had directed a right-wing propaganda piece. *BFL* – László Cserépy, B. 19334/1949, 21–6. László Cserépy, 'Gyanusított kihallgatásáról Jgykv', MÁBFPRO report 3298./1946, 27 February 1946. The accounts of Bánky and Cserépy were more-or-less confirmed by Ernő Gottesmann, although all are at odds with the general behavior of the president of the OMB Alfréd Szöllőssy, who as Keresztes-Fischer's delegate generally pursued a policy of muting, not raising, antisemitic sentiment. Gottesmann testified that the Interior Ministry, specifically through Szöllőssy, interceded in the making of the film. Szöllőssy supposedly told the producers that the film would be permitted on the condition that it make a totally antisemitic statement. Thus, Gottesmann labeled the film a 'directed film' [irányfilm], one prepared under government pressure whereby its original form and original script were distant from the final outcome. *BFL* – László Cserépy, B. 19334/1949. Ernő Gottesmann, 'Tanúvallomási Jgykv', MÁBFPRO report 430sz./1946, 25 February 1946.
75. *BFL* – Robert Bánky 17670/1949, 39. Script for *The Thirtieth*, scene 20, 4.
76. Gy. Wlassics Jr., the head of the ONFB, hints at this in 'Nyilatozatok egy film betiltása körül', *MF* V/4 (3 February 1943), 5.
77. *BFL* -MFSzSz, XVII 647/6 - 287 B. Testimony of Dr. István Erdélyi, Jgykv felvétetett a Magyar Filmalkalmazottak Szabad Szakszervezetéhez a NB által kiküldött IB XVII.ülésről, 14 August 1945.
78. *MOL-Ó* - Hunnia, Z1124-r4-d44. Dr. Bingert János vezérigazgató levelése, 1942–4. Bingert to L. Temesváry, 7 February 1943. Quoted in Sándor, *Őrségváltás után,* 179–80.
79. Further evidence that the government had given up on celluloid populism includes the ONFB's 1943 decision to reject the populist script *Precipice* [*Szakadék*], written by J. Darvas and L. Ranódy. Nemeskürty, *A képpé varázsolt idő,* 639. On Bingert, see Frey, interview with Imre Hecht, 18 June 2007.

80. *MOL* - MFI, K675-cs2-t13. Hiradó mozira vonatkozó iratok/MTI, MFI végrehajtóbizottsági jgykv-einek. 1938–42, 1125 - back. 32.sz jgykv - 1942.25 August - 1 September-ig. Run length from *Interfilm* Sonderheft 1 (Spielfilm-Produktion 1942), 54–5.
81. Ernő Gyimesy Kásás, 'Őrségváltás', *MF* IV/33 (18 August 1942), 8.
82. Bánky's assertion of reluctance might have been specious, but he claimed that he was at risk of losing his military service exemption and jeopardizing his wife, whom he said was legally Jewish, if he refused. The Third Jewish Law, issued in August 1941, outlawed Christian-Jewish marriage, placing Bánky in a vulnerable position. *BFL* - Victor Bánky [Bánki], Nb. 2540/1945, 81–2. Testimony of Bánky in 'Bánki Gyula Viktor, Filmrendező', Nü 4959/1945, 8 August 1945. In other testimony, Bánky ironically claimed that having worked with Károly Nóti, the Jewish screenwriter, on other films and in violation of the Jewish Laws, he was at the mercy of radical government workers who threatened to blackball him. This assertion was confirmed by other film professionals. *BFL* - Victor Bánky [Bánki], Nb. 2540/1945, 198–9, 206. 'Jgykv.', Nb.VI.2.540/1945/16, 25 June 1947.
83. *BFL* - Victor Bánky [Bánki], Nb. 2540/1945, 103. Testimony of Dr. L. Balogh, Népügyészség 5050/1 - 45, 17 July 1945, from file Dr. L. Barla, Mester Film co-owner.
84. Sándor, *Őrségváltás után,* 177–9.
85. *BFL* - Dr. Nagy Sándor - Nb 3482/1946, Testimony of A. Bródy, 14 September 1945 and I. Radó, 12 September 1945. German and Hungarian sources reported that a Tiszaeszlár film, re-written by *Függetlenség* publicist Lajos Marschalko and György Patkos, almost made it to production during the Nazi occupation. *Film Hiradó* II/22 (1–7 June 1944) & *Deutsch Zeitung in Ungarn,* from *BADH* - R63, Fach 353, 'Pressestelle. Vertrauliche Mitteilung'. 140. Kb. aus SOE, 29.5.–4.6.44, 12.
86. It would be wrong to interpret my comment to mean that Jews lived well under Kállay. While until March 1944 they were protected from deportation to Nazi extermination camps, most were deprived of basic rights. Tens of thousands were forced to do labor service, many of whom died from overwork, abuse by Hungarians, exposure to the elements, and exposure to the Red Army's weapons.
87. While Szőts may have denied the incorporation of antisemitism in his film, the works of Nyirő were certainly anti-Jewish. A popular author, Nyirő also edited the rightwing paper *Hungarian Power* [*Magyar Erő*], later became a figure in the Arrow Cross government, and, accused of war crimes, fled to Franco's Spain after the war.
88. *MOL* - MFI, K675-cs3-t16, 1942–3. 'Jelentés a Jenes Nándor olaszországi utjáról', 3 November 1942.

89. *MOL-Ó* – Hunnia, Z1123-r1-d2. Igazg. jgyek, 1940–2. Report on Italian press response summarized in 23 September 1942 Jgykv – Hunnia Igazg-i ülésről, minutes 2. See also reprints of review clippings in Italian newspapers in September 1942 issues of *Magyar Film*. For film historians, see Cunningham, *Hungarian Cinema*, 56; David Cook, *History of Narrative Film* (New York, 1990), 700; J. Pinter & M. Záhonyi-Ábel (eds), *Ember a havason - Szőts István 100* (Budapest, 2013).
90. 'Spanish, ungarische und schwedische Spielfilme – Starker Beifall für deutschen Kulturfilm', *Film-Kurier* 24/211 (9 September 1942), 1.
91. 'Negyediziglen', 'Szirius' és 'Emberek a havason' képviselik Magyarországot a velencei Biennalen', *MF* IV/33 (18 August 1942), 1.
92. 'A magyar filmgyártás ügye a képviselőház előtt', *MF* IV/47 (23 November 1942), 1–2. Oláh was one of the leading figures of the populist right.
93. *MOL-Ó* – Hunnia, Z1123-r1-d2. Igazg. jgyek, 1940–2. 20 June 1941 Jgykv – Hunnia Igazg-i ülésről, minutes 4.
94. In his book, Szőts explains authorities criticized his film as 'too Catholic' and that he resisted state efforts to make the film overtly antisemitic. 'They [state censors] tried to smuggle trendy propaganda into films prepared with state support. I, however, rigidly opposed all of these sorts of controlling experiments, although it would have been easy to create a repulsive Jewish figure out of the evil tree-cultivator, and with this stir up antisemitic hatred'. Szőts, *Szilánkok és gyaluforgácsok*, 27–8, 37.
95. Sándor, *Őrségváltás után*, 180.
96. 'Padányi Gulyás Jenő nyilatkozik a magyar filmgyártás gondjairól és a jövő útjáról!' *MF* V/7 (17 February 1943), 2. Padányi-Gulyás went on to cite the example of Germany: 'Symptomatic of this is the case of Germany. The Germans have not produced a film analyzing domestic social questions since *Jud Süß*. They say: because there is a war, more entertainment films must be brought before the audiences'.
97. Langer, 'Fejezetek', 221.
98. *AAA* – GsB, Fach 27, Kult 12, Nr. 4a, 'Filmwesen Ungarn'. Folder 'Ignaz Semmelweiss'. Hunnia labeled *Semmelweis* a 'moral success', differentiating it from The *Ball is On* [*Áll a bál*] and *5:40* [*5 óra 40*], which were 'business successes'. *BFL* – Hunnia Filmgyár Rt. – File # Cg29830 – 3949. 'Közgyülés', 25 June 1940. DEFA, ironically, made the film *Semmelweis, Retter der Mütter* in 1950.
99. *MOL* – ME – Tájékoztatási Osztály, K30-cs4, év nélkül, B/45.t: A magyar-német társaságra vonatkozó és egyéb német vonatkozásu iratok, 737–8. 'Feljegyzés', 3 March 1941, re; Ufa.
100. *BFL* – László Balogh, Nb. 1699/1945. ONF Ikt. Sz. 291/1941, note from Wlassics, Jr., to Balogh, 16 October 1941. Re: *Béla Kun (133 vörös nap* vagy *Vérbiztos),* written by L. Nagykéry-Babura.

101. Sándor, *Őrségváltás után*, 200.
102. Langer estimates that between 1 September 1942 and 31 October 1943, 40 films were made in Hunnia's studios. Six were Hunnia's own productions, and they averaged 40 percent higher costs than films made by private companies. Langer, 'Fejezetek', 278–9.
103. Quotes from Jenő Szinyei Merse's speeches before the Lower House (11 November 1942) and Upper House (17 December 1942) of the Hungarian Parliament. In *Magyarország Évkönyve - 1942.* IX.év. (Budapest, 1943), 92.
104. *MOL-Ó* - Hunnia, Z1123-r1-d3, 1943–5. 9 June 1943 Jgykv - Hunnia Igazg-i ülésről, minutes 5.
105. *MOL-Ó* - Hunnia, Z1123-r1-d3, 1943–5. 11 June 1943 Jgykv - Hunnia Igazg-i ülésről, minutes 2.
106. 'Amerikai méretekben készül az első magyar történelmi film', *MF* V/26 (30 June 1943), 3.
107. *MOL-Ó* - Hunnia, Z1123-r1-d3, 1943–5. 16 February 1944 Jgykv - Hunnia Igazg-i ülésről, minutes 9. The cost estimate of HP 714,000 comes from Langer, 'Fejezetek', 316.
108. Géza Matolay, 'Filmkritika és szakértelem', *MF* V/26 (30 June 1943), 1–2; and Géza Matolay, 'Miért nem gyártunk történelmi filmet?' *MF* V/31 (4 August 1943), 1–2.
109. 'Vaszary János komoly nyilatkozata a vidám filmről', *Mozi Ujság* III/30 (28 July– 3 August 1943), 2.
110. *MOL-Ó* - Hunnia, Z1124-r1-d14. 'Széchenyi film forgatókönyv'. 'Pro memoria' [1943]. Hunnia officials dismissed the Széchenyi text as a poor 'series of symbolic pictures', absent of history and not a feature film.
111. *MOL-Ó* - Hunnia, Z1123-r1-d3, 1943–5. 19 January 1944 Jgykv - Hunnia Igazg-i ülésről, minutes 12–13; and 29 March 1944 Jgykv - Hunnia Igazg-i ülésről, minutes 6. Zilahy was selected in spite of his known anti-fascist opinions.
112. Cunningham, *Hungarian Cinema*, 53.
113. Györgyi Vajdovich, citing Károly Szalay, argues kinds of heroes in melodramas changed as did the types of melodramas, but I would argue these were transnational trends not specific to Hungary. Vajdovich, 'A Magyar film 1939 és 1945 között'.
114. 'Magyar mozi, magyar név', *A Filmszinházak Közlönye* I/1 (4 August 1942), 3.
115. Theodor W. Adorno, 'Culture industry reconsidered', in J.M. Bernstein (ed), *The Culture Industry. Selected Essays on Mass Culture* (London, 1991), 89.
116. 'Vaszary János komoly nyilatkozata a vidám filmről', *Mozi Ujság* III/ 30 (28 July–3 August 1943), 2. Italics in original.
117. *BFL* - László Balogh, Nb. 1699/1945, 663. Balogh, quoted in Jgykv felvétetett az ONF 1943. december 10-iki üléséről', 5.
118. Homi K. Bhabha, 'The Commitment to Theory', in J. Pines and P. Willemen (eds), *Questions of Third Cinema* (London, 1989), 127–8.

119. Jon E. Fox & Cynthia Miller-Idriss, 'Everyday Nationhood', *Ethnicities* 8/4 (2008), 536–7.
120. The concept of film elites as an 'imaginative community' comes from Randall Halle, *The Europeanization of Cinema. Interzones and Imaginative Communities* (Urbana, IL, 2014).
121. Nemeskürty, *Word and Image*, 124.
122. 'Filmankét', *Mozi Ujság* III/35 (1–7 September 1943), 3.
123. 'Happy-end?' *Mozi Ujság* II/43 (28 October – 3 November 1942). This sentiment became even more widespread after the debacle at Stalingrad.
124. 'Nagyváradi moziélet', *MF* V/49 (8 December 1943), 6–7.
125. *MOL* – MFI, K675-cs3-t16, 1942–3. 'Jelentés a Jenes Nándor olaszországi utjáról', 3 November 1942.
126. Dr. Molnár Károly, 'Film és közvélemény', *MF* VI/18–19 (1 October 1944), 2–3. That comedies were preferred was surprising. Very few of the respondents were from Budapest and its surroundings (16 of 525 respondents). The official survey results were: ≈39 percent (203/525) wanted more films dealing with the Hungarian spirit and life-or-death questions; ≈24 percent (125/525) wanted higher quality films; 18 percent (93/525) were satisfied with present situation; 11 percent (59/525) wanted more comedies; 4 percent (20/525) wanted more populist/everyday life type movies; 2.5 percent (13/525) wanted historical films; and less than 1 percent wanted more war movies (4/525), agriculture/land matter movies (3/525), or erotic films (2/525). The survey was skewed by the vagueness of the first category, 'films dealing with the Hungarian spirit and life-or-death questions', which was essentially code for everything not in pre-1939 films. The results also indicate a popular belief in the lack of success in the search for a national essence.
127. Ibid., 4. Comedies, both historians and film historians agree, have the most subversive potential of any genre. See Leach, *British Film*, 144–7.
128. Imre Bencsik, 'A nemzeti filmstilus kialakitása', *Filmszolgálat. Az 'Oroszlánkölyök' magyar filmművészeti mozgalom tájékoztató levele*, 2 (1 December 1943), 5. Underline in original.

Conclusion

The Nazi occupation of Hungary instantly changed the Hungarian film world, accelerating the descent of an industry already in the midst of a downward spiral.[1] The end of independent Italy in 1943 meant that Hungary had become totally reliant upon Nazi raw film stock, a commodity doled out sparsely. German and IFK restrictions limited the number of films Hungary could produce to 45 during the 1943/44 season and severely hampered exports. When the *Wehrmacht* took control of Hungary's lands on 19 March 1944, film production had already slowed considerably. While at least thirty new Hungarian films opened in 1944, most of them had begun or even finished filming before German troops arrived. As Hungarian sovereignty disappeared, the film world experienced yet another paroxysm of purge, redefinition, and self-destruction.

American intelligence agents assumed that Germany would incorporate Hungary's movie industry into its own, using the blueprint of Prague Film, the Czech motion picture firm 'cleansed' and reorganized in 1939–40.[2] Instead, Nazi officials, for lack of time, energy, and resources, aided the Hungarians as they obliterated themselves. To the more extreme film industry elements, the German occupation presented an opportunity to carry through the transformation that Hungary's conservative government had been unwilling to complete. The first step was to depose those considered the greatest impediments to a truly Christian national industry. Nazi puppet Prime Minister Döme Sztójay, his allies, and his German backers forced out Interior Minister Ferenc Keresztes-Fischer, Hunnia head János Bingert, and ONFB chief Baron Gyula Wlassics Jr., by mid-1944.[3] All were relative moderates, some with anti-German proclivities, whom the new regime cast as protectors of Jewish interests. Into their offices crept a cohort of more rightist, generally more antisemitic men, aspiring to streamline the industry under more direct state control. István Antal took over the Ministry of Religion and Education. He gave speeches touting the 'epoch-making…and worldview shaping power of film', and named Miklós Mester his State Secretary in charge of film matters.[4] Mester promptly called for a reorganization of the movie business, promising his office would restart production and bring state-induced 'unity' to the industry.[5] He began with Hunnia, where Bingert's replacement Sándor Nagy immediately restructured the company. The government also named Ágoston Tari, probably the most radical of the new appointees, as Government Commissioner for Movie Theaters in an attempt to force the recalcitrant theater owners' association, the MMOE, into submission. Antal Páger became a government commissioner, empowered to ideologically align the acting corps and prepare an industry blacklist.[6] László Balogh not only assumed leadership of the ONFB when authorities shoved Wlassics Jr. aside, but also became the Censorship Committee's chief, finally unifying Hungary's pre- and post-production censorship authorities.

With these and other appointments came renewed hope in radical circles that the Film Chamber would be resuscitated, re-designed as the single association within which all questions related to movies were decided.[7] Through the summer, film professionals revisited all the old arguments, promising 'serious' film and structural reorganization. This, naturally, would lead to 'rebirth', a 'complete reorganization of the film industry', and the 'spiritual and ethical purification of film production'.[8] Central to these visions was an obsessive craving to finally 'solve' the industry's Jewish question. *The Hungarian Word* [*Magyar Szó*] called for a review of the rental agreements for every one of Hungary's 1,400-plus small and normal theaters, charging that in almost every case rents continued to be paid to Jewish enterprises.[9] The first post-occupation issue of *Magyar Film* announced 'No more Jews in the Film Chamber', heralding new antisemitic legislation which required the immediate purge of all Jews.[10] Along with 14 known

Jewish film professionals who were Chamber members, *Magyar Film* listed as expelled approximately 250 stage and screen actors and actresses, and 32 additional film workers recently identified as Jewish. As of 31 March 1944, all were fired from their positions.[11] In official announcements that followed, authorities spoke of how 'Hungarian blood greeted the new direction as a beginning to a new age of national film production'.[12]

German authorities forced exhibitors to reprise *Jud Süß* and Hungarian equivalents such as *Changing of the Guard*. The Nazis appropriated all American films still in circulation. They proposed joint productions, such as a Géza von Bolváry creation about the Hungarian composer and pianist Franz Liszt, and through the summer and fall initiated planning for 18 additional films.[13] Many Hungarian professionals welcomed these changes. Pál Benyovszky, the editor of the nascent journal *Film News* [*Film Hiradó*], expressed the hope that the occupiers' actions would finally liberate the Hungarian industry and crush the 'conspiracy' between Hungarian authorities and 'film Jews' that had for so long suppressed the emergence of the legitimate national film style:

> In the course of the past few years, with Christian money and Christian directors, several such Hungarian films were directed which not only fit our conceptions of art but stood their ground internationally as well. However, what happened during this time is that Hungarian film production was not able to totally sterilize itself of the Jewish influence and, in business matters, for example, a weak Karádi [the starlet Katalin Karády] film received greater, louder advertisements than *The Thirtieth*, the Antal Páger film.[14]

Hungarian filmmakers had, in fact, created a legitimate, unadulterated national style, claimed Benyovszky. However, it had been trampled in dark rooms by cigar-chomping Jews and government bureaucrats:

> ...spiritual pioneers wanted to put the racial and social problems before audiences....[A]fter being filtered by the film censorship committee it ended up that many great Hungarian films were removed from the program as a result of narrow-minded and compromised political viewpoints. Here such secret powers worked together, whose many tentacles it would be worthwhile to track down in the future.

> Finally, a forceful and comprehensive government order recently came into being, which clarified and stabilized the film front, and 100 per cent Aryanized the film profession....[T]he road is paved which will lead to a true artistic renaissance in Hungarian film.[15]

Whether Benyovszky foresaw that the paving he advocated would actually end in the devastation of Jewish lives is not clear. Many Jewish intellectuals did survive the Holocaust, and that was the case for many of Hungary's Jewish film professionals. However, some of Hungary's prominent film figures were unable to escape their eliminationist compatriots and their Nazi cohorts. Béla Gaál, one the most influential pre-1938 directors, was probably the most renowned film personality killed in the Hungarian Holocaust. Béla Karla, a screenwriter and the director of the Artistica Film Company, died in 1944, likely in a Nazi extermination camp. Fülöp Engel was an American-born Hungarian who returned to Hungary in the twenties and worked as head of distribution for several film companies, including Warner Brothers-First National, before founding his own import-export distribution company. He too died in the camps, as did several of the men employed as top managers in the Budapest branch of MGM during the 1930s, such as Pál Faragó and Miklós Salamon. László Ádám, one of Hungary's leading film journalists whose column appeared in the newspaper *Esti Kurir* through the 1930s, was among the hundreds of thousands of Hungarian Jews slaughtered at Auschwitz. Several former editors of the influential private film journals of the 1930s, including Ferenc Endrei, Andor Garami, and Imre Somló, were murdered in camps or during forced marches in 1944–5. Also among the dead were the producer Sándor Ungár and the theater owner and distributor Jenő Gertler. Scores of other lesser-known actors, actresses, and film professionals perished in the Hungarian Holocaust, while hundreds of others suffered severe depredations in camps or labor service.[16]

Some Christian film celebrities possessed the courage to defend their colleagues. Hungary's top female star of the war years, Katalin Karády, whose *femme fatale* roles and sultry singing voice won her legions of loyal fans, walked out mid-production in late 1943 and ultimately quit acting, in part to protest the industry's offensive treatment of its Jews.[17] The Gestapo arrested her for her pro-Allied leanings and she was later incarcerated by

Figure 29 Katalin Karády, one of Hungary's most beloved, and controversial, wartime film starlets.

the Arrow Cross for her 'defeatism' and lack of patriotism.[18] Pál Jávor, the industry's leading man from 1931 through 1944 and Karády's frequent co-star during the war, was expelled from the Film Chamber, banned from production, and eventually jailed for his leftist leanings and his defense of the Jews, particularly his wife. Miklós Mester, the Christian producer in the 1940s whose company had been funded by German monies since the late 1930s, epitomized the contradictory politics of the time. Seeking to consolidate the movie industry while serving as Film Commissioner during the Nazi occupation, he also made substantial and risky efforts to save Jews.[19] Surely there were others who spoke out or acted covertly, but the evidence is limited.[20]

In the end, however, few of the actions of Hungary's film elites, whether in defense of Jews or in favor of their elimination, had much of a bearing on the fate of the industry. The Hungarian motion picture business was not rejuvenated with the final solution to its Jewish problem, as Benyovszky had hoped. Rather, capital investment disappeared, feature filmmaking slowed, and by July 1944 stopped altogether. Even the radical right paper *Új*

Magyarság reluctantly admitted that mistrust reigned among film producers and the desire to make movies had vanished.[21] By the time of the Arrow Cross takeover in October 1944, conscription of male film professionals, film stock shortages, skyrocketing production costs, domestic instability, Allied bombing, and industry turmoil had all taken their toll. Hungary produced newsreels, a smattering of short films, and nothing more. The appointment of an entire slate of rabid antisemitic fascists headed by Ferenc Kiss, Dezső Bariss, and Ferenc Rajniss could not revive Hungary's defunct industry. Beyond forcing Hungary's theaters to headline the Nazi film *Horst Wessel,* the new regime's largest film project consisted of confiscating all filmmaking equipment and shipping it to Germany. Arrow Cross men did this in the desperate hope that against all odds, the Reich would triumph and they would rebuild Hungary's feature capacity on a Danubian plain purified in blood.[22]

Conclusion

However one defines culture and its nature, accepting the perception that cultures do exist and have some concreteness is an essential precondition for imagining and building any nation state.[23] Hungary's political and cultural elites understood this, granting great significance to the role of culture in framing the national space. Culture and the related discourses about character became the means of asserting boundaries, evoking emotions, and making claims to recognition, privilege, and sovereignty. They provided platforms for the articulation of ideas of nation, arenas where groups asserted difference by discriminating, evaluating, and encoding inequality in a battle for power.[24]

When governments and politicians invoke 'the national culture', as Hungary's did, they often do so with prejudicial intentions, hoping to demean and weaken any alternative visions. In many cases, they frame their discussions in terms of protecting something sacred from foreign attack or intra-territorial betrayal. The national culture, the argument goes, is *sui generis,* 'unique', 'authentic', and the principle fundament supporting a set of local and national 'values'. The national culture and its producers must be safeguarded, for if they are not, national specificity and national worth crumble. The dominoes will continue to fall until eventually the political establishment's worst fear, the pollution or dilution of national identity, is realized. And in the age of the nation-state, once the allegiance to the nation is lost, the state becomes insecure, if not untenable.[25] Whether the opponent of this national culture is cosmopolitanism, internationalism, or national indifference, the threat is existential.

Since its fall from great power status after post-World War I, Hungary's leadership has had difficulty coming to terms with its place in Europe and the world, difficulty negotiating a path between former glory and relative impotence. Hungary's political and cultural elites have long discussed the issue of Hungary's national self-image. While some have offered sophisticated, tolerant, malleable, and inclusive answers, many have not.[26] The latter group, which dominated the 1930s and early 1940s and whose views continue to be expressed in current Hungarian public discourse, hews to a fundamentally ethnonationalist view. Despite being debunked by scholars, this ethnonationalism insists that ethnicities exist, that they beget discrete cultures, and that these cultural differences justify political sovereignty.[27] This circular reasoning, which presumes the existence of a culture always under attack and thus in need of purification and protection, becomes the basis for arguing that a unified national identity must exist and that the proper role of government is to control the processes of culture construction. Hungarian Prime Minister Viktor Orbán remarked in 1999 that:

> we have been … trying … to sum up exactly, what it means to be Hungarian, what exactly we are and what exactly our ambitions and aims are. This image, however, is emerging only very slowly from the pieces of the mosaic (of history), which we have occasionally managed to clutch on to. …[I]f we ourselves cannot produce an image that gets straight to the heart of what we are, if we cannot produce an image that both captures the essence (of the Hungarian nation) and is convincing, then others will produce it in our stead. …[We will be] squeezed irrevocably into the mould of that image of us as an industrious, ambitious, but otherwise not particularly exciting, small East-Central European people.
>
> …[We do] not envision the government inventing and creating the national image.…[Yet] the role of the state will continue to be indispensable even in the centuries that lie ahead of

> us. There is simply no other institution capable of harmonising, fitting together, and coordinating this work. This is the government's task.[28]

Orbán's statements, made nearly 80 years after the Treaty of Trianon and a full 70 since sound film appeared in Hungary, reveal both fundamental insecurities and utopian nationalist aspirations for recognition. Worries about the lack of a uniform Hungarian identity, about foreign agents unfairly pigeon-holing Hungary and stereotyping its people, and about the government's obligation to distill the essence of 'Hungarianness' – these could easily be concerns of a prime minister from the 1930s or 1940s. With increasing stridency since the late 1990s through the refugee crisis of 2015, Prime Minister Orbán has repeated the mantra that the Hungarian 'nation' is a 'primordial community of blood' rooted in Christianity.[29] He has claimed that Hungary is comprised of an ethnic community of Hungarians and that the 'preservation and cultivation of this distinctive ethnocultural [Christian-nationalist] quality is the foremost duty of [the] government'.[30]

Orbán's rhetoric illustrates that conceptualization of national identity based on cultural, and even racial, distinction remains salient for Hungary's leaders. Recent Hungarian debates, particularly over 2011 constitutional changes; statues, monuments, education curricula, and museums; the rise in overt antisemitism and its corollary, the ascent of the far-right Jobbik party; and the government's response to the influx of refugees since 2015 underscore this point. Use of the definite article *the* in the new constitution's preamble referencing '*the* Hungarian culture', '*the* spiritual and intellectual unity of our nation', '*the* role of Christianity in preserving our nationhood', not to mention the explicit notation of 'our national culture' and its contribution to European diversity, show significant continuity in thought from the interwar and wartime periods.[31] Public arguments over the contents of the House of Terror and the Budapest Holocaust Memorial demonstrate that Hungary is still working out its current and historical relationship with the Holocaust and the roles Jews have and will play in Hungarian society. Construction of statues of Miklós Horthy, Bálint Hóman, and others who collaborated in the destruction of the Jews, exacerbated by divisive attempts to rehabilitate these individuals, are but one expression of a palpable rise in antisemitism in Hungary. Debates over these actions have drawn virtually all of Hungarian civil society, as well as international organizations such as the World Jewish Congress and the US Congress, into Hungary's on-going struggle for a usable past.[32] Even commemorative stamps featuring the antisemitic wartime starlet Zita Szeleczky illustrate that clashes over Hungary's historical narratives of the 1930s and 1940s continue. Disturbingly, justifications for affording individuals like Szeleczky and Hóman public space indicate that Hungarian society is either ignorant of the past or willing to overlook unsavory elements of it. In terms of tolerating difference, the picture is not a positive one. Polls conducted in 2013 and 2016 by the Medián Public Opinion and Market Research Institute, as well as follow-on studies, have found that some 30 to 40 percent of Hungarian society holds strong or moderately antisemitic views.[33] These attitudes appear to resonate among many of Hungary's leaders, whose exclusivist definition of 'Hungarian' now discriminates against all non-Christians. 'We welcome non-Christian investors, artists, scientists', announced Prime Minister Orbán in June 2015, 'but we don't want to mix on a mass scale.'[34] Papers all over the globe continue to publish Orbán's assertions, made with growing bellicosity since 2015, that by taking a harsh stance toward most Muslim refugees, Hungary was and continues to defend 'Christian Europe'.[35] Multiculturalism, he and his more radical compatriots have argued, destroys national cultures and has no place in Hungary.[36]

This rhetoric and these battles over history are more than critical. They can be, in the words of László Lőrinc, 'life-threatening' when they mobilize myths and falsehoods about 'enemies' in order to justify heavy-handed actions in supposed defense of the homeland.[37] This mix of Christian-nationalist populism and multiple forms of xenophobia – encapsulating a longing for an essential, transcendent, and empowering identity, protected from mis- or re-interpretation by outsiders or contamination from insiders, stemming from an agreed upon cultural, emotional, and historical canon – simultaneously reveals the persistent power and tenuous nature of the national imaginary. Orbán and his allies, even as they sometimes acknowledge the contributions of other nationalities and ethnic groups to the well-being of the Hungarian nation, resolutely tilt against the central reality of national identity: its lack of concreteness, existing only as a subjective and contingent discursive phenomenon. Contested and reformulated according to political need and historical change, nationness is simultaneously a local, regional, and global phenomenon. It is a process

Figure 30 Szeleczky stamp. Stamp commemorating the 100th anniversary of the birth of the actress Zita Szeleczky, issued by the Hungarian Post Office in April 2015, part of a broader effort to rehabilitate Horthy-era figures. Source: Magyar Posta, graphics by Orsolya Kara.

that depends upon who mobilizes it, who consumes it, where and when. Like all but a few of their predecessors who governed in the interwar and wartime periods, today's nationalists continue to will their subject into existence and to disregard the intrinsic flaw in the call for a timeless core essence. If such a thing were possible to forge, it would already exist and need no fine-tuning or clarification.

Film offered interwar and wartime elites new vistas to define and visualize the nation and its character, to produce authenticity, to gain recognition, and to strive for transcendence. It promised new forms of empowerment in a chaotic world, new ways of overcoming the trauma of Hungary's Great War disaster. This study of the formation and evolution of the Hungarian national sound film industry and its place in the broader Hungarian culture has endeavored to show how difficult and continuous the practice of producing the nation is. It demonstrates just how much nation-formation is influenced by contradictory forces emanating from multiple directions. We should not presume, as most nationalists did and do, that groups have pre-existing identities waiting to be found. But neither should we dismiss their searches by declaring that matters of culture and identity, while intricately linked, are indefinable due to their inherent fluidity, subjectivity, and temporal specificity. We must explore the multiple valences of nation construction. We must interrogate the internal inconsistencies in the ways nationalists construct their nations, public resistance or indifference to their constructions, and alternatives to their visions. Doing so reveals why, for example, Nazis can create conditions which encourage Jewish cultural success, cosmopolitan Jews can create Christian national cultural products, and ardent Hungarian nationalists often come to loggerheads when working out nationalist practice. To wit, those characteristics of Hungary and its film industry that *prima facie* appeared to contribute to national unity were sometimes detriments to the nationalizing mission. Hungary's unique language conferred a degree of protection for domestic film. However, it made production for export a challenge. Hungary's supposed ethnic homogeneity seemed to make the nation ripe for a coherent national cinema, but its urban/rural split and its elites' fear of its own people and of democracy proved intractable impediments or convenient excuses for the film and political masters.[38] The process worked in reverse too – detriments acted as attributes. The Hungarian state's inability to make financial commitments to the film industry in the late 1920s and early 1930s forced the movie business to develop a sounder, market-based footing. However, Hungary's tortured relationship with capitalism in turn resulted in a reliance on Jewish capital that then became a fundamental conundrum when concepts of the national changed and few 'nationalists' stepped up to replace the unwanted minority. Hungary's culture of étatism hindered the development of the movie business yet slowed the spread of antisemitism. The basic divisions between the three branches of the movie business – production, distribution, and exhibition – had exactly the same effects. Janus appeared everywhere in the Hungarian national sound film saga.

International relations, namely transactions with the United States, France, Austria, the Balkan states, and especially Germany all had substantial influences on the evolution of the Hungarian motion picture business as it simultaneously strove for homogeneity and universality. In some cases these relations encouraged unity and coherence in the Hungarian industry, but often in unintended and domestically undesirable ways. In other cases, geopolitical events or interactions with foreign countries exacerbated rifts already present in the film world and the wider Hungarian society, altering conceptions of what was national and in the national interest. Suppositions about national sovereignty, international recognition, and a special Hungarian destiny in Europe often had as much influence on shaping the film industry as the craving for cultural cohesion and authenticity.

Ultimately, film proved to be ill-suited as a purveyor of an ideology of purity or as a cultural homogenizer. Much of this failure is due to the material, political, cultural, and transnational nature of the medium itself. Filmmaking is a commercial art that demands profits to survive. It is a mass medium with unparalleled persuasive power which most political authorities both coveted and felt compelled to regulate. Film is also a consumable good, watched by audiences with different frames of reference. The experience of Hungary in the 1930s and 1940s shows us that the purposeful construction of a cultural industry on a presumed national basis reveals more about what divides a society than what unites it. This episode demonstrates why in a small country with a limiting language, restricted market, and a finite group of elites, no consensus emerged concerning what their industry's goals were. Priorities and identities in celluloid are shaped by far too many creative individuals, regulators, material forces, and even influences from outside the culture to transfigure into stable, harmonious, and substantial forms. When something resembling coherence emerged, the 1930s visions of Jewish middle-class-led modernization and Jewish comedic forms, political and cultural leaders alike accepted it for consumption beyond Hungary's borders. Yet instead of embracing this vision domestically, the large majority of the governing class and the vociferous right rejected it as anational and alien, illustrating that what was Hungarian abroad was not the same as what was Hungarian at home. Their counterparts in film responded by attempting to craft something explicitly Christian and national, to affix the reality of wartime Hungary to the imagined ideal, and to teach Hungarians who they were. Most of these attempts to nationalize and demarcate the national culture failed.

Competing imperatives undermined all efforts to produce a coherent particularism. The result was an industry that could neither fulfill the dreams most nationalists had for it, nor survive the war that had propelled it to such unprecedented heights.

Notes

1. The impact on Hungarian radio was more immediate, exposing Nazi plans. No Jewish-written or composed works could be played, nor could any performers of Jewish origin have any role. See Miklós Szabó, '1944. március 19. történetéhez', *Levéltári Közvelemények* 53/2 (1982), 300.
2. *NARA* - RG208, Box 275, Entry 367, OWI Overseas Branch, Bureau of Overseas Intelligence Central Files 1941–5. File: 'E: Balkans (Hungary) 6. 'Response to Special Request' 27 December 1943, OWI BOI to Mr. Robert Riskin, re: Interest in American Motion Pictures in Hungary.
3. Tim Kirk, 'Film and Politics in South-east Europe: Germany as "Leading Cultural Nation", 1933–45', in Vande Winkel & Welch (eds), *Cinema and the Swastika,* 246.
4. Antal speech translation in *BADH* - R 63, Fach 354, 'Pressestelle'. 147. Kb. SOE, 17.7–23.7.44, 19–20.
5. 'Mester Miklós államtitkár a filmügyek átszervezéséről', *Film Hiradó* II/29 (20–26 July 1944).
6. *BADH* - R 63, Fach 353, 'Pressestelle'. 137. Kb. SOE, 8.5-14.5.44, 18; 144. Kb. SOE, 26.6.–2.7.44, 1.
7. Géza Matolay, 'Kezdődjék meg a teljes átrendezés!' *MF* VI/8 (20 April 1944), 1–2.
8. Reprints in *BADH* – R 63, Fach 353 and 354. 'Pressestelle'.
9. *BADH* – R 63, Fach 353, 'Pressestelle'. 141. Kb. SOE, 5.6.-11.6.44, 14–15.
10. Law 1220/1944 M.E. sz.
11. 'Nincs többé zsidó a Filmkamarában!' *MF* VI/7 (1 April 1944), 2–3.
12. *BADH* – R 63, Fach 354, 'Pressestelle'. 148. Kb. SOE, 17.7–23.7.44, 1.
13. *BADH* – R 63, Fach 354, 'Pressestelle'. 149. Kb. SOE, 31.7–20.8.44, 35. Hungary premiered only 22 films in 1944 before production ceased in June/July. Most of the 22 films were manufactured in late 1943 or pre-March 1944. Hungarian producers made a mere handful during the Nazi occupation. *BADH* – R 63, Fach 354, 'Pressestelle'. 149. Kb. SOE, 31.7–20.8.44, 35; 154. Kb. SOE, 25.9–1.10.44, 14–15.
14. Pál Benyovszky, 'A magyar filmgyártás és a keresztény tőke', *Film Hiradó* II/19 (11–18 May 1944), 1.
15. Ibid.
16. The above (and below) list of Jewish film professionals killed during the Holocaust is compiled from Nemeskürty, *A magyar film 1939–1944,* 19 and

Ábel, 'Magyar filmgyártó'. Many more important film professionals were murdered after the Arrow Cross takeover in October 1944. Gaál, the man who made the blockbusters *The Dream Car* and *The New Landlord*, was specifically targeted and gunned down by Arrow Cross men. Rezső Faragó, a life-long professional distributor and producer known for funding one of Hungary's most successful interwar films, *Young Noszty*, was luckier. He survived an Arrow Cross murder attempt. In addition to Gaál, lost in late 1944 and early 1945 were: Izzó Feldmann, a film establishment stalwart who had led several professional associations in the 1920s and 1930s and was the founder/owner of Cito Cinema; Jenő Deák, the last owner of the Deák distribution company and a well-known distributor/producer; Miklós Kormos, the co-owner or owner of several distribution companies, including Express and the Kormos Miklós Distribution company; and György Engel, the head of the Focus Film Company and an experienced professional who had worked extensively with Joe Pasternak and Hermann Kosterlitz.

17. Karády was posthumously named 'Righteous among the Nations' for her wartime actions. She spoke out publicly in defense of other Jews, was involved in secret negotiations with the Allies to withdraw Hungary from the war, and hid Jews from the Nazis and Arrow Cross. She and her films experienced a renaissance between the 1980s and early 2000s. See Frey, 'Mata Hari or the body of the nation? Interpretations of Katalin Karády', *Hungarian Studies Review* XLII/1–2 (2014), 89–106; László Kelecsenyi, *Karády Katalin* (Budapest, 1982), 40; and Sándor, *Őrségváltás után*, 212.
18. Cunningham, *Hungarian Cinema*, 42.
19. Tibor Sándor references Randolph Braham, who called Mester a former 'ultra-rightist' who became 'an admirable man'. Mester switched to the anti-German side even before becoming Film Commissioner. As Undersecretary of State in the Ministry of Religion and Education during the summer of 1944, he helped save hundreds of Jews by arranging for exemptions from deportations and/or the requirement to wear the yellow star, or by helping to arrange passage from Budapest to safety. Braham, *The Politics of Genocide, Condensed Edition*, 160; Braham, *Studies on the Holocaust in Hungary* (Boulder, CO, 1990), 67, 74; Sándor, *Őrségváltás után*, 218; Leslie Laszlo, 'The Role of the Christian Churches in the Rescue of the Budapest Jews', *Hungarian Studies Review* XI/1 (Spring 1984), 28, 41 ftn 28.
20. One of the reasons for the house arrest and later Arrow Cross jailing of Former Army Chief of Staff Ferenc Szombathelyi was his opposition to the wide reprise of the antisemitic *Changing of the Guard*. The Arrow Cross also arrested other actors with Jewish spouses or pro-Jewish proclivities, such as the actor Zoltán Egyed and actresses Gizi Bájor and Ilona Titkos. Bános, *Jávor Pál*, 223, 228–9.

21. *BADH* – R 63, Fach 354, 'Pressestelle. Vertrauliche Mitteilung'. 148. Kb. SOE, 17.7–23.7.44, pp. II–III.
22. *BFL* – Testimony in hearings of István Kauser, B.23186/1949 and Dezső Bariss, B. 20500/1949. See also *TH* – István Kauser (OMME), testimony in Szűcs Demeter és tsa., ÁVH Vizsgálati d., III/3.c osztály, sz. 93363.
23. G. Eley and R.G. Suny, 'Introduction: From the Moment of Social History to the Work of Cultural Representation', in Eley & Suny (eds), *Becoming National: A Reader* (New York, 1996), 8.
24. Edward Said, *The Word, the Text, and the Critic* (London, 1984), 11.
25. This exact sentiment was expressed in historian Mária Schmidt's interview for the August-October 2012 'Mi a Magyar? Kortárs Válaszok' exhibit at Budapest's Műcsarnok.
26. Examples of the tolerant approach in the 1930s and more recently are Mihály Babits' 'On the Characteristics of Hungarians', originally written in the late-1930s, quoted in George Bisztray, 'Hungarian National Character: Specific or Interactive?' *Hungarian Studies Review* XXVII/1–2 (2000), 101–11; and George Konrád, *The Melancholy of Rebirth: Essays from Postcommunist Central Europe, 1989–94*. Trans. H.M. Heim (San Diego, 1995).
27. Magdolna Marsovszky, 'Antiszemitizmus és nemzeti áldozati mítosz', *Beszélő Online* (20 May 2013), http://beszelo.c3.hu/onlinecikk/antiszemitizmus-es-nemzeti-aldozati-mitosz.
28. Quotes from the speeches of Prime Minister Viktor Orbán, translated in Gusztáv Kosztolányi, 'Hungarian Identity, Globalisation and EU Accession' *Central European Review* 2/6 (14 February 2000), http://www.ce-review.org/00/6/essay6.html. In an interview on 4 March 2012 the *Frankfurter Allgemeine Zeitung*, Orbán declared that 'nations without ambitions will not be able to make the European Community great', in which he conflated nation, family, and Hungary's special place in Europe with Christian cultural unity.
29. See, for example, Orbán's 29 September 2012 speech dedicating a sculpture in Ópusztaszer, in which he refers to Hungary as a Christian, blood-based community. Video embedded in Zsófia Mihancsik, 'A "vérközösség" országáról és a miniszterelnökéről', *Galamus-csoport* (1 October 2012), http://galamus.hu/index.php?option=com_content&view=article&id=165077:a-verkoezoesseg-orszagarol-es-miniszterelnoekerl-164091&catid=51&Itemid=129. On Orbán's recent use of the term 'Christian-national' see 'Viktor Orbán and the "Christian-National Idea"', *Hungarian Spectrum* (22 September 2015), http://hungarianspectrum.org/2015/09/22/viktor-orban-and-the-christian-national-idea/.
30. György Péteri, 'Should We Really Be Surprised by Where Viktor Orbán's Hungary is Heading?' *Transit Europäisches Revue/Tr@nsit Online* (28 October 2014) http://www.iwm.at/read-listen-watch/transit-online/really-surprised-viktor-orbans-hungary-heading/; and Liz Fekete, 'Hungary:

Power, punishment and the "Christian-national Idea"', *Race & Class* 57/4 (2016), 40, 43–4.

31. 'Magyarország Alaptörvénye', *Magyar Közlöny* 43 (25 April 2011), 10656. http://www.kozlonyok.hu/nkonline/MKPDF/hiteles/mk11043.pdf. Debates between historians over latent antisemitic rhetoric in books and articles, and broader public arguments over Hungarian culture, its folk traditions, the new national unity holiday, and the construction of the innovation museum – these are just some indicators of Hungarians' continued and multivalent fixation with national identity.
32. The rehabilitation and proposal to build a statue of Bálint Hóman in Székesfehérvár provoked a visceral public debate which played out in the major newspapers, particularly on the pages of *Magyar Nemzet,* in Hungary's most popular and most important intellectual journals, such as *HVG* and *Élet és Irodalom,* and in the Hungarian Academy of Sciences. It spilled over into street protests, in which those opposed to the Hóman statue donned yellow stars. A group of American Congressmen sent a letter of protest to Hungary, and World Jewish Congress President Ronald Lauder met directly with Prime Minister Órban to urge him not to commit state funds to the construction of the Hóman monument. Paul Shapiro, the Director for Advanced Holocaust Studies at the United States Holocaust Museum, testified before Congress about Hungarian antisemitism and the multipronged Fidesz and Jobbik-led 'assault on memory of the Holocaust' on 19 March 2013.
33. 'A Magyarországi antiszemitizmus számokban', *Szombat* (24 January 2014), http://www.szombat.org/politika/a-magyarorszagi-antiszemitizmus-szamokban; Gábor T. Szántó, 'A magyar antiszemitizmus nem kirívó, de nincs ok a megnyugvásra', *Szombat* (21 January 2014), http://www.szombat.org/politika/a-magyar-antiszemitizmus-nem-kirivo-de-nincs-ok-a-megnyugvasra; 'Medián: Eeyhén [sic] csökkent az antiszemitizmus Magyarországon', *Világgazdaság Online* (31 March 2015), http://www.vg.hu/kozelet/tarsadalom/median-eeyhen-csokkent-az-antiszemitizmus-magyarorszagon-447254; Marton Dunai, 'Anti-Semitism Taboo under Threat in Hungary', *Reuters* (20 May 2014), Available at http://www.reuters.com/article/us-hungary-antisemitism-idUSKBN0E10E420140521. For late 2016 statistics, see 'Enyhén nőtt az antiszemitizmus Magyarországon', *Medián Közvélemény- és Piackutató Intézet* (19 April 2017) http://www.webaudit.hu/object.4c0cb8a1-00ce-4223-aeee-e64ac72dbec5.ivy. On antisemitism in Hungary since the fall of Communism, see the works of András Kovács, most recently 'Antisemitic Prejudices and Dynamics of Antisemitism in Post-Communist Hungary', *Proceedings/'Antisemitism in Europe Today: the Phenomena, the Conflicts'* (Berlin, 8–9 November 2013) and András Kovács, 'Zsidóellenes előítéletesség és az antiszemitizmus dinamikája a mai Magyarországon', *Társadalmi Riport* (2014), 486–508.

34. 'Multiculturalism doesn't work in Hungary, says Orbán', *Reuters* (3 June 2015), http://www.reuters.com/article/us-hungary-orban-idUSKBN0OJ0T920150603.
35. E.g. 'Orbán, "EU's Christian Identity under Threat from Muslim Migrants"', *Financial Times* (30 March 2017), https://www.ft.com/content/7ecde2c2-af12-329a-9133-29a7bee08e31?mhq5j=e1; Ian Traynor, 'Migration crisis: Hungarian PM says Europe in the grip of madness', The *Guardian* (3 September 2015), http://www.theguardian.com/world/2015/sep/03/migration-crisis-hungary-pm-victor-orban-europe-response-madness; Rick Noack, 'Muslims threaten Europe's Christian identity, Hungary's Leader says', *The Washington Post* (3 September 2015), https://www.washingtonpost.com/news/worldviews/wp/2015/09/03/muslims-threaten-europes-christian-identity-hungarys-leader-says/.
36. 'Multiculturalism Doesn't Work in Hungary, Says Orbán', op. cit. and Szabolcs Dull, 'Orbán Viktor mindenkit óv a multikulturális Európától'. Index.hu (2 June 2015). http://index.hu/belfold/2015/06/02/orban_viktor_mindenkit_ov_a_multikulturalis_europatol/.
37. László Lőrinc, 'Most akkor kik támadják Magyarországot?' *HVG* (12 June 2015), http://hvg.hu/velemeny/20150612_Most_akkor_kik_tamadjak_Magyarorszagot.
38. On 'performative ethnicity' and the persistence of 'ethnicism' in Central European history, see Rogers Brubaker and Jeremy King. Brubaker, *Ethnicity without Groups* (Cambridge, MA, 2004), 7–27, esp. 9–12 and King, 'The Nationalization of East Central Europe. Ethnicism, Ethnicity, and Beyond', in M. Bucur & N. Wingfield (eds), *Staging the Past* (West Lafayette, IN, 2001), 112–52, esp. 113–14, 124–30, 136–42.

Bibliography

All web sources accessed 30 July 2017 unless specified.

ARCHIVAL MATERIALS

Academy of Motion Picture Arts and Sciences, Margaret Herrick Library – Beverly Hills, CA

Adolf Zukor Collection

Library Special Collections

Motion Picture Association of America General Correspondence Files

Motion Picture Association of America Production Code Administration Files

Oral History Collection: Hans Salter, M.L. Gunzburg, Eugene Zukor, Adolph & Eugene Zukor

Auswärtiges Amt Archiv – Berlin

Büro Reichministers

Deutsche Kulturpolitik

Gesandtschaft in Budapest

Handakten VLR Clodius

Handelspolitische Abteilung

Kult – Generalia

Kult. Pol.- Generalia

Kulturpolitische [Abt] G (Geheim)

Politische Abteilung II

Politische Abteilung II – Wirtschaft

Presseabteilung, Protokoll, Sonderreferat Wirtschaft und Abteilung W IX [Schifffahrt]

Rundfunkpolitische Abteilung

Bayerisches Hauptstaatsarchiv – Munich

MK (Kultusministerium)
ML (Landwirtschaftsministerium)
MWi (Wirtschaftministerium)
Reichsstatthalter (Epps)
StK (Staatskanzlei)

Budapest Fővárosi Levéltár – Budapest

Files relating to People's Courts [*Népbíróság*] trials and investigations of:
Antal, István
Balogh, László
Banás[s], József
Bánky [von Bánky], Robert
Bánky, Viktor
Bariss, Dezső
Barla, László
Bornemissza, Gábor
Bornemissza, Géza
Cselle, Lajos
Cserépy, László
Cziffra, András
Göttesmann, Ernő
Haász, Aladár
Hidvégi, Ernő
Kádas, Károly
Kauser, István
Kibedy Varga, Sándor
Kiss, Ferenc
Kőszeghy, Iván
Kulcsár, István
Lohr, Ferenc
Mester, Miklós
Muráti, Lili

Nagy, Iván
Nagy, Sándor
Páger, Antal
Pásztóy, Ámon
Podmanicky, Félix
Schlick, István
Szalontai Kiss, Miklós
Szeleczky, Zita
Tomcsányi, Kálmán
Tőrey, Zoltán
Träger, Ernő
Váczi, Dezső

Postwar Vetting [*Igazoló bizottság*] Records of:
Hunnia Filmgyár Rt
Magyar Filmalkalmazottak Szabad Szakszervezet
Magyar Filmiroda Rt.
Magyar Hiradó
Magyar Nemzeti Apolló Mozgófénykép Rt
Magyar Szinészek Szabad Szakszervezet
Star Filmgyár és Filmkereskedelmi Rt

Business Committee [*Cégbizottság*] Records of:
Auróra Film Kft
Budapest Film Rt
City Film Rt
Continental Film
Délibáb
Dr. Baik Mihály Filmkölcsönző
Eco Filmkereskedelmi és Kölcsönző Rt
Genius Film
Glóbus Film Kft
Hajdu Film Kft
Harmónia Filmipari és Filmforgalmi Vállalat Kft
Hausz Mária Filmkölcsönző és Filmgyártó

Hirsch Lajos és Tsuk Imre Filmkölcsönző Vállalata
Hunnia Filmgyár Rt
Kárpát Film
Kárpát Filmkereskedelmi Kft
Kino Filmipari és Kereskedelmi Kft
Kovács Emil és Társa Cég
Magyar Film Iroda
Magyar Írók Filmje
Mester Film
MGM Rt
Minerva Filmgyátó Kft
Mozgóképipari Kft
Mozgóképüzemi Rt
Müvészfilm Kölcsönző Vállalat Kft
Objektiv Film Kft
Osso Film Rt
Palatinus
Palatinus Filmterjesztő Kft
Pallas Filmgyártó és Filmforgalmi Kft
Paramount Filmforgalmi Kft
Pásztor Film Kft
Pátria Filmkereskedelmi és Filmkölcsönző Vállalat
Pegazus Filmgyártó és Filmkölcsönző Kft
Pictura Film Kft
Prizma FilmRádius Filmipari Rt
Reflektor Film
Rex Filmgyártó és Filmkereskedelmi Kft
Sláger Filmeketkölcsönző és Filmforgalmi Kft
Stylus Filmipari Rt
Takács Film
Tricolor Filmkereskedelmi és Filmforgalmi Rt
Ufa Rt
Universal Film Rt
Walter Tibor Filmkölcsönző Vállalat
Warner Brothers-First National

Bundesarchiv – Berlin

Auslandsorganisation der NSDAP
Auswärtiges Amt Nachrichten- und Presseabteilung
BDC Documents – Individual files
Der Beauftragte des Führers für die Überwachung der gesamten geistigen und weltanschaulichen Schulung und Erziehung der NSDAP
Deutsche Arbeitsfront Arbeitswissenschaftliches Institut
Deutsches Konsulat in Temesvar
Deutsches Nachrichtenbüro
Dienststellen Rosenberg – Kanzlei Rosenberg
Kampfbund für deutsche Kultur – Reichsleitung
Persönlicher Stab Reichsführer-SS
Reichsfilmkammer
Reichsfinanzministerium
Reichskreditgesellschaft
Reichskulturkammer
Reichslandbund-Pressearchiv
Reichsministerium für Volksaufklärung und Propaganda
Reichsministerium für Wissenschaft, Erziehung, und Volksbildung
Reichspropagandaamt Ausland
Reichsanstalt für Film und Bild in Wissenschaft und Unterricht
Reichssicherheitshauptamt
Reichswirtschaftministerium
Reichswirtschaftskammer
Reichswirtschaftsministerium
Statistisches Reichsamt

Bundesarchiv (Zwischenarchiv) Dahlwitz – Hoppegarten

Deutsche Akademie
Deutsches Auslandswissenschaftliches Institut Zeitungsausschnitssammlung
Reichssicherheitshauptamt personnel records – Individual files
Südosteuropäische Gesellschaft

Bundesfilmarchiv – Berlin

Individual film and censor files pertaining to Hungarian films shown in Germany or Hungarian-German co-productions
Individual personnel files of Hungarian film figures living in Germany
Film and *Reichsministerium für Volksaufklärung und Propaganda*-related documents

Hadtörténeti Levéltár – Budapest

Vezérkari Főnökség Iratai

Institut für Zeitgeschichte – Munich

Records of Alfred Rosenberg
Records of Hans Hinkel

Magyar Film Intézet/Magyar Film Archivum – Budapest

Film collection
Pamphlet and film journal collections
Photo collection
Unpublished manuscript collection

Magyar Országos Levéltár – Budapest

Belügyminisztériumi Levéltár
Iparügyi Minisztériumi Levéltár
Kereskedelem- és Közlekedésügyi Minisztériumi Levéltár
Kozma Miklós iratai
Külügyminisztériumi Levéltár
Magyar Filmiroda Rt
Miniszterelnökségi Levéltár, Minisztertanács
Vallás-és Közoktatásügyi Minisztériumi Levéltár

Magyar Országos Levéltár – Óbuda

Hunnia Filmgyár Rt
Kisebb takarékpénztári és bankfondok
Magyar Általános Hitelbank Rt
Microfilm Records of Captured German Documents
Pénzügyminisztérium/Pesti Magyar Kereskedelmi Bank Rt

Magyar Zsidó Levéltár – Budapest

Miscellaneous collections

National Archives and Records Administration – College Park, MD

US Government Document Research:
RG 44, *Office of Government Reports*
RG 208, *Office of War Information*
RG 169, *Records of the Foreign Economic Administration*
RG 59, *State Department*
RG 59, *State Department* (Microfilm)

Captured German Document Research:
T-77, *Army High Command Records*
T-580, Captured German Records Filmed at Berlin
T-120, *Records of the German Foreign Office Received by the Department of State*
T-175, *Records of the Reich Leader of the SS and Chief of the German Police*
T-70, *Records of the Reich Ministry for Public Enlightenment and Propaganda, 1936–44*
T-71, *Reich Ministry of Economics Records*

'Schuchmacher Records'
BDC [Berlin Document Center] Documents, Individual files

Reich Chamber of Culture Collection
Reich Ministry of Economics Collection
Reich Propaganda Ministry Collection

Párttörténeti Levéltár – Budapest

MSzMP KB Párttörténeti Intézete Archivum
Révész Mihály gyüjteménye

Országos Széchényi Könyvtár, Special Collections – Budapest

Letter collections of:
Balogh, László
Hóman, Bálint
Jávor, Pál
Kádár, Levente
Keresztes-Fischer, Ferenc
Klebelsberg, Kuno
Kozma, Miklós
Lajta, Andor
Scitovszky, Béla
Szeleczky, Zita
Wlassics, Gyula, Jr.

Small Imprint and Poster Collection: period posters and advertisements
Journal and Newspaper Collections: various newspapers and film/culture journals

Történeti Hivatal – Budapest (currently the Állambiztonsági Szolgálatok Történeti Levéltára)

Records pertaining to:
Barla, László
Hives, Henrik

Kádár, L. Levente
Kauser, István
Páger, Antal
Szeleczky, Zita
Tomcsányi, Kálmán

United States Holocaust Memorial Museum – Washington, DC

RG-39.004M Magyar Kiràlyi Honvèdelmi Minisztèrium, Elnöki A osztàly-elnöki sorozat 1938–41

FILM SCREENINGS, INTERVIEWS, CONFERENCES & EXHBITIONS

Film Screenings:

Bundesfilmarchiv – Berlin
Magyar Filmintézet – Budapest
Stiftung deutsche Kinematek – Berlin

Interviews:

'Élő Filmtörténet: Hamza D. Ákos a Magyar Filmintézetben'. Budapest: 2 June 1987. Transcript provided by Gyönyi Balogh.
Personal interview with Imre Hecht. Pound Ridge, New York, 18 June 2007.
Personal interview with Péter Bokór. Budapest: Országos Széchenyi Könyvtár, 9 March 1999.

Conferences and Exhibitions:

'Mi a Magyar? Kortárs Válaszok' Budapest, Műcsarnok, August-October 2012
Sándor, Tibor. 'Őrségváltás elött – őrségváltás után'. Budapest, Puskin mozi, 13 December 1998.

PUBLISHED SOURCES

Statistical/Documentary Collections and Period Writings

A Magyar Film Iroda [M.F.I.] hiradóinak cenzuraszövegei. Vols III–VI. Budapest, Magyar Filmtudományi Intézet és Filmarchivum, 1969.

Ádám, Magda, Gyula Juhász & Lajos Kerekes (eds). *Diplomáciai iratok Magyarország külpolitikájához, 1936–45*. Budapest, 1962.

—— (eds) *Allianz Hitler-Horthy-Mussolini: Dokumente zur ungarischen Außenpolitik (1933–1944)*. Trans. Johanna Till. Budapest, 1966.

Anonymous. *A második zsidótörvény célja, indokolása, következményei. Néhány szó a magyar zsidó értelmiség nevében a magyar közvéleményhez*. Budapest, 1939.

Ávédik, Félix. *A mozi és a közönsége*. Budapest, 1942.

Az 1935. évi április hó 27-ére hirdetett Országgyülés Képviselőházának naplója, Vol.3. Budapest, 1935.

Az Országos Mozgóképvizsgáló Bizottság határozatai: A Budapesti Közlönyben megjelent cenzuradöntések másolatai. Budapest, 1969.

Bencs, Zoltán. *Magyar világnézet*. Budapest, 1939.

Bingert, János. *A mozgófényképüzemek és filmügyek rendészete*. Compiled by Mihály Baik. Budapest, 1928.

Dáloky, János. *Igy készül a magyar film. A magyar film múltja és jelene, kulisszatitkai és problémái. A könyv többszáz érdekes képpel világítja meg a magyar filmek születését*. Budapest, 1942.

Fejér, Tamás. *A film gyártása, esztétikája és jövője*. Budapest, 1943.

Foreign Ministry. *Külpolitikai adatok az 1939.évről*. Budapest, 1941.

Gertler, Viktor. *Az én filmem*. Budapest, 1942.

Gyárfás, Gy., S. Nagy, R. Vári, & F. Lohr (eds). *Mozgóképszinházak üzemvitele, magánjogi, jogszabályai, hangtechnikáj*. Budapest, 1937.

Illyés, Gyula. *Magyarok-Naplójegyzetek*. Budapest, 1938.

Kecskeméti, Vilmos (ed). *Zsidó évkönyv az 5692–3.bibliai évre*. Budapest, 1932–3.

Klebelsberg, Kuno. *Neonacionalizmus*. Budapest, 1928.

Kovács, Alajos. *A zsidóság térfoglalása Magyarországon*. Budapest, 1922.

Kozma, Miklós. 'A legélesebb magyar fegyverek egyike a magyar rádió', *Cikkek, nyilatkozatok, 1921–39*. Budapest, 1939.

Lajta, Andor. *A tizéves magyar hangosfilm 1931–1941*. Budapest, 1942.

Lukács, György (ed). *Magyarok a kultúráért*. Budapest, 1929.

Magyar Zsidók naptára 1941–5701. Budapest, 1941.

Magyarország Évkönyve - 1942. IX.évfolyam. Budapest, 1943.

Nagy, Sz. Péter (ed). *A népi-urbánus vita dokumentumai 1932–1947*. Budapest, 1990.

Smolka, János. *Mesegép a valóságban*. Budapest:, n.d.[1938?].

Stern, Samu. *A zsidókérdés Magyarországon*. Budapest, 1938.

Székely, Sándor István. *A Második Zsidótörvény és a Végrehajtási Utasitás. Teljes, hiteles szövege részletes magyarázatokkal*. Budapest, 1939.

———. *Mit akar tudni a filmről?* Budapest, 1943.

Szekfű, Gyula (ed). *Három nemzedék és ami utána következik*. Budapest, 1935.

———. *Mi a magyar?* Budapest, 1939.

Szűcs, Ádám. *Az igazi filmművészetről*. Budapest, n.d.

Traub, Hans. *Der Film als politisches Machtmittel*. München, 1933.

von Boehmer, Henning & Helmut Reitz. *Der Film in Wirtschaft und Recht*. Berlin, 1933.

Wersényi, György. *A mozgófényképüzemek rendészete*. Debrecen, 1943.

Primary Sources – Selected Periodicals

A Film. A Magyar Mozi és Filmszakma Lapja, 1935–8

Bajtárs, 1936–7

Berlini Visszhang [Germany], 1944

Deutsche Filmkunst [Germany], 1943–4

Egyenlőség, 1934, 1936–8

Film Daily Year Book of Motion Pictures, 1933–44

Film Élet, 1933

Film Hiradó, 1943–4

Film und Bild [Germany], Occasional issues 1936–43

Film, Szinház, Irodalom, Occasional issues

Filmcompass, 1935, 1939, 1941, 1945.

Filmkultúra, 1929–38

Filmművészeti Évkönyv, 1929–45

Filmszinházak Közlönye, 1942–4

Filmtechnika, 1932

Filmtisztviselők Lapja, 1938–9

Filmújság. Zsolnai László Hétilapja, 1932–7

Interfilm [Germany], 1936–7, 1942–4

International Motion Picture Almanac, 1935–42

Jahrbuch der Reichsfilmkammer [Germany], 1937–9

Magyar Film, 1939–44

Magyar Filmkurír, 1929–32; 1935–8

Magyar Múzsa, 1943–4

Magyarság, Occasional issues 1940–2

Mozgó Élet, 1937–8
Mozi- és Filmvilág. Guthy György Moziszaklapja, 1937–9
Mozi Ujság, 1942–4
Mozivilág - Kino Welt, 1935–8
Mvwwozivilág, 1930–3
Nemzetőr, 1940
Szinházi Magazin, 1939–40

Secondary Sources – Unpublished Papers and Theses

Ábel, Péter. 'Magyar filmgyártó, kölcsönző vállalatok és filmlaboratóriumok törzslapjai /1928–38. Budapest: Unpublished MFI manuscript, 1972.

Gati, Charles. 'The Populist Current in Hungarian Politics, 1935–44'. Indiana University: Unpublished Ph.D. dissertation, 1965.

Hirsch, Eric. 'Pure Sources, Pure Souls: Folk Nationalism and Folk Music in Hungary in the 1930s'. University of California at Berkeley, CA: Unpublished Ph.D. dissertation, 1995.

Lajta, Andor. 'A magyar film története. V. A magyar hangosfilm korszak első 16 éve, 1929–44'. Budapest: Szinháztudományi és Filmtudományi Intézet kézirat, 1958.

Langer, István. 'Fejezetek a filmgyár történetéből, I–II.rész, 1919–48'. Parts 1–2, 1919–1948. Budapest: MFI kézirat, 1980.

Manchin, Anna. 'Fables of Modernity. Entertainment Films and the Social Imaginary in Interwar Hungary', Brown University: Unpublished Ph.D. dissertation, 2008.

Nagy, Zsolt. 'Grand Delusions: Interwar Hungarian Cultural Diplomacy, 1918–1941'. University of North Carolina: Unpublished Ph.D. dissertation, 2012.

Ránki, Vera. 'The Politics of Inclusion and Exclusion: The Dynamics of Nationalism, Assimilation and Antisemitism. The Case of Hungary'. Sydney, Australia: Unpublished Ph.D. dissertation, University of New South Wales, 1996.

Strausz, Péter. 'Kamarák a két világháború közötti Magyarországon'. ELTE, Budapest: Unpublished Ph.D. dissertation, 2007.

Selected Secondary Sources

'25 éves a mozi'. *Filmspirál* VI/22 (2000), 43–67. See http://epa.oszk.hu/00300/00336/00006/25eves.htm.

'25 éves a mozi'. *Filmspirál* VI/23 (2000), 75–100. See http://epa.oszk.hu/00300/00336/00007/amozi25.htm.

'A Magyarországi antiszemitizmus számokban'. *Szombat* (24 January 2014), http://www.szombat.org/politika/a-magyarorszagi-antiszemitizmus-szamokban

'A mozirevízió'. *Filmspirál* VII/27-3 (2001), http://epa.oszk.hu/00300/00336/00011/torv.htm

'Budapesti mozgóképszínházak, 1903–1923'. *Filmspirál* VII/25-1 (2001), 83–97.

'Magyarország Alaptörvénye' *Magyar Közlöny* 43 (25 April 2011), 10656. http://www.kozlonyok.hu/nkonline/MKPDF/hiteles/mk11043.pdf.

'Sites of indifference' articles. *Austrian History Yearbook* 43 (2012), 21–137.

'Special Issue: Cosmopolitanism, Nationalism and the Jews of East Central Europe', *European Review of History/Revue Européenne d'Histoire* 17/3 (June 2010).

'Színészkönyvtár'. http://www.szineszkonyvtar.hu/.

Abel, Richard (ed). *Silent Film*. New Brunswick, NJ, 1996.

———. *The Red Rooster Scare: Making Cinema American, 1900–1910*. Berkeley, CA & London, 1999.

Ablonczy, Balázs. *Pál Teleki: The Life of a Controversial Hungarian Politician*. Trans. T & H. DeKornfeld. Boulder, CO, 2006.

Ádám, Magda. *A kisantant (1920–38)*. Budapest, 1981.

Albrecht, Gerd. *Film im Dritten Reich*. Karlsruhe, Germany, 1979.

Almog, Shmuel. *Nationalism and Antisemitism in Modern Europe 1815–1945*. New York, 1990.

Altman, Rick (ed). *Sound Theory, Sound Practice*. New York, 1992.

———. *Cinema/Genre*. London, 1999.

Anderson, Benedict. *Imagined Communities: Reflections on the Origins and Spread of Nationalism*. London & New York, 1983.

———. *The Spectre of Comparison: Nationalism, Southeast Asia, and the World*. London, 1998.

Applegate, Celia. *A Nation of Provincials: The German Idea of Heimat*. Berkeley, CA, 1990.

Arnold, Matthew. *Culture and Anarchy: An Essay in Political and Social Criticism*. Indianapolis, IN & New York, 1971 [1882].

Aschheim, Steven E. 'German History and German Jewry: Boundaries, Functions and Interdependence', *Leo Baeck Institute Year Book*, 43 (1998), 315–22.

Asper, Helmut G. 'Filmavantgardisten im Exil'. In *Exil und Avant Garden. Exilforschung • Ein internationales Jahrbuch • Band 16*. Munich, 1998: 174–93.

———. *Filmexilanten im Universal Studio*. Berlin, 2005.

Bakker, Gerben. *Entertainment Industrialised: The Emergence of the International Film Industry, 1890–1940*. Cambridge, UK, 2008.

Ballinger, Pamela. *History in Exile: Memory and Identity at the Borders of the Balkans*. Princeton, NJ, 2003.

Balog, Margit. *A KALOT és a katolikus társadalompolitika 1935–1946*. Budapest, 1998.

Balogh, Gyöngyi and Jenő Király. *'Czak egy nap a világ...' A magyar film műfaj- és stílustörténete, 1929–1936*. Budapest, 2000.

Balogh, Sandor, et al. *Magyarország a XX. Században, 2-edik kiadás*. Budapest, 1986.

Balogh, Gy., V. Gyürey, & P. Honffy. *A magyar játékfilm történte a kezdetektől 1990-ig*. Budapest, 2004.

Bamford, Kenton. *Distorted Images: British National Identity and Film in the 1920s*. London, 1999.

Banac, I. & K. Verdery (eds). *National Character and National Ideology in Interwar Eastern Europe*. New Haven, CT, 1995.

Bános, Tibor. *Jávor Pal szemtől szemben*. Budapest, 1978.

——. *A pesti kabaré 100 éve*. Budapest, 2008.

Barbian, Jan-Pieter. '"Kulturwerte in Zeitkampf". Die Kulturabkommen des "Dritten Reiches" als Instrument nationalsozialistischer Auβenpolitik'. *Archiv für Kulturgeschichte* 74/2 (1992), 415–59.

Barbu, Bogdan. 'Hollywood Movies, American Music and Cultural Policies behind the Iron Curtain. Case Study: Cold War Romania, 1945–1971'. In S. Jakelic and J. Varsoke (eds). *Crossing Boundaries: From Syria to Slovakia*. Vienna: *Junior Visiting Fellows' Conference* XIV/2 (2003), 1–18. http://www.iwm.at/publ-jvc/jc-14-02.pdf

Barron, Stephanie & Sabine Eckmann. *Exil. Flucht und Emigration europäischer Künstler 1933–45*. München, 1998.

Bartov, Omer. *The 'Jew' In Cinema: From the Golem to Don't Touch My Holocaust*. Bloomington, IN, 2005.

Becker, Wolfgang. *Film und Herrschaft: Organisationsprinzipien und Organisationsstrukturen der nationalisozialisten Filmpropaganda*. Berlin, 1973.

Benjamin, Walter. 'The Work of Art in the Age of Mechanical Reproduction'. In Leo Braudy and Marshall Cohen (eds). *Film Theory and Criticism: Introductory Readings*. 5th edition. New York & Oxford, UK, 1999: 731–51.

Berend, Iván T. *Decades of Crisis*. Berkeley, CA, 2001.

Berend, Iván T. & György Ránki. *Magyarország a fasiszta Németország 'életterében' 1933–1939*. Budapest, 1960.

Bernstein, J.M. (ed). *The Culture Industry. Selected Essays on Mass Culture. Theodor W. Adorno*. London, 1991.

Bethlen, István and I. Bolza. *Hungarian Politics during World War Two: Treatise and Indictment*. Munich, 1985.

Bhabha, Homi K. *Nation and Narration*. London, 1990.

Bibó, István. 'A kelet-európai kisállamok nyomorúsága', (1946). In Bibó, *Valogatott tanulmányok,* vol. II. Budapest, 1986.

——. 'The Jewish Question in Hungary after 1944' (1948). In Károly Nagy (ed). *István Bibó. Democracy, Revolution and Self-Determination: Selected Writings*. Trans. A. Boros-Kazai. Highland Lakes, NJ, 1991.

Billig, Michael. *Banal Nationalism*. London, 1995.

Biltereyst, Daniel, Richard Maltby, & Philippe Meers (eds). *Cinema, Audiences and Modernity: New perspectives on European cinema history*. New York, 2012.

Birch, Brian. 'Hungarian Cinema and Fascism'. In J. Davis (ed). *Film, History, and the Jewish Experience: A Reader*. London, 1986: 79–95.

Bischoff, Ralph F. *Nazi Conquest through German Culture*. Cambridge, MA, 1942.

Bisztray, George. 'Hungarian National Character: Specific or Interactive?' *Hungarian Studies Review* XXVII/1–2 (2000), 101–11.

Bjork, James. *Neither German nor Pole: Catholicism and National Indifference in a Central European Borderland*. Ann Arbor, MI, 2008.

Black, Gregory D. *Hollywood Censored. Morality Codes, Catholics and the Movies*. New York, 1994.

Bláhová, Jindřiška, 'A Tough Job for Donald Duck: Czechoslovakia and Hollywood 1945–1969'. *Iluminace* 19/1 (2007), 215–18.

Bock, Hans-Michael & M. Tötenberg (eds). *Das Ufa-Buch. Kunst und Krisen, Stars und Regisseure, Wirtschaft und Politik*. Frankfurt-am-main, Germany, 1992.

Bock, Hans-Michael & W. Jacobsen (eds). *Recherche: Film. Quellen und Methoden der Filmforschung*. Munich, 1997.

Bolykovszky, Béla. 'A magyar filmművészet szolgálatában. Visszaemlékezések I'. *Filmkultúra* 27/4 (1991), 5–16.

———. 'Visszaemlékezések II'. *Filmkultúra* 27/5 (1991), 32–42.

Bonnell, V. E. & L. Hunt (eds). *Beyond the Cultural Turn*. Berkeley, CA, 1999.

Borbándi, Gyula. *Der ungarische Populismus*. Mainz, 1976.

Bordwell, David. *On the History of Film Style*. Cambridge, MA, 1997.

Braham, Randolph L. *Anti-Semitism and the Treatment of the Holocaust in Postcommunist Eastern Europe*. Boulder, CO, 1994.

Braham, R. L. (ed). *The Tragedy of Hungarian Jewry. Essays, Documents, Depositions*. Boulder, CO, 1986.

———. *The Politics of Genocide: The Holocaust in Hungary*, Vol. 1. Boulder, CO, 1994.

Braham, R. L. & A. Pók (eds). *The Holocaust in Hungary: Fifty Years Later*. Boulder, CO, 1997.

Braham, R. L. & R. Vagó (eds). *The Holocaust in Hungary: Forty Years Later*. New York, 1985.

Broszat, Martin. 'Faschismus und Kollaboration in Ostmitteleuropa zwischen den Weltkriegen'. *Vierteljahrshefte für Zeitgeschichte* 14/3 (Juli 1966), 225–51.

———. 'Deutschland – Ungarn – Rumänien. Entwicklung und Grundfaktoren nationalsozialistischer Hegemonial- und Bündnispolitik 1938–1941'. *Historische Zeitschrift* 206 (1968), 45–96.

Brubaker, Rogers. *Nationalism Reframed. Nationhood and the National Question in the New Europe*. Cambridge, UK, 1996.

———. *Ethnicity without Groups*. Cambridge, MA, 2004.

Bryant, Chad. *Prague in Black*. Cambridge, MA, 2007.

Bucur, Maria. *Eugenics and Modernisation in Interwar Romania*. Pittsburgh, PA 2002.

Burke, Peter. *Varieties of Cultural History*. Ithaca, NY, 1997.
Burns, Bryan. *World Cinema: Hungary*. Trowbridge, UK, 1996.
Buscombe, Edward. 'Film History and the Idea of National Cinema'. *Australian Journal of Screen Theory* 9/10 (1981), 141–53.
Butzel, M. & A. M. López (eds). Special issue on Nation and Film. *Quarterly Review of Film and Video* 14/3 (1993).
Carr, Steven. *Hollywood & Anti-Semitism: A Cultural History up to World War II*. Cambridge, UK, 2001.
Case, Holly. *Between States*. Stanford, CA, 2009.
Cenner, Mihály, et al. (eds). *Kabos Gyula 1887–1941*. Budapest, 1987.
Cesarini, David (ed). *Genocide and Rescue*. Oxford, UK, 1997.
Chakravarty, Sumita S. *National Identity in Indian Popular Cinema, 1947–1987*. Austin, TX, 1993.
Chambers, J. W. II & D. Culbert (eds). *World War II, Film and History*. New York & Oxford, UK, 1996.
Chapman, James & Nicholas Cull. *Projecting Empire: Imperialism and Popular Cinema*. London, 2009.
Chatterjee, Partha. *The Nation and Its Fragments. Colonial and Postcolonial Histories*. Princeton, NJ, 1993.
Cherchi Usai, Paolo. *Silent Cinema: An Introduction*. London, 2000.
Cohen, Asher. 'La politique antijuive en Europe (Allemagne exclue), 1938–41'. *Guerre Mondaile et Conflits Contemporains* 150 (April 1988), 45–59.
Cole, Tim. 'Constructing the "Jew", Writing the Holocaust: Hungary 1920–45'. *Patterns of Prejudice* 33/3 (1999), 19–27.
Confino, Alon. *The Nation as a Local Metaphor*. Chapel Hill, NC & London, 1997.
——. 'Edgar Reitz's *Heimat* and German Nationhood: Film, Memory, and Understandings of the Past'. *German History* 16/2 (1998), 185–208.
Connelly, John. 'Nazis and Slavs: From Racial Theory to Racist Practice'. *Central European History* 32/1 (1999), 1–33.
Connor, Walker. *Ethnonationalism: The Quest for Understanding*. Princeton, NJ, 1994.
Cook, David. *History of Narrative Film*. New York, 1990.
Cooper, Frederick & R. Brubaker. 'Identity'. In F. Cooper (ed). *Colonialism in Question: Theory, Knowledge, History*. Berkeley, CA, 2005: 59–90.
Cosandey, R. & A. Gaudreault (eds). *Images across Borders*. Lausanne, 1995.
Crofts, Stephen. 'Concepts of national cinema'. In John Hill and P. Church Gibson (eds). *The Oxford Guide to Film Studies*. Oxford, UK, 1998: 385–94.
——. 'Concepts of national cinema'. In John Hill and P. C. Gibson (eds). *World Cinema: Critical Approaches*. Oxford, UK, 2000: 1–10.
Csepeli, György. *National Identity in Contemporary Hungary*. Trans. M. D. Fenyo. Boulder, CO, 1997.

——. *Structures and Contents of Hungarian National Identity*. Trans. Chris Tennant. Frankfurt-am-main, Germany, 1989.

Cultural Policy in Hungary. Paris, 1974. (http://unesdoc.unesco.org/images/0000/000090/009050eo.pdf)

Cunningham, John. *Hungarian Cinema: From Coffeehouse to Multiplex*. London, 2004.

——. 'From Arcadia to Collective Farm and Beyond. The Rural in Hungarian Cinema'. In C. Fowler & G. Helfield (eds). *Representing the Rural. Space, Place and Identity in Films about the Land*. Detroit, MI, 2006: 292–306.

Curran, J. & V. Porter (eds). *British Cinema History*. London, 1983.

Czettler, Anton. *Pál Graf Teleki und die Außenpolitik Ungarns 1939–1941*. München, 1996.

Dahm, Volker. 'Die Reichskulturkammer als Instrument kulturpolitischer Steuerung und sozialer Reglementierung'. *Vierteljahrshefte für Zeitgeschichte* 34/1 (January 1986), 53–84.

Dallos, Szilvia. *Magyar hangja…a szikronizálás története*. Budapest, 1999.

de Grazia, Victoria. *The Culture of Consent: Mass Organization of Leisure in Fascist Italy*. Cambridge, UK, 1981.

——. 'Mass Culture and Sovereignty: The American Challenge to European Cinema, 1920–1960'. *Journal of Modern History* 61 (March 1989), 53–87.

——. 'European Cinema and the Idea of Europe, 1925–1995'. In G. Nowell-Smith (ed). *Hollywood and Europe: Economics, Culture, National Identity 1945–1995*. London, 1998: 19–33.

——. *Irresistible Empire. America's Advance through Twentieth-Century Europe*. Cambridge, MA, 2005.

Deák, István. 'Hungary from 1918 to 1945'. New York, 1989.

Deák, Tamás & József Mudrák. *Magyar hangosfilm lexikon 1931–1944*. Budapest, 2006.

Dennison, Stephanie & Lisa Shaw. *Popular Cinema in Brazil*. Manchester, UK, 2004.

Dickinson, Margaret & Sarah Street. *Cinema and State: The Film Industry and the British Government, 1927–84*. London, 1985.

Dirks, Nicholas B. 'In Near Ruins: Cultural Theory at the End of the Century'. In Nicholas B. Dirks (ed). *In Near Ruins: Cultural Theory at the End of the Century*. Minneapolis, MN, 1998.

Don, Yehuda. 'The Economic Dimensions of Antisemitism: Anti-Jewish Legislation in Hungary, 1938–44'. *East European Quarterly* 20/4 (January 1987), 447–65.

Drewniak, Bogusław. *Der deutsche Film 1938–1945. Ein Gesamtüberblick*. Düsseldorf, Germany 1987.

Duara, Prasenjit. *Rescuing History from the Nation: Questioning Narratives of Modern China*. Chicago, IL, 1995.

Düwell, Kurt. *Deutschlands auswärtige Kulturpolitik 1918–32*. Köln, Germany, 1976.

Dyer, Richard. *The Matter of Images: Essays on the Role of Representation*. London, 1993.

Dyer, R. & G. Vincendeau (eds). *Popular European Cinema*. London & New York, 1992.

Edensor, Tim. *National Identity, Popular Culture and Everyday Life*. Oxford, UK, 2002.

Ehrlich, Evelyn. *Cinema of Paradox: French Filmmaking under the German Occupation*. New York, 1985.

Eisenstein, Sergei. *The Film Sense*. Ed. & Trans. Jay Leyda. New York, 1947.

Eley, G. & R. G. Suny (eds). *Becoming National: A Reader*. New York & Oxford, UK, 1996.

Ellwood, D. & R. Kroes (eds). *Hollywood in Europe: Experiences of Cultural Hegemony*. Amsterdam, 1994.

Elsaesser, Thomas. *New German Cinema: A History*. London, 1989.

——. (ed). *Early Cinema: Space, Frame, Narrative*. London, 1990.

——. 'European Cinema. Conditions of Impossibility?' In *European Cinema: Face to Face with Hollywood*. Amsterdam, 2007: 13–34.

Erdélyi, Z. Ágnes (ed). *Kabos Gyula 1887–1941. Születésének századik évfordulójára*. Budapest, 1987.

Erényi, Tibor. 'Zsidók és a magyar politikai élet 1848–1938'. *Múltunk* 39/4 (1994), 3–30.

Eriksen, T.H. *Ethnicity and Nationalism: Anthroplogical Perspectives*. London, 1993.

Esbenshade, Richard S. 'The Radical Assimilated: Hungarian "Urbanists" and Jewish Identity in the 1930s'. In L. Greenspoon, R. Simkins & B. Horowitz (eds). *The Jews of Eastern Europe, Studies in Jewish Civilization,* v. 16. Omaha, NE, 2005: 117–42.

Everett, Wendy. 'Introduction: European film and the quest for identity'. In Wendy Everett, (ed). *European Identity in Cinema*. Bristol, UK, 2005: 8–14.

Eyman, Scott. *The Speed of Sound: Hollywood and the Talkie Revolution, 1926–1930*. New York, 1997.

Ezra, Elizabeth (ed). *European Cinema*. Oxford, UK, 2004.

Fabó, Beáta. 'A moziépítészet és a város'. *Budapesti Negyed* V/2–3 (Summer/Fall 1997). http://epa.oszk.hu/00000/00003/00014/fabo.htm

Fehervári, Zoltán (ed). *A Budapest mozi 100 éve*. Budapest, 2001.

Fehrenbach, Heide. *Cinema in Democratizing Germany*. Chapel Hill, NC & London, 1995.

Fein, Seth. 'Hollywood, U.S.-Mexican Relations, and the Devolution of the Golden Age of Mexican Cinema'. *Filmhistoria* [*Spain*] 4/2 (June 1994), 103–35.

——. 'Transnationalization and cultural collaboration'. In A. M. López (ed). *Studies in Latin American Popular Culture*, vol. 17. Las Cruces, NM, 1998: 105–28.

——. 'From Collaboration to Containment'. In J. Hershfield and D.R. Maciel (eds). *Mexico's Cinema: A Century of Film and Filmmakers*. Wilmington, DE, 1999: 123–63.

Fejes, Judit. 'A magyar-német gazdasági és politikai kapcsolatok kérdéséhez az 1920-as - 1930-as évek fordulóján'. *Történelmi Szemle* 76/3 (1976), 361–84.

Fejtö, François. *Hongrois et Juifs*. Paris: Balland, 1997.

Fekete, Bálint. 'A Horthy-kori filmhíradók kutatása és használata az oktatásban'. *Mediakutató* XV/4 (Winter 2014), 13–23. http://www.mediakutato.hu/cikk/2014_04_tel/02_horthy_filmhirado.pdf

Fekete, Liz. 'Hungary: Power, punishment and the "Christian-national Idea"'. *Race & Class* 57/4 (2016), 39–53.

Fenyő, Mario D. *Hitler, Horthy and Hungary. German-Hungarian Relations, 1941–44*. New Haven, CT, 1972.

Ferro, Marc. 'Film as an Agent, Product and Source of History'. *Journal of Contemporary History* 18 (1983), 357–64.

——. *Cinema and History*. Trans. Naomi Greene. Detroit, MI, 1988.

Fitzpatrick, Sheila. *The Cultural Front: Power and Culture in Revolutionary Russia*. Ithaca, NY, 1992.

Ford, Charles & Robert Hammond. *Polish Film: A Twentieth Century History*. Jefferson, NC, 2005.

Formisano, Ronald P. 'The Concept of Political Culture'. *Journal of Interdisciplinary History* XXXI/3 (Winter 2001), 393–426.

Fox, J. E. & C. Miller-Idriss, 'Everyday Nationhood', *Ethnicities* 8/4 (2008), 536–63.

Frank, Tibor. *Double Exile: Migrations of Jewish-Hungarian Professionals through Germany to the United States, 1919–1945*. Oxford, UK, 2009.

Freidlander, Saul (ed). *Probing the Limits of Representation: Nazism and the 'Final Solution'*. Cambridge, MA, 1992.

Frey, David S. 'Aristocrats, Gypsies, and Cowboys All: Film Stereotypes and Hungarian National Identity in the 1930s'. *Nationalities Papers* 30/2 (2002), 383–401.

——. '"Just what is Hungarian?" Concepts of national identity in the Hungarian film industry, 1931–44'. In Marsha Rozenblit & Pieter Judson (eds). *Reconstructing Nationalities in East Central Europe*. New York, 2004, 203–23.

——. '"Why We Fight" Hungarian Style: War, Civil War, and the Red Menace in Hungarian Wartime Feature Film.' *KinoKultura* Special Edition (February 2008). http://www.kinokultura.com/specials/7/frey.shtml

——. 'Mata Hari or the body of the nation? Interpretations of Katalin Karády'. *Hungarian Studies Review* XLII/1–2 (2014), 89–106.

——. 'A Smashing Success? The Paradox of Hungarian Cultural Imperialism in Nazi New Order Europe, 1939–1942'. *Journal of Contemporary History* 51/3 (2016), 577–605.

Fritz, Walter. *Kino in Österreich 1929–1945. Der Tonfilm*. Vienna, 1991.

Fröhlich, Elke (ed). *Die Tagebücher von Dr. Goebbels. Sämtliche Fragamente*. Munich, 1996.

Fülöp, Mihály & Péter Sipos. *Magyarország külpolitikája a XX.században*. Budapest, 1998.

Gaál, Zsuzsanna G. '*Magyar Múzsa* (1943–44)'. *Magyar Könyvszemle* 107/4 (1991), 399–407.

Gál Molnár, Péter. *A Páger-ügy*. Budapest, 1988.

Gastony, Endre B. 'Hungarian Foreign Minister Kálmán Kánya's Grand Design, 1933–36. *East European Quarterly* XIX/2 (June 1985), 175–89.

Geertz, Clifford. *The Interpretation of Cultures*. New York, 1973.

Gellner, Ernest. *Nations and Nationalism*. Ithaca, NY, 1983.

Gentile, Emilio. *Politics as Religion*. Princeton, NJ, 2006.

Geoffroy, Rene. *Ungarn als Zufluchtsort und Wirkungsstätte deutschsprachiger Emigranten (1933–1938–9)*. Frankfurt-am-main, Germany, 2001.

Georgescu, Tudor. 'Ethnic minorities and the eugenic promise: the Transylvanian Saxon experiment with national renewal in inter-war Romania'. *European Review of History/Revue europeenne d'histoire,* 17/6 (December 2010), 861–80.

Gergely, Jenő & Pál Pritz. *A Trianoni Magyarország*. Budapest, 1998.

——. *Gömbös Gyula. Politikai Pályakép*. Budapest, 2001.

Gerlach, C. & C. Dieckmann (eds). *Besatzung und Bündnis: deutsche Herrschaftstrategien in Ost- und Südosteuropa*. Berlin, 1995.

Gerő, András. *Modern Hungarian Society in the Making: The Unfinished Experience*. Trans. James Patterson and Enikő Koncz. Budapest, 1995.

Gerő, András & J. Poór (eds). *Budapest. A History from Its Beginnings to 1998*. Trans. J. Zinner, C. D. Eby & N. Arató. Highland Lakes, NJ, 1997.

Gerrits, A. & N. Adler (eds). *Vampires Unstaked: National Images, Stereotypes and Myths in East Central Europe*. Amsterdam, 1995.

Gheorghiu-Cernat, Manuela. *A concise history of the Romanian film*. Bucharest, 1982.

Gittings, Christopher E. *Canadian National Cinema: Ideology, Difference and Representation*. London & New York, 2002.

Gluck, Mary. 'The Budapest Flaneur: Urban Modernity, Popular Culture, and the "Jewish Question" in *Fin-De-Siècle* Hungary'. *Jewish Social Studies* 10/3 (2004), 1–22.

Gombár, Csaba, Elemér Hankiss, László Lengyel, & Györgyi Várnai (eds). *The Appeal of Sovereignty: Hungary, Austria and Russia*. Boulder, CO, 1998.

Gomery, Douglas. 'Economic Struggle and Hollywood Imperialism: Europe Converts to Sound'. *Yale French Studies* 60 (1980), 80–93.

Götz, Eszter & Tibor Sándor. *Ennyi/Cut...Budapest-Bécs-Hollywood: A magyar filmes emigráció*. Budapest, 1999.

Goulding, Daniel J. *Liberated Cinema: The Yugoslav Experience, 1945–2001*. Bloomington, IN, 2002.

Grant, Barry K. (ed). *Film Genre: Theory and Criticism*. Metuchen, NJ, 1977.

Greene, Naomi. *Landscapes of Loss: The National Past in Postwar French Cinema*. Princeton, NJ, 1999.

Gregor, Ulrich & Enno Patalas. *Geschichte des Films*. Gütersloh, Germany, 1962.

Grenzebach Jr., William S. *Germany's Informal Empire in East-Central Europe: German Economic Policy toward Yugoslavia and Romania, 1933–39*. Stuttgart, Germany, 1988.

Grieveson, Lee. 'On governmentality and screens'. *Screen* 50/1 (Spring 2009), 180–7.

Grindon, Leger. *Shadows on the Past: Studies in the Historical Fiction Film*. Philadelphia, PA, 1994.

Grossberg, L., C. Nelson & P.A. Treichler (eds). *Cultural Studies*. New York, 1992.

Gundle, Stephen. *Mussolini's Dream Factory: Film Stardom in Fascist Italy*. New York, 2013.

Gyáni, Gábor. 'Polgárosodás mint zsidó identitás,' *BUKSz* 9/3 (Fall 1997), 266–78.

———. 'Forráskritika és bizonyítás. Viszontválasz Karády Viktornak'. *BUKSz* 10/1 (Spring 1998), 20–7.

Gyurgyák, János. *A zsidókérdés magyarországon. Politikai Eszmetörténet*. Budapest, 2001.

Haase, Christine. *When Heimat Meets Hollywood: German Filmmakers and America, 1985–2005*. Rochester, NY, 2007.

Hake, Sabine. *German National Cinema*. London & New York, 2002.

———. *Popular Cinema of the Third Reich*. Austin, TX, 2002.

Hall, John R. 'Cultural meanings and cultural structures in historical explanation'. *History and Theory* 39 (October 2000), 331–47.

Halle, Randall. *The Europeanization of Cinema. Interzones and Imaginative Communities*. Urbana, IL, 2014.

Haltof, Marek. *Polish National Cinema*. New York, 2002.

Hames, Peter (ed). *The Cinema of Central Europe*. London, 2004.

———. *Czech and Slovak cinema: theme and tradition*. Edinburgh, 2009.

Hanák, Peter. *The Garden and the Workshop. Essays on the Cultural History of Vienna and Budapest*. Princeton, NJ, 1998.

———. 'The Image of the Germans and the Jews in Nineteenth Century Hungary'. *CEU History Department, Working Paper Series 2*. Budapest, 1995: 67–87.

Hanebrink, Paul. *In Defense of Christian Hungary*. Ithaca, NY, 2006.

Hansen, Miriam. 'The Mass Production of the Senses: Classical Cinema as Vernacular Modernism'. *Modernism/Modernity* 6/2 (1999), 59–77.

Hay, James. *Popular Film Culture in Fascist Italy: The Passing of the Rex*. Bloomington, IN, 1987.

Hayward, Susan. *French National Cinema*. New York, 1993.

Heiss, G. & I. Klimeš (eds). *Obrazy času/Bilder der Zeit*. Prague, 2003.

Higson, A. 'The Idea of National Cinema'. *Screen* 30/4 (Autumn 1989), 36–46.

——. *Waving the Flag: Constructing a National Cinema in Britain*. Oxford, UK: Clarendon Press, 1995.

Higson, Andrew. & R. Maltby (eds). *'Film Europe' and 'Film America': Cinema, Commerce and Cultural Exchange 1920–39*. Exeter, UK, 1999.

Hill, John. 'The Issue of National Cinema and British Film Production'. in Duncan Petrie (ed). *New Questions of British Cinema*. London, 1992, 10–21.

Hill, J. P. Gibson, R. Dyer, E. Ann Kaplan, & P. Willemen (eds). *The Oxford Guide to Film Studies*. New York, 1998.

Hill, J. & K. Rockett (eds). *Film History and National Cinema*. Dublin, 2005.

Hirsch, Eric. 'Voices from the Black Box: Folk Song, Boy Scouts and the Construction of Folk Nationalist Hegemony in Hungary, 1930–44'. *Antipode* 29/2 (1997), 197–215.

Hirschfeld G. & P. Marsh (eds). *Collaboration in France: Politics and Culture during the Nazi Occupation, 1940–44*. Oxford, UK, 1989.

Hitchins, Keith. 'Interwar Southeastern Europe Confronts the West'. *Angelaki* 15/3 (December 2010), 9–26.

Hjort, Mette. *Small Nation, Global Cinema: The New Danish Cinema*. Minneapolis, MN, 2005.

Hjort, M. & S. MacKenzie (eds). *Cinema & Nation*. London, 2000.

Hobsbawm, E. *Nations and Nationalism since 1780: Programme, Myth, Reality*, 2nd edition. Cambridge, UK, 1993.

Hochscherf, Tobias. '"Kennen Wir Uns nicht aus Wien?" Émigré Film-makers from Austria in London, 1928–1945'. *The Yearbook of the Research Centre for German and Austrian Exile Studies* (2006), 133–47.

Hoensch, Joerg. *A History of Modern Hungary 1867–1994*, 2nd edn. Trans. Kim Traynor. London, 1996.

Hoffmann, Hilmar. *The Triumph of Propaganda. Film and National Socialism 1933–45*. Trans. J. A. Broadwin & V.R. Berghahn. Providence, RI, 1996.

Holloway, Ronald. *The Bulgarian Cinema*. Rutherford, NJ, 1986.

Horak, Jan-Christopher. 'German Exile Cinema, 1933–1945'. *Film History* 4 (1996), 373–89.

Horak, J-C. and Helmut Asper. 'Three Smart Guys: How a Few Penniless German Émigrés Saved Universal Studios'. *Film History* 11/2 (1999), 134–53.

Hroch, Miroslav. *The Social Preconditions of National Revival in Europe: A Comparative Analysis of the Social Composition of Patriotic Groups among the Smaller European Nations*. Trans. Ben Fowkes. Cambridge, UK, 1985.

Hull, David S. *Film in the Third Reich: A Study of the German Cinema 1933–45*. Berkeley, CA, 1969.

Huseby-Darvas, Éva. 'The Search for Hungarian National Identity'. In L. Romanucci-Ross & G. A. De Vos (eds). *Ethnic Identity: Creation, Conflict, and Accommodation*. London, 1995.

Ignotus, Paul. *Hungary*. London, 1972.

Imre, Anikó (ed). *East European Cinemas*. New York, 2005.

Ingebrigsten, Erik. 'Privileged Origins. 'National Models' and the Reform of Public Health in Hungary'. In György Péteri (ed). *Imagining the West in Eastern Europe and the Soviet Union*. Pittsburgh, PA, 2010, 36–58.

Iordanova, D., D. Martin-Jones, & B. Vidal (eds). *Cinema at the Periphery*. Detroit, MI, 2010.

Jackel, Anne. 'France and Romanian Cinema, 1896–1999'. *French Cultural Studies* 11/3 (October 2000), 409–24.

James, Beverly. 'Hungary'. In G. Kindem (ed). *The International Movie Industry*. Carbondale, IL, 2000: 165–77.

Janos, Andrew C. *The Politics of Backwardness in Hungary 1825–1945*. Princeton, NJ, 1982.

———. Janos, Andrew C. *East Central Europe in the Modern World. The Politics of the Borderlands from Pre- to Postcommunism*. Stanford, CA, 2000.

Jansen, Sue Curry. *Censorship: The Knot That Binds Power and Knowledge*. New York, 1988.

Jarvie, Ian. 'Dollars and Ideology: Will Hays' Economic Foreign Policy 1922–45'. *Film History* 2/3 (September 1988), 207–21.

———. *Hollywood's Overseas Campaign: the North Atlantic Movie Trade, 1920–1950*. Cambridge, UK, 1992.

Jávor, Pál. *Egy szinész elmondja*. Budapest, [1946] 1989.

Jeancolas, Jean-Pierre. *Cinéma hongrois 1963–1988*. Paris, 1989.

Johnson, Kevin. 'Kulturelle (nicht-)Vermischung. Nation, Ort und Identität in tschechisch-deutschen Mehrsprachenversionen'. In Johannes Roschlau (ed). *Zwischen Barrandov und Babelsberg. Deutsch-tschechische Filmbeziehungen im 20. Jahrhundert*. Hamburg, Germany, 2008: 71–83.

Judson, Pieter. *Guardians of the Nation*. Cambridge, MA, 2006.

Juhász, Gyula. *A Teleki-kormány külpolitikája 1939–41*. Budapest, 1964.

———. *Magyarország külpolitikája 1919–45*. Budapest, 1969.

———. *Uralkodó eszmék Magyarországon 1939–1944* Budapest, 1983.

———. 'Some Aspects of Relations between Hungary and Germany during the Second World War'. In György Ránki (ed). *Hungarian History, World History*. Budapest, 1984: 209–20.

———. *Magyarország nemzetközi helyzete és a magyar szellemi élet 1938–44*. Budapest, 1987.

Juhász, István. *Szinház a moziban 1931–1944*. Budapest, 2002.

Kaes, Anton. *From Hitler to Heimat: The Return of History as Film*. Cambridge, MA, 1989.

Kállay, Nicholas. *Hungarian Premier*. New York, 1954.

Karabel, Jerome. 'Towards a theory of intellectuals and politics'. *Theory and Society* 25 (1996), 205–33.

Karády, Katalin. *Hogyan lettem szinésznő*. Budapest, [1941] 1989.

Karády, Viktor. *Zsidóság, polgárosodás, asszimiláció. Tanulmányok*. Budapest, 1997.

———. 'Zsidó vagy nem zsidó polgárosodás? Válasz Gyáni Gábornak'. *BUKSz* 10/1 (Spring 1998), 9–19.

———. 'Zsidó ídentitás és modernizáció, avagy az asszimiláció paradoxonjai: A Csoportazonosság mint sórs és mint választás'. *Világosság* 40/7 (1999), 15–31.

———. 'Different Experiences of Modernization and the Rise of Anti-Semitism. Social-political Foundations of the *numerus clausus* (1920) and the "Christian Course" in Post World War I Hungary'. *Transversal* 4/2 (2003), 3–34.

Karcsai Kulcsár, István. 'Irodalom, színház, és film kapcsolata Magyarországon a harmincas években'. *Filmtudományi Szemle* 2 (1973), 7–42.

———. '"Őrségváltás" után. "Népi film" 1939–1944'. *Filmkultúra* XXI/11 (1985), 14–24.

Karsai, László (ed). *Kirekesztők*. Budapest, 1992.

———. 'Searchlight into Darkness. The Diary of Ferenc Szálasi'. *Yad Vashem Studies* 37/2 (2009), 93–134.

Katzburg, Nathaniel. *Hungary and the Jews*. Ramat-Gan, Israel, 1981.

Kelbecheva, Evelina. 'Creation of a New National Style in Bulgarian Art after the First World War'. *Bulgarian Historical Review* 21/2–3 (June 1993), 108–24.

Kelecsényi, László. *A magyar hangosfilm hét évtizede, 1931–2000. Hyppolittól Werckmeisterig*. Budapest, 2003.

———. *Mozizó Budapest*. Budapest, 2008.

Kenez, Peter. *The Birth of the Propaganda State: Soviet Methods of Mass Mobilization, 1917–1929*. Cambridge, UK, 1985.

Kenyeres, Zoltán (ed). *Nyugat, 1930–1941: Válogatás – Viták, programok, kritikák*. Budapest, 1988.

Kepley Jr., Vance. 'The origins of Soviet cinema: a study in industry development'. In Richard Taylor and Ian Christie (eds). *Inside the Film Factory: New Approaches to Russian and Soviet Cinema*. London, 1991: 60–79.

Kézdi-Kovács, Zsolt. *Magyar filmesek a világban/Hungarians in Film*. Budapest, 1996.

———. 'Film als kultureller Integrationsfaktor in Mitteleuropa'. In H. Pflügl (ed). *Mein 20. Jahrhundert: Der Traum vom Glück/Az Én XX. Századom: Visszaszámlálás*. Wien, 1999: 15–18.

Kinder, Marsha. *Blood Cinema: The Reconstruction of National Identity in Spain*. Berkeley, CA, 1993.

King, Jeremy. 'The Nationalization of East Central Europe: Ethnicism, Ethnicity, and Beyond'. In M. Bucur & N. M. Wingfield (eds). *Staging the Past: The Politics of Commemoration in Habsburg Central*. Purdue, IN, 2001, 112–52.

———. *Budweisers into Czechs and Germans*. Princeton, NJ, 2002.

Király, Jenő. *Karády mítosza és mágiája*. Budapest, 1989.

Kleβmann, Christoph. 'Osteuropaforschung und Lebensraumpolitik in Dritten Reich'. In P. Lundgreen (ed). *Wissenschaft im Dritten Reich*. Frankfurt-am-main, Germany, 1985, 350–85.

Klimeš, Ivan. 'Multiple-Language Versions of Czech Films and the Film Industry in Czechoslovakia in the 1930s'. *Cinema & Cie* 4 (Spring 2004), 89–101.

Klinger, Barbara. 'Film history terminable and interminable: recovering the past in reception studies'. *Screen* 38/2 (1997), 107–28.

Koerber, Esther-Beate. 'Wie interpretiert man eine Wochenschau? Überlegungen an Beispielen aus der Nachkriegszeit'. *Geschichte, Wissenschaft und Unterricht* 45 (1994), 137–50.

Kőháti, Zsolt. 'Magyar film hangot keres (1931–1938)'. *Filmspiral* II/2-1 (1996), 67–131.

——. 'Magyar film hangot keres (1931–1938)'. *Filmspiral* II/3-2 (1996), 1–52.

——. *Tovamozduló ember, tovamozduló világban. A magyar némafilm 1896–1931 között*. Budapest, 1996.

—— (ed). *A Magyar Film Olvasókönyve (1908–1943)*. Budapest, 2001.

Komoróczy, Géza, K. Frojimovics, et al. (eds). *Jewish Budapest: Monuments, Rites, History*. Budapest, 1999.

Konrád, George. *The Melancholy of Rebirth: Essays from Postcommunist Central Europe, 1989–94*. Trans. H.M. Heim. San Diego, 1995.

Kontler, László. *A History of Hungary*. New York, 2002.

Kosanović, Dejan. *History of Cinema in Bosnia and Herzegovina 1897–1945*. Belgrade, 2005.

——. *A Short History of Cinema in the Vojvodina. Part one 1896–1941*. Belgrade, 2012.

Kosztolányi, Gusztáv. 'Hungarian Identity, Globalisation and EU Accession'. *Central European Review* 2/6 (14 February 2000). http://www.ce-review.org/00/6/essay6.html

Kovács, András. 'Zsidóellenes előítéletesség és az antiszemitizmus dinamikája a mai Magyarországon'. *Társadalmi Riport* (2014), 486–508.

Kovács, Attila S. 'National Symbols in Hungarian Folk Art'. *Acta Ethnographica Hungarica* 41/1–4 (1996), 155–78.

Kovács, Mária M. *Liberal Professions and Illiberal Politics: Hungary from the Habsburgs to the Holocaust*. Washington, DC, 1994.

Kreimeier, Klaus. *Die UFA Story. Geschichte Eines Filmkonzerns*. Munich, 1992.

Kroes, Rob. 'America and the European Sense of History'. *Journal of American History* 86/3 (December 1999), 1135–55.

Kühne, Olaf. 'Geographie der nationalen Stereotype. Fallbeispiel Mittel- und Osteuropa'. *Osteuropa* 51/11–12 (November/December 2001), 1416–34.

Lackó, Miklós. *Korszellem és tudomány 1910–1945*. Budapest, 1988.

——. 'Populism in Hungary: Yesterday and Today'. In J. Held (ed). *Populism in Eastern Europe*. Boulder, CO, 1996: 107–27.

——. 'A két világháború között'. *Budapesti Negyed* (1998/2–3), 20–1. http://epa.oszk.hu/00000/00003/00016/lacko.htm

Lajta, Andor. *A Magyar oktatófilm története 1912–1951*. Budapest, 1960.

Lampland, Martha. 'Death of a hero: Hungarian National Identity and the Funeral of Lajos Kossuth'. *Hungarian Studies* 8/1 (1993), 29–35.

Landy, Marcia. *Fascism in Film: The Italian Commercial Cinema, 1931–1943*. Princeton, NJ, 1986.

——. 'Frames or Frame Ups of National Cinema?' *KinoKultura* #6 (October 2004). http://www.kinokultura.com/articles/oct04-natcine-landy.html

Langer, István. 'A két 'Lila akác'. Beszélgetés Székely István filmrendezővel'. *Filmspirál* II/2-1 (1996), 132–54.

Langman, Larry. *Destination Hollywood: The Influence of Europeans on American Filmmaking*. Jefferson, NC, 2000.

Leach, Jim. *British Film*. Cambridge, UK, 2004.

Lendvai, F. L., A. Sohár & P. Horváth (eds). *Hét évtized a hazai zsidóság életében*, vols I–II, Budapest, 1990.

Lendvai, Paul. *The Hungarians: A Thousand Years of Victory in Defeat*. Trans. A. Major. Princeton, NJ, 2003.

Liehm, Mira & A.J. Liehm. *The Most Important Art: Eastern European Film after 1945*. Berkeley, CA, 1977.

Linde-Laursen, Anders. 'Taking the National Family to the Movies: Changing Frameworks for the Formation of Danish Identity, 1930–90'. *Anthropological Quarterly* 72/1 (January 1999), 18–33.

Livezeanu, Irina. *Cultural Politics in Greater Romania*. Ithaca, NY, 1995.

Loacker, A. & M. Prucha (eds). *Unerwünschtes Kino. Der deutschesprachige Emigrantenfilm 1934–1937*. Vienna, 2000.

Lorman, Thomas. *Counter-Revolutionary Hungary, 1920–25: István Bethlen and the Politics of Consolidation*. Boulder, CO, 2006.

Lowry, Stephen. *Pathos und Politik. Ideologie in Spielfilmen des Nationalsozialismus*. Tübingen, 1991.

——. 'Ideology and Excess in Nazi Melodrama: *The Golden City*'. *New German Critique* 74 (Spring-Summer 1998), 125–50.

Lukacs, John. *Budapest 1900: A Historical Portrait of a City & Its Culture*. New York, 1988.

Macartney, C.A. *Hungary and Her Successors: The Treaty of Trianon and Its Consequences 1919–37*. London, 1937.

——. *October Fifteenth: A History of Modern Hungary, 1929–44*, vol. I. Edinburgh, 1957.

Macrae, C. N., C. Stangor & M. Hewstone (eds). *Stereotypes and Stereotyping*. New York, 1996.

Manchin, Anna. 'Interwar Hungarian Entertainment Films and the Reinvention of Rural Modernity'. *Rural History* 21/2 (October 2010), 195–212.

——. 'Jewish Humor and the Cabaret Tradition in Interwar Hungarian Entertainment Films'. In L. Khatib (ed). *Story Telling in World Cinema: Forms* v.1. New York, 2012, 32–47.

——. 'Gyula Kabos and "Jewish Difference": Reconstructing Interwar Jewish History through Film'. *The Hungarian Quarterly* 54/209 (2014), 172–90.

Mandler, Peter. *The English National Character: The History of an Idea from Edmund Burke to Tony Blair*. New Haven, CT, 2006.

Margry, Karel. 'Newsreels in Nazi-Occupied Czechoslovakia'. *Historical Journal of Film, Radio and Television* 24/1 (2004), 69–116.

Marsovszky, Magdolna. 'Antiszemitizmus és nemzeti áldozati mítosz'. *Beszélő Online* (20 May 2013), http://beszelo.c3.hu/onlinecikk/antiszemitizmus-es-nemzeti-aldozati-mitosz

Marton, Kati. *The Great Escape: Nine Jews Who Fled Hitler and Changed the World*. New York, 2006.

Mathijs, Ernest (ed). *The Cinema of the Low Countries*. London, 2004.

Mellencamp, P. & P. Rosen (eds). *Cinema Histories, Cinema Practices*. Frederick, MD, 1984.

Mihancsik, Zsófia. 'A "vérközösség" országáról és a miniszterelnökéről'. *Galamus-csoport* (1 October 2012), http://galamus.hu/index.php?option=com_content&view=article&id=165077:a-verkoezoesseg-orszagarol-es-miniszterel-noekerl-164091&catid=51&Itemid=129.

Miron, Guy. 'History, Remembrance, and a 'Useful Past' in the Public Thought of Hungarian Jewry, 1938–1939'. *Yad Vashem Studies* 32 (2004), 131–70.

Moeller, Felix. *Der Filmminister. Goebbels und der Film im Dritten Reich*. Berlin, 1998.

Molnár, Miklós. 'The Hungarian Intellectual and the Choice of Commitment or Neutrality'. In I. Maclean, A. Montefiore & P. Winch (eds). *The Political Responsibility of Intellectuals*. Cambridge, UK, 1990, 189–200.

Mosse, George L. *The Culture of Western Europe: The Nineteenth and Twentieth Centuries*, 3rd edn. Boulder, CO, 1988 [1961].

Mozgóképtár. Magyar filmtörténeti sorozat [*Moving picture gallery. A Hungarian Film History Series*], Compact Disk, volume 1: Játékfilmek a kezdetektől 1944-ig. Budapest, 1996.

Mühl-Benninghaus, Wolfgang. *Das Ringen um den Tonfilm. Strategien der Elektro- und der Filmindustrie in den 20er und 30er Jahren*. Düsseldorf, 1999.

Murai, András. *Film és kollektív emlékezet: magyar múltfilmek a rendszerváltozás után*. Szombathely, Hungary, 2008.

Nairn, Tom. *Faces of Nationalism: Janus Revisited*. London, 1997.

Nelson, Barbara A. 'Hollywood's Struggle for Romania, 1938–1945'. *Historical Journal of Film, Radio and Television* 29/3 (September 2009), 295–319.

Nemes, Károly. *Miért jók a magyar filmek?* Budapest, 1968.

Nemes, Károly & Sándor Papp. *A magyar film 1945–1956 között*. Budapest, 1980.

Nemeskürty, István. *A meseautó utasai*. Budapest: Corvina, 1965.

——. *Word and Image. History of the Hungarian Cinema*, 2nd edition. Trans. Z. Horn and F. Macnicol. Budapest, 1974.

——. (ed). *A magyar film 1939–1944: Egész müsort betöltö játékfilmek*. Budapest, 1980.

——. *A képpé varázsolt idő*. Budapest, 1983.

——. *Abriss der Kulturgeschichte Ungarns*. Trans. Hans Skirecki. Budapest, 1994.

——. *Nemeskürty István összes műve. A mozgókép varázsa*. Budapest, 2002.

Nemeskürty, I, I. Karcsai Kulcsár & M. Kovács (eds). *A magyar hangosfilm története a kezdetektöl 1939-ig*. Budapest, 1975.

Nolan, Mary. *Visions of Modernity. American Business and the Modernization of Germany*. New York, 1994.

Nowell-Smith, G. & S. Ricci (eds). *Hollywood and Europe: Economics, Culture, National Identity: 1945–95*. London, 1998.

O'Regan, Tom. *Australian National Cinema*. London, 1996.

O'Regan, T. & B. Shoesmith (eds). *History on/and/in Film: Selected Papers from the 3rd Austrialian History and Film Conference, Perth*. Perth, Australia, 1987.

Ormos, Mária. *Egy magyar médiavezér: Miklós Kozma*, 2 vols Budapest, 2000.

Orr, John. 'The Art of National Identity: Peter Greenaway and Derek Jarman'. In J. Ashby & A. Higson (eds). *British Cinema, Past and Present*. London, 2000, 327–38.

Orzoff, Andrea. *Battle for the Castle: The Myth of Czechoslovakia in Europe, 1914–1948*. Oxford, UK, 2009.

Paces, Cynthia. *Prague Panoramas*. Pittsburgh, PA, 2009.

Paikert, G.C. 'Hungarian Foreign Policy in Intercultural Relations, 1919–1944'. *American Slavic and East European Review*, 11/1 (February 1952), 42–65.

Pánczél, Lajos. *Az ötvenéves mozi*. Budapest, 1946.

Pápai Zsolt & B. Varga. 'Hollywood innen és túl'. *Metropolis* 14/1 (2010). http://metropolis.org.hu/?pid=16&aid=329

Passek, Jean-Loup (ed). *Le cinéma hongrois*. Paris, 1979.

Pasternak, Joe & D. Chandler, *Easy the Hard Way*. New York, 1953

Patai, Raphael. *The Jews of Hungary: History, Culture, Psychology*. Detroit, MI, 1996.

Pelle, János. *Sowing the Seeds of Hatred*. Boulder, CO, 2004.

Perneki, Mihály (ed). *Shvoy Kálmán titkos naplója és emlékirata 1918–1945*. Budapest, 1983.

Péter, László & L. Czigány. 'Urbánusok és népi(es)ek nálunk'. In László Péter (ed). *Az Elbától keletre: Tanulmányok a magyar és kelet-európai történelemből*. Budapest, 1998: 379–84.

Petrie, Duncan (ed). *New Questions of British Cinema*. London, 1992.

Petrie, Graham. 'Hungarian Silent Cinema Rediscovered'. *The Hungarian Quarterly*, XXXVIII/147 (Autumn 1997), 152–60.

Petropoulos, Jonathan. *Art as Politics in the Third Reich*. Chapel Hill, NC, 1996.

Pines, Jim & Paul Willemen (eds). *Questions of Third Cinema*. London, 1989.

Pók, Attila. 'Atonement and Sacrifice: Scapegoats in Modern Eastern and Central Europe'. *East European Quarterly* XXXII/4 (January 1999), 531–48.

——. 'Scapegoating and Antisemitism after World War I'. *Jewish Studies at the CEU* II (2002–3), 125–34.

——. *The Politics of Hatred in the Middle of Europe. Scapegoating in Twentieth Century Hungary: History and Historiography*. Szombathely, Hungary, 2009.

Polony, Csaba. 'Globalization and Cultural Nationalism. Remarks on the populist/urbanist debate in the arts in Hungary.' New York, 2000.

Pór, Katalin. *De Budapest à Hollywood: le theater hongrois à Hollywood, 1930–43*. Rennes, France, 2010.

Porter-Szűcs, Brian. 'Beyond the Study of Nationalism'. In T. Kamusella & K. Jaskułowski (eds). *Nationalism Today*. Oxford, UK, 2009, 3–15.

Possony, Stefan T. 'National Socialistic Economics: The Contradictions of the New Order'. *The Journal of Politics* 4/2 (May 1942), 149–82.

Pritz, Pál. 'Magyarország külpolitikája a formálódó Berlin-Róma tengely árnyékában'. *Századok* 115/5 (1981), 924–57.

——. *Magyarország külpolitikája Gömbös Gyula miniszterelnöksége idején 1932–36*. Budapest, 1982.

——. *Magyar diplomácia a két háború között*. Budapest, 1995.

Pusztaszeri, László. *Karády és Ujszászy*. Budapest, 2008.

——. *Szeleczky Zita pályaképe*. Budapest, 2011.

Puttnam, David. *The Undeclared War: The Struggle for Control of the World's Film Industry*. London, 1997.

Ramirez, Bruno. 'Clio in Words and in Motion: Practices of Narrating the Past'. *Journal of American History* 86/3 (December 1999), 987–1014.

Ránki, György. *Economy and Foreign Policy: The Struggle of the Great Powers for Hegemony in the Danube Valley, 1919–39*. Boulder, CO, 1983.

Ránki, György, et al. (eds). *Magyarország története*, vol. 8 [1918–1945]. Budapest, 1976.

Ranki, Vera. *The Politics of Inclusion and Exclusion: Jews and Nationalism in Hungary*. Teaneck, NJ, 1999.

Reeves, Nicholas. *The Power of Film Propaganda: Myth or Reality?* London, 1999.

Reicher, S., N. Hopkins, & S. Condor, 'Stereotype construction as a strategy of influence'. In R. Spears and P. J. Oakes, et al. (eds). *The Social Psychology of Stereotyping and Group Life*. Malden, MA, 1997, 94–118.

Reményi Gyenes, István (ed). *Ismerjük őket? Zsidó származású nevezetes magyarok arcképcsarnoka*. Budapest, 1997.

Rentschler, Eric. *The Ministry of Illusion: Nazi Cinema and Its Afterlife*. Cambridge, MA, 1996.

Réti, György. 'Az 1938.évi Anschluss a magyar-olasz kapcsolatok tükrében'. *Múltunk* XLIII/1 (1998), 71–93.

Ricci, Steven. *Cinema and Fascism: Italian Film and Society, 1922–1943*. Berkeley, CA, 2008.

Richards, Jeffery. *Films and British National Identity: From Dickens to* Dad's Army. Manchester, UK, 1997.

——. *The Unknown 1930s: An Alternative History of the British Cinema*. London, 1998.

Riemenschneider, Michael. *Die deutsche Wirtschaftpolitik gegenüber Ungarn 1933–1944. Ein Beitrag zur Interdependenz von Wirtschaft und Politik unter dem Nationalsozialismus*. Frankfurt-am-main, Germany, 1987.

Ritschl, A. O. 'Nazi Economic Imperialism and the Exploitation of the Small: Evidence from Germany's Secret Foreign Exchange Balances, 1938–40'. *The Economic History Review*, New Series 54/2 (May 2001), 324–45.

Rogger, Hans & E. Weber (eds). *The European Radical Right: A Historical Profile*. Berkeley, CA, 1966.

Romány, Pál. 'Tanyavilág és falukutatás. Tanirodalom Magyarországon', *Magyar Tudomány* 163/9 (September 2002), 1187–95.

Romsics, I. *István Bethlen: A Great Conservative Statesman of Hungary, 1874–1946*. Trans. M. D. Fenyő. Boulder, CO, 1995.

——.*Hungary in the Twentieth Century*. Trans. T. Wilkinson. Budapest, 1999.

——. 'From Christian Shield to EU Member'. *The Hungarian Quarterly* 48/188 (Winter, 2007), 3–27.

——. *Magyar Sorsfordulók*. Budapest, 2012.

Rosen, Philip. 'History, Textuality, Nation: Kracauer, Burch and Some Problems in the Study of National Cinemas'. *Iris* 2/2 (1984), 69–83.

Rosenstone, Robert A. *Visions of the Past: The Challenge of Film to Our Idea of History*. Cambridge, MA, 1995.

——. (ed). *Revisioning History. Film and the Construction of a New Past*. Princeton, NJ, 1995.

Roshwald, Aviel. *The Endurance of Nationalism*. Cambridge, UK, 2006.

Roth, Joseph. *Flight without End*. New York, 2003 [1927].

Said, Edward. *The Word, the Text, and the Critic*. London, 1984.

Sajti, Enikő A. *Délvidék 1941–44. A magyar kormányok délszláv politikája*. Budapest, 1987.

Sakmyster, Thomas. *Hungary's Admiral on Horseback. Miklós Horthy, 1918–1944*. Boulder, CO, 1994.

Salmi, Hannu. 'Film as historical narrative'. *Film-Historia* [*Spain*] V/1 (1995), 45–54.

Sándor, Tibor. *Őrségváltás. A magyar film és a szélsőjobboldal a harmincas-negyvenes években*. Budapest, 1992.

——.*Őrségváltás után: zsidókérdés és filmpolitika 1938–44*. Budapest, 1997.

——. 'Urak és parasztok – viták és mitoszok'. *Budapesti negyed* 5/2-3 [16–17] (1997), http://epa.oszk.hu/00000/00003/00014/sandor.htm.

Saunders, Thomas. *Hollywood in Berlin: American Cinema and Weimar Germany*. Berkeley, CA, 1994.

Schlüpmann, Heide. 'Politik als Schuld'. *Frauen und Film* 38 (1985), 46–57.

Schorske, Carl. *Fin-de-siècle Vienna: politics and culture*. New York, 1980.

Schweck, Ingo. *'[...] weil wir lieber im Kino sitzen als in Sack und Asche'*. Muenster, 2002.

Sedgwick, John & Mike Pokorny. 'Hollywood's foreign earnings during the 1930s'. *Transnational Cinemas* 1/1 (2010), 83–97.

Shafik, Viola. *Popular Egyptian Cinema: Gender, Class, and Nation*. Cairo & New York, 2006.

——. *Arab Cinema: History and Cultural Identity*. Cairo & New York, 2007.

Short, K.R.M. (ed). *Feature Films as History*. Knoxville, TN, 1981.

Silber, Michael K. *Jews in the Hungarian Economy 1760–1945*. Jerusalem, 1992.

Sipos, András. 'Who is a "True Hungarian"? The Movement of "Spiritual Defence of the Fatherland" and the Image of the Enemy'. *CEU History Department, Working Paper Series 2*. Budapest, 1995, 119–31.

Sipos, Péter. *Imrédy Béla a vádlottak padján*. Budapest, 1999.

Skaff, Shiela. *The Law of the Looking Glass: Cinema in Poland 1896–1939*. Athens, OH, 2008.

Skaper, Brigitta. 'Magyar filmsztárok a két világháború közötti Magyarországon'. *Médiakutató* (Fall 2008), http://www.mediakutato.hu/cikk/2008_03_osz/08_regi_magyar_filmsztarok

Sklar, Robert. *Movie-made America: A Social History of American Movies*. New York, 1975.

Sklar, Robert & C. Musser (eds). *Resisting Images: Essays on Cinema and History*. Philadelphia, PA, 1990.

Smith, A. D. *The Ethnic Origins of Nations*. Oxford, UK, 1986.

——. *National Identity*. London, 1991.

Sobchack, Vivian (ed). *The Persistence of History: Cinema, Television and the Modern Event*. New York, 1996.

Soila, T., A. Söderbergh-Widding, and G. Iverson. *Nordic National Cinemas*. London, 1998.

Sorlin, Pierre. *The Film in History: Restaging the Past*. Totowa, NJ, 1980.

——. *Italian National Cinema 1896–1996*. London, 1996.

Spears, R. & P. J. Oakes, et al. (eds). *The Social Psychology of Stereotyping and Group Life*. Malden, MA, 1997.

Special Issue on Nationalism. *The Philosophical Forum* XXVIII/1–2 (Fall-Winter 1996–7).

Spector, Scott. 'Was the Third Reich Movie-Made? Interdisciplinarity and the Reframing of "Ideology"'. *AHR* 106/2 (April 2001), 460–84.

Spiker, Jürgen. *Film und Kapital*. Berlin, 1975.

Spira, Thomas. *German-Hungarian Relations and the Swabian Problem: From Károlyi to Gömbös 1919–1936*. Boulder, CO, 1977.

Stachura, Peter (ed). *The Shaping of the Nazi State*. London, 1978.

Steinweis, Alan E. 'German Cultural Imperialism in Czechoslovakia and Poland, 1938–1945'. *The International History Review* XIII/3 (August 1991), 466–80.

——. *Art, Ideology, & Economics in Nazi Germany: The Reich Chambers of Music, Theater, and the Visual Arts*. Chapel Hill, NC, 1993.

Stoil, Michael Jon. *Cinema beyond the Danube: The Camera and Politics*. Metuchen, NJ, 1974.

Stradomski, Wieslaw. 'The Jewish Cinema in Inter-war Poland'. Trans. Aleksandra Rodzińska. *Polish Art Studies* 10 (1989), 167–77.

Street, Sarah. *British National Cinema*. London, 1997.

Sugar, Peter & I. J. Lederer (eds). *Nationalism in Eastern Europe*. Seattle, WA & London, 1994 [1969].

Sugar, P., et al. (eds). *A History of Hungary*. Bloomington, IN, 1994.

Suny Ronald & Michael Kennedy (eds). *Intellectuals and the Articulation of Nation*. Ann Arbor, MI, 2001.

Suranyi, Vera. *Minarik, Sonnenschein, és a többiek. Zsidó sorsok magyar filmen*. Budapest, 2001.

Szabó, Miklós. '1944. március 19. történetéhez'. *Levéltári Közvelemények* 53/2 (1982), 295–302.

Szakall, S.K. *The Story of Cuddles: My Life under the Emperor Francis Joseph, Adolf Hitler and the Warner Brothers*. Trans. Paul Tabori. London, 1954.

Szalai, Anna. 'Will the Past Protect Hungarian Jewry? The Response of Jewish Intellectuals to Anti-Jewish Legislation'. *Yad Vashem Studies* 32 (2004), 171–208.

Szántó, Gábor T. 'A magyar antiszemitizmus nem kirívó, de nincs ok a megnyugvásra'. *Szombat* (21 January 2014), http://www.szombat.org/politika/a-magyar-antiszemitizmus-nem-kirivo-de-nincs-ok-a-megnyugvasra

Szczepanczik, Petr. 'Undoing the National: Representing National Space in 1930s Czechoslovak Multiple-Language Versions'. *Cinema & Cie* 4 (Spring 2004), 55–65.

——. '"Tief in einem deutschen Einflussbereich." Die Aufführung und Rezeption deutschsprachiger Filme in der Tschechoslowakei in den frühen 1930er Jahren'. In Jan Distelmeyer (ed). *Babylon in FilmEuropa. Mehrspachen-Versionen der 1930er Jahre* (Hamburg, Germany, 2006), 89–102.

Szegedy-Maszák, Mihály (ed). *National Heritage, National Canon*. Budapest, 2001.

Székely, Gyula, M. Cenner & I. Szilágyi (eds). *A magyar színészet nagy képeskönyve*. Budapest, 1984.

Székely, István. *A Hyppolittól a Lila Akácig*. Budapest, 1978.

Szinai, Miklós. 'A magyar kormányok politikája a zsidókérdésben (1936–44. március 19): Hozzászólás a felelősség kérdéséhez'. *Magyar Tudomány* 38/12 (1994), 1476–87.

Szőts, István. *Szilánkok és gyaluforgácsok*. Budapest, 1999.

Szűcs, László. 'Kormányzói audiencia az első Zsidótörvény előtt, (1937. November 11)'. *Levéltári Közlemények* 1–2 (1993), 143–64.

Taylor, Richard & D. Spring (eds). *Stalinism and Soviet Cinema*. London, 1993.

Tegel, Susan. *Nazis and the Cinema*. London, 2007.

Thompson, Kristin. *Exporting Entertainment: America in the World Film Market 1907–34*. London, 1985.

——. 'National or International Films?' *Film History* 8/3 (1996), 281–96.

Tihanov, Galin. 'Why did modern literary theory originate in Central and Eastern Europe? (And why is it now dead?)'. *Common Knowledge* 10/1 (Winter 2004), 61–81.

Tilkovszky, Loránt. *Ungarn und die deutsche Volksgruppenpolitik 1938–45*. Köln, Germany, 1981.

——. *Teufelskreis: Die Minderheitenfrage in den deutsch-ungarischen Beziehungen 1933–38*. Trans. Johanna Till. Budapest, 1989.

Todorov, Tzevtan. 'The Co-Existence of Culture', *Oxford Literary Review* 19 (1997), 3–17.

Todorova, Maria. *Imagining the Balkans*. New York, 1997.

Trencsényi, Balázs. '"Imposed Authenticity": Approaching Eastern European National Characterologies in the Inter-war Period', *Central Europe* 8/1 (May 2010), 20–47.

——. *The Politics of 'National Character': A Study in Interwar East European Thought*. Oxford, UK, 2012.

Triana-Toribio, Núria. *Spanish National Cinema*. London, 2003.

Tsivian, Yuri. *Early Cinema in Russia and its Cultural Reception*. Trans. Alan Bodger. London, 1994.

Turda, Marius. 'The Biology of War: Eugenics in Hungary, 1914–1918'. *Austrian History Yearbook* XL (2009), 238–64.

Turda, M. & P. Weindling (eds). *'Blood and Homeland': Eugenics and Racial Nationalism in Central and Southeast Europe*. Budapest, 2007.

Ujváry, Gábor. *A harmincharmadik nemzedék*. Budapest, 2010.

Ungvári, Tamás. *The 'Jewish Question' in Europe: The Case of Hungary*. Boulder, CO, 2000.

Vadas, József. *A magyar art deco*. Budapest, 2005.

Vajdovich, Györgyi. 'A magyar film 1939 és 1945 között'. *Metropolis* 17/2 (2013). http://metropolis.org.hu/?pid=16&aid=480

——. 'Ideologiai üzenet az 1939–1944 közötti magyar filmekben. *Metropolis* 17/2 (2013). http://metropolis.org.hu/?pid=16&aid=484

——. 'Művészetek: Filmtörténete. A magyar film 1918-tól 1945-ig'. In *Dunánál. A magyarok a 20. században*, (2001) http://mek.oszk.hu/01900/01906/html/index4.html.

——, 'Vígjátékváltozatok az 1931–1944 közötti magyar fimben'. 18/3 (2014). http://www.metropolis.org.hu/?pid=16&aid=514.

van der Let, Petrus & A. Loacker. *Unerwünschtes Kino*. DVD. Wien/Budapest, 2005.

Varga, Balázs, et al. (eds). *Játékfilmek/Hungarian Feature Films 1931–1997*. Budapest, 1998.

Vargyai, Gyula. *A vezérkar szerepe a katonai és politikai döntésekben Magyarországon, 1919–44*. Budapest, 1994.

Vasey, Ruth. 'Foreign Parts: Hollywood's Global Distribution and the Representations of Ethnicity'. *American Quarterly* (December 1992), 617–42.

——. *The World According to Hollywood, 1918–39*. Madison, WI, 1997.

Vasudevan, Ravi S. 'Addressing the spectator of a "third world" national cinema: the Bombay "social" film of the 1940s and 1950s'. *Screen* 36/4 (Winter 1995), 305–24.

Veress, József. *A magyar film története*. Budapest, 2006.

Vincendeau, Ginette. 'Hollywood Babel: The Coming of Sound and the Multiple Language Version'. *Screen* 29/2 (Spring 1988), 24–39.

Vitali, Valentina & P. Willemen (eds). *Theorising National Cinema*. London, 2006.

von Dassanowsky, Robert. *Austrian Cinema: A History*. Jefferson, NC, 2005.

Vörös, Kati. 'How Jewish is Jewish Budapest?' *Jewish Social Studies* 8/1 (2001), 88–125.

Votruba, Martin. 'Historical and National Background of Slovak Filmmaking'. *KinoKultura* Special #3 (December 2005), http://www.kinokultura.com/specials/3/votruba.pdf

Wahl, Chris. 'Paprika in the Blood'. *Spectator* 27/2 (Fall 2007), 11–20.

Walicki, Andrzej. 'Intellectual Elites and the Vicissitudes of Imagined Nation in Poland'. *East European Politics and Societies* 11/3 (Fall 1997), 227–53.

Walsh, Michael. 'National Cinema, National Imaginary'. *Film History* 8/1 (1996), 5–17.

Wasel, Ernst & E. Korani (eds). *Film posters 1912–1945. A magyar filmplakát aranykora/the golden age of Hungarian film posters*. Budapest, 2004.

Weber, Eugen. *Peasants into Frenchmen: The Modernization of Rural France, 1870–1914*. Stanford, CA, 1976.

Weis, Elisabeth & J. Belton (eds). *Film Sound: Theory and Practice*. New York, 1985.

Weiss, Gernot. 'Aus gemeinsamen Wurzeln. Filmproduktion zwischen Prag un Wien bis 1938'. In J. Roschlau (ed). *Zwischen Barrandov und Babelsberg. Deutsch-tschechische Filmbeziehungen im 20. Jahrhundert*. Hamburg, Germany, 2008: 36–40.

Welch, David. (ed). *Nazi Propaganda: The Power and the Limitations*. Totowa, NJ, 1983.

——. *Propaganda and the German Cinema 1933–45*. Oxford, UK, 1983.

Welch, D. & R. Vande Winkel (eds). *Cinema and the Swastika*. London, 2011.

White, G. W. *Nationalism and Territory: Constructing Group Identity in Southeastern Europe*. New York, 2000.

Williams, Alan. *Republic of Images: A History of French Filmmaking*. Cambridge, MA, 1992.

—— (ed). *Film and Nationalism*. New Brunswick, NJ, 2002.

Williams, Raymond. *Culture and Society: 1780–1950*. New York, 1983 [1958].

Wingfield, Nancy. 'When Film Became National: 'Talkies' and the Anti-German Demonstrations in Prague'. *Austrian History Yearbook* XXIX/1 (1998), 113–38.

——(ed). *Creating the Other: Ethnic Conflict and Nationalism in Habsburg Central Europe*. New York, 2004.

——. 'The Problem with "Backwardness": Ivan T. Berend's Central and Eastern Europe in the Nineteenth and Twentieth Centuries'. *European History Quarterly* 34/4 (2004), 535–51.

——. *Flag Wars and Stone Saints*. Cambridge, MA, 2007.

Witte, Karsten. 'Film im Nationalsozialismus: Blendung und Überblendung'. in W. Jacobsen, A. Kaes, and H. H. Prinzler (eds). *Geschichte des deutschen Films*. Stuttgart, Germany, 1993: 119–70.

Wulf, Joseph. *Theater and Film im Dritten Reich. Eine Dokumentation*. Frankfurt-am-main, Germany, 1989.

Youngblood, Denise J. *Movies for the Masses. Popular Cinema and Soviet Society in the 1920s*. Cambridge, UK, 1992.

Záhonyi-Ábel, M. 'Filmcenzúra Magyarországon a Horthy-korszak első évtizedében'. *Médiakutató* (Summer 2012), 91–101. http://www.mediakutato.hu/cikk/2012_02_nyar/12_filmcenzura_magyarorszagon/

——. 'A magyar filmes intézményrendszer 1938–1944'. *Metropolis* 17/2 (2013), 12–27. http://metropolis.org.hu/?pid=16&aid=503

Zahra, Tara. *Kidnapped Souls*. Ithaca, NY, 2008.

——. 'Imagined Noncommunities: National Indifference as a Category of Analysis'. *Slavic Review* 69/1 (Spring 2010), 93–119.

Zhang, Yingjin. *Chinese National Cinema*. London & New York, 2004.

Index